Design and Analysis
of Experiments

STATISTICS: Textbooks and Monographs

A SERIES EDITED BY

D. B. OWEN, Coordinating Editor

Department of Statistics
Southern Methodist University
Dallas, Texas

Vol. 1: The Generalized Jackknife Statistic, *H. L. Gray and W. R. Schucany*
Vol. 2: Multivariate Analysis, *Anant M. Kshirsagar*
Vol. 3: Statistics and Society, *Walter T. Federer*
Vol. 4: Multivariate Analysis: A Selected and Abstracted Bibliography, 1957-1972, *Kocherlakota Subrahmaniam and Kathleen Subrahmaniam* (out of print)
Vol. 5: Design of Experiments: A Realistic Approach, *Virgil L. Anderson and Robert A. McLean*
Vol. 6: Statistical and Mathematical Aspects of Pollution Problems, *John W. Pratt*
Vol. 7: Introduction to Probability and Statistics (in two parts), Part I: Probability; Part II: Statistics, *Narayan C. Giri*
Vol. 8: Statistical Theory of the Analysis of Experimental Designs, *J. Ogawa*
Vol. 9: Statistical Techniques in Simulation (in two parts), *Jack P. C. Kleijnen*
Vol. 10: Data Quality Control and Editing, *Joseph I. Naus* (out of print)
Vol. 11: Cost of Living Index Numbers: Practice, Precision, and Theory, *Kali S. Banerjee*
Vol. 12: Weighing Designs: For Chemistry, Medicine, Economics, Operations Research, Statistics, *Kali S. Banerjee*
Vol. 13: The Search for Oil: Some Statistical Methods and Techniques, *edited by D. B. Owen*
Vol. 14: Sample Size Choice: Charts for Experiments with Linear Models, *Robert E. Odeh and Martin Fox*
Vol. 15: Statistical Methods for Engineers and Scientists, *Robert M. Bethea, Benjamin S. Duran, and Thomas L. Boullion*
Vol. 16: Statistical Quality Control Methods, *Irving W. Burr*
Vol. 17: On the History of Statistics and Probability, *edited by D. B. Owen*
Vol. 18: Econometrics, *Peter Schmidt*
Vol. 19: Sufficient Statistics: Selected Contributions, *Vasant S. Huzurbazar (edited by Anant M. Kshirsagar)*
Vol. 20: Handbook of Statistical Distributions, *Jagdish K. Patel, C. H. Kapadia, and D. B. Owen*
Vol. 21: Case Studies in Sample Design, *A. C. Rosander*
Vol. 22: Pocket Book of Statistical Tables, *compiled by R. E. Odeh, D. B. Owen, Z. W. Birnbaum, and L. Fisher*
Vol. 23: The Information in Contingency Tables, *D. V. Gokhale and Solomon Kullback*
Vol. 24: Statistical Analysis of Reliability and Life-Testing Models: Theory and Methods, *Lee J. Bain*
Vol. 25: Elementary Statistical Quality Control, *Irving W. Burr*
Vol. 26: An Introduction to Probability and Statistics Using BASIC, *Richard A. Groeneveld*
Vol. 27: Basic Applied Statistics, *B. L. Raktoe and J. J. Hubert*
Vol. 28: A Primer in Probability, *Kathleen Subrahmaniam*
Vol. 29: Random Processes: A First Look, *R. Syski*

Vol. 30: Regression Methods: A Tool for Data Analysis, *Rudolf J. Freund and Paul D. Minton*

Vol. 31: Randomization Tests, *Eugene S. Edgington*

Vol. 32: Tables for Normal Tolerance Limits, Sampling Plans, and Screening, *Robert E. Odeh and D. B. Owen*

Vol. 33: Statistical Computing, *William J. Kennedy, Jr. and James E. Gentle*

Vol. 34: Regression Analysis and Its Application: A Data-Oriented Approach, *Richard F. Gunst and Robert L. Mason*

Vol. 35: Scientific Strategies to Save Your Life, *I. D. J. Bross*

Vol. 36: Statistics in the Pharmaceutical Industry, *edited by C. Ralph Buncher and Jia-Yeong Tsay*

Vol. 37: Sampling from a Finite Population, *J. Hajek*

Vol. 38: Statistical Modeling Techniques, *S. S. Shapiro*

Vol. 39: Statistical Theory and Inference in Research, *T. A. Bancroft and C.-P. Han*

Vol. 40: Handbook of the Normal Distribution, *Jagdish K. Patel and Campbell B. Read*

Vol. 41: Recent Advances in Regression Methods, *Hrishikesh D. Vinod and Aman Ullah*

Vol. 42: Acceptance Sampling in Quality Control, *Edward G. Schilling*

Vol. 43: The Randomized Clinical Trial and Therapeutic Decisions, *edited by Niels Tygstrup, John M. Lachin, and Erik Juhl*

Vol. 44: Regression Analysis of Survival Data in Cancer Chemotherapy, *Walter H. Carter, Jr., Galen L. Wampler, and Donald M. Stablein*

Vol. 45: A Course in Linear Models, *Anant M. Kshirsagar*

Vol. 46: Clinical Trials: Issues and Approaches, *edited by Stanley H. Shapiro and Thomas H. Louis*

Vol. 47: Statistical Analysis of DNA Sequence Data, *edited by B. S. Weir*

Vol. 48: Nonlinear Regression Modeling: A Unified Practical Approach, *David A. Ratkowsky*

Vol. 49: Attribute Sampling Plans, Tables of Tests and Confidence Limits for Proportions, *Robert E. Odeh and D. B. Owen*

Vol. 50: Experimental Design, Statistical Models, and Genetic Statistics, *edited by Klaus Hinkelmann*

Vol. 51: Statistical Methods for Cancer Studies, *edited by Richard G. Cornell*

Vol. 52: Practical Statistical Sampling for Auditors, *Arthur J. Wilburn*

Vol. 53: Statistical Signal Processing, *edited by Edward J. Wegman and James G. Smith*

Vol. 54: Self-Organizing Methods in Modeling: GMDH Type Algorithms, *edited by Stanley J. Farlow*

Vol. 55: Applied Factorial and Fractional Designs, *Robert A. McLean and Virgil L. Anderson*

Vol. 56: Design of Experiments: Ranking and Selection, *edited by Thomas J. Santner and Ajit C. Tamhane*

Vol. 57: Statistical Methods for Engineers and Scientists. Second Edition, Revised and Expanded, *Robert M. Bethea, Benjamin S. Duran, and Thomas L. Boullion*

Vol. 58: Ensemble Modeling: Inference from Small-Scale Properties to Large-Scale Systems, *Alan E. Gelfand and Crayton C. Walker*

Vol. 59: Computer Modeling for Business and Industry, *Bruce L. Bowerman and Richard T. O'Connell*

Vol. 60: Bayesian Analysis of Linear Models, *Lyle D. Broemeling*

Vol. 61: Methodological Issues for Health Care Surveys, *Brenda Cox and Steven Cohen*

Vol. 62: Applied Regression Analysis and Experimental Design, *Richard J. Brook and Gregory C. Arnold*

Vol. 63: Statpal: A Statistical Package for Microcomputers – PC-DOS Version for the IBM PC and Compatibles, *Bruce J. Chalmer and David G. Whitmore*

Vol. 64: Statpal: A Statistical Package for Microcomputers – Apple Version for the II, II+, and IIe, *David G. Whitmore and Bruce J. Chalmer*

Vol. 65: Nonparametric Statistical Inference, Second Edition, Revised and Expanded, *Jean Dickinson Gibbons*

Vol. 66: Design and Analysis of Experiments, *Roger G. Petersen*

OTHER VOLUMES IN PREPARATION

Design and Analysis of Experiments

ROGER G. PETERSEN

Oregon State University
Corvallis, Oregon

MARCEL DEKKER, INC.   NEW YORK · BASEL · HONG KONG

Library of Congress Cataloging in Publication Data

Petersen, Roger G., [date]
  Design and analysis of experiments.

  (Statistics, textbooks and monographs; v. 66)
  Bibliography: p.
  Includes index.
  1. Experimental design. I. Title. II. Series.
QA279.P48   1985   001.4'34   85-20745
ISBN 0-8247-7340-3

MARCEL DEKKER, INC.
270 Madison Avenue, New York, New York 10016

Current printing (last digit):
10

PRINTED IN THE UNITED STATES OF AMERICA

*Dedicated to the memory of
Henry Laurence Lucas, Jr.,
a scholar, teacher, and friend*

Preface

The intent of this book is to present a variety of experimental designs, to look at the advantages, disadvantages, and uses of each type of design, to outline the procedure for constructing the designs, and to consider the analysis and interpretation of data from each type. The purpose of the book is twofold: to provide a textbook in modern experimental design for students both in statistics and in fields of applied science, and to serve as a reference book for research scientists faced with the necessity of obtaining and interpreting precise data in an efficient way.

It is assumed that the reader has had an introduction to statistical inference including estimation, significance tests, some analysis of variance, and an exposure to simple and multiple regression. Except for Chapter 11, the required mathematical background is college algebra. In Chapter 11 the presentation and analysis of response surface designs is developed with the use of elementary matrix algebra. Response surface designs, however, may be used and analyzed without a knowledge of matrix algebra. The procedures are given in Chapter 11.

Except for fractional replication and confounding in the factorial experiments, little is presented on the theory of design construction. Further, no attempt has been made to include all possible experimental designs. Rather, in addition to the standard basic designs, I have included the types of design that I have found to be useful during more than 25 years as an agricultural experiment station statistician and consulting biometrician. Except for the change-over designs in Chapter 12 and the lattice designs in Chapter 13, catalogs of the various kinds of design are not included. Plans for most of the designs can be constructed by using the principles presented here, or may be found in other experimental design books.

It is my feeling that many books on experimental design place too much emphasis on the details of design construction and data analysis, and too little emphasis on interpretation. They overlook the fact that the purpose of a designed experiment is to obtain data which will provide answers to a research problem. I have attempted to avoid this by including a "Report of Statistical Analysis" with each numerical example as a guide to how the data might be interpreted by the research scientist. It is also my feeling that contrasts, particularly orthogonal sets of contrasts, provide one of the most powerful tools available to the research data analyst. For this reason I have included a fairly extensive section on contrasts in Chapter 5. I have also attempted to put multiple comparison procedures into their proper perspective in Chapter 5. In addition, Chapter 7 is included to illustrate the detailed analysis of data from a number of typical experiments.

The material in this book forms the basis of two courses at Oregon State University. The material in Chapters 1 through 7 forms a one-term course in experimental design and analysis for students in the applied sciences. The remainder of the book serves as the basis of an intermediate level one-term course for statistics majors and graduate students in applied sciences.

Some of the tables in the book are reproduced with the kind permission of the original authors and publishers. I am grateful to the literary executor of the late Sir Ronald A. Fisher, F.R.S., to Dr. Frank Yates, F.R.S., and to Longman Group Ltd., London, for permission to reprint Table III from their book *Statistical Tables for Biological, Agricultural and Medical Research* (6th edition, 1974), which appears as Table A2 in the Appendix of this book. The table of Percentage Points of the *F* Distribution (Table A3) and the table of Percentage Points of the Studentized Range (Table A4) appear as Table 18 and Table 29, respectively, in *Biometrika Tables for Statisticians,* Vol. 1 (3rd edition, 1966). These are reproduced with the permission of the Biometrika Trustees. The table of Minimum-Average-Risk *t* Values (Table A5) was originally published in R. A. Waller and D. B. Duncan (1969), A Bayes Rule for the Symmetric Multiple Comparison Problem, *J. Amer. Stat. Assoc. 64*:1485–1503, with "corregenda" in *J. Amer. Stat. Assoc. 67*:253–255 (1972). This table is reproduced with the permission of the American Statistical Association. Plans for the lattice designs (Sections 13.5.1–13.5.8) in Chapter 13 were published as plans 10.1–10.8 in W. G. Cochran and G. M. Cox, *Experimental Designs* (2nd edition, 1957). These are reproduced with the permission of the publisher, John Wiley & Sons, Inc., New York.

I am indebted to a number of people for pointing me in the direction of an interest in experimental design. In particular, I am grateful to Gertrude Cox and H. L. "Curly" Lucas, who were on the faculty of North Carolina State University while I was there. The change-over designs in Chapter 12 were developed largely by Curly Lucas, for which I am grateful. I would also like to express my appreciation to scientists at the agricultural experiment stations of North Carolina and Oregon and those at the International Center for Agricultural Research in the Dry Areas (ICARDA) in Aleppo, Syria, for providing interesting problems to work on and for assisting in the development of new designs and new analytical techniques. I would also like to thank the many students whose comments and suggestions helped shape the notes that led to this book. I am grateful to my wife, Jean, who typed the original draft of this manuscript and who provided encouragement throughout the whole project. Finally, I would like to thank Ms. Clover Redfern, who typed the final copy of the manuscript and served as my final authority when it came to questions of style.

Roger G. Petersen

Contents

Preface   v

1 Introduction   1
  1.1 Preliminaries   1
  1.2 Replication   2
  1.3 Randomization   2
  1.4 Blocking   3
  1.5 An Overview   4
  1.6 Definitions   5
     Further Reading   6

2 Completely Randomized Design   7
  2.1 Introduction   7
  2.2 Randomization   8
  2.3 Data Analysis   9
  2.4 Equal Replication   10
  2.5 Unequal Replication   16
  2.6 Fixed and Random Effects Models   20
  2.7 Subsampling (Nested Data, Hierarchical Classification)   24
     Further Reading   33

3 Randomized Block Design   34
  3.1 Blocking   34
  3.2 Randomized Block Design   35
  3.3 Model and Assumptions   42
  3.4 Missing Values   43
  3.5 Relative Efficiency   44
  3.6 Numerical Example: Randomized Block Design   44
     Further Reading   47

4 More Restrictive Designs 48
   4.1 Introduction 48
   4.2 Latin Square 48
   4.3 Missing Observations 56
   4.4 Relative Efficiency 56
   4.5 Numerical Example: Latin Square Design 57
   4.6 Restrictions 59
   4.7 Numerical Example: Multiple Latin Squares 60
   4.8 Further Restriction: The Graeco-Latin Square 64
   4.9 Summary of Basic Experimental Designs 68
       Further Reading 71

5 Separation of Means 72
   5.1 Introduction 72
   5.2 Selected Pairwise Comparisons 73
   5.3 Multiple Comparisons (Data Snooping) 75
   5.4 Comments on Multiple Comparison Procedures 84
   5.5 Contrasts 85
   5.6 Orthogonal Contrasts 89
   5.7 Contrasts of Totals 91
   5.8 Summary of Rules for Contrasts 94
   5.9 Numerical Example: Design and Analysis 96
   5.10 Regression Components of the Treatment SS 100
       Further Reading 111
       References 111

6 Factorial Experiments 112
   6.1 Introduction 112
   6.2 Factorial Experiments 114
   6.3 Two-Factor Experiments 117
   6.4 Three-Factor Experiments 124
   6.5 Extension 134
   6.6 Split-Plot Designs 134
       Further Reading 145
       References 145

7 Data Interpretation: Some Examples 146
   7.1 Introduction 146
   7.2 Example I: Washday Products 147
   7.3 Example II: Frozen Orange Juice 149
   7.4 Example III: Bush Bean Spacing 154
   7.5 Example IV: Wheat Yield Trial 159
   7.6 Statistical Computing Packages 162
       Further Reading 165
       References 165

8 Multifactor Experiments   **166**
   8.1 Introduction   166
   8.2 The 2^k Factorial Series   166
   8.3 Single Replication   175
   8.4 Fractional Replication   179
   8.5 Further Fractionation   186
   8.6 Sequences of Fractional Replications   187
   8.7 Formal Consideration of Effects   189
   8.8 Factors at Three Levels   192
   8.9 More Than Two Factors at Three Levels   198
   8.10 Single and Fractional Replication: Factors at Three Levels   200
      Further Reading   202
      References   202

9 Confounding   **203**
   9.1 Introduction   203
   9.2 Principles of Confounding   204
   9.3 Analysis   205
   9.4 Further Confounding   207
   9.5 Partial Confounding   208
   9.6 Confounding in the 3^k Series   213
   9.7 Further Subdivision   213
   9.8 Extension   219
   9.9 Mixed Level Confounding   219
   9.10 Blocking in Fractional Replication   222
      Further Reading   229
      References   229

10 Split-Plot Design: Variations   **230**
   10.1 Introduction   230
   10.2 Alternatives   232
   10.3 Further Subdivision   235
   10.4 Further Variations   239
   10.5 Extension   244
   10.6 Compact Designs for Exploratory Research   247
      Further Reading   250
      References   250

11 Response Surfaces   **252**
   11.1 Introduction   252
   11.2 Basic Ideas   253
   11.3 Design Selection   264
   11.4 Bias   267
   11.5 Augmented 2^k Factorials   271

11.6	Second-Order Surfaces	278
11.7	Numerical Example	290
11.8	Searching for Optima	297
11.9	List of Symbols	300
	Further Reading	301
	References	301

12 Change-Over Trials — **302**

12.1	Introduction	302
12.2	Carry-Over (Residual) Effects	302
12.3	Cross-Over, or Simple Reversal, Design	303
12.4	Switch-Back, or Double Reversal, Design	306
12.5	Incomplete Block Switch-Back Design	310
12.6	Latin Square Change-Over, or Round Robin, Design	315
12.7	Extra-Period Latin Square Change-Over Design	324
12.8	Incomplete Latin Square Change-Over Design	330
12.9	Catalog of Balanced Latin Square Designs	340
12.10	Selected Incomplete Latin Square Designs	344
	Further Reading	350

13 Incomplete Block Designs — **351**

13.1	Introduction	351
13.2	Lattice Designs	353
13.3	Balanced Lattices	354
13.4	Partially Balanced Lattices	361
13.5	Plans for Lattice Designs	386
	Further Reading	400
	References	400

Bibliography	401

Appendix		
Table A1	2500 Randomly Assorted Digits	405
Table A2	Student's *t* Distribution	406
Table A3	Percentage Points of the *F* Distribution	407
Table A4	Percentage Points of the Studentized Range	409
Table A5	Minimum-Average-Risk *t* Values	411

Author Index	417
Subject Index	419

Design and Analysis of Experiments

1

Introduction

1.1 PRELIMINARIES

Suppose a wheat breeder wants to compare the yield of a newly developed variety with that of a well-known, standard variety. He has two basic objectives in making this comparison. The first is to answer the question, "Is there a real difference in yield between the two varieties?" The second, related to the first, is to estimate the size of the difference.

As might be expected, the first objective involves a test of hypothesis that there is no real difference in yield and hope that the data lead us to reject this hypothesis. We meet the second objective by computing an estimate, preferably an interval estimate, of the true difference in yields. Almost all experimentation is done for one or both of these purposes: Testing hypotheses about, and estimating differences among, the effects of different treatments. An additional purpose is to obtain information on why treatments affect the experimental material as they do. The role of experimental design is to provide efficient and precise information to meet these objectives.

More particularly, experiments are designed for the following purposes:

1. To provide estimates of treatment effects or differences among treatment effects
2. To provide an efficient way of confirming or denying conjectures about the response to treatment
3. To assess the reliability of estimates and conjectures
4. To estimate the variability of the experimental material
5. To increase precision by eliminating extraneous variation from the comparisons of interest
6. To provide a systematic, efficient pattern for conducting an experiment

Suppose the wheat breeder decided to conduct an experiment to compare the new variety with the standard. The simplest procedure would be to plant one variety in one plot and the other variety in another plot. At harvest he could then observe the difference in yield between the two plots. There is one obvious drawback to this procedure, however. The breeder would have no way to determine how much of his observed difference is due to a real difference between varieties and how much is due to the natural variation found in all biological material.

In the field of experimental design the random variation among experimental units subjected to the same treatment is called *experimental error*. This does not mean that mistakes have been made in conducting the experiment. Rather, it is due to biological variation, soil variation, variation in technique, etc. If the breeder wants to test the significance of the difference between varieties, or to compute an interval estimate of the difference, he must have a measure of experimental error.

1.2 REPLICATION

Although estimates of error may sometimes be obtained from previous experiments, the preferred procedure is to obtain the estimate from the experiment itself. To do this the breeder must repeat, or *replicate*, each variety more than once in his experiment. Replication serves a number of purposes in a designed experiment:

1. It provides an *estimate of experimental error* because it provides several observations on experimental units receiving the same treatment.
2. It *increases precision* by reducing standard errors. Recall that the standard error of a mean, $s_{\bar{y}}$, is given by

$$s_{\bar{y}} = \sqrt{s^2/r}$$

   where s^2 is the sample variance and r is the number of observations (replications) upon which the mean is based. Clearly, as r increases $s_{\bar{y}}$, decreases.
3. It can *broaden the base for making inference*. As replication is increased a wider variety of units can be brought into the experiment. This, in turn, can lead to a greater range of conditions over which the experimental results will apply.

1.3 RANDOMIZATION

Replication is not the only factor which must be considered in designing an experiment. Suppose the breeder had arranged his experiment according to Fig. 1.1. Now, assume that there was a fertility gradient in the field ranging from

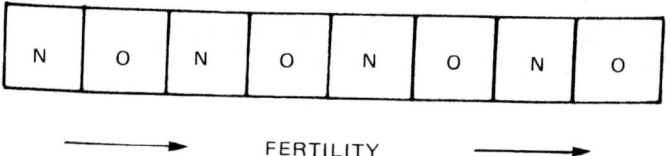

FIG. 1.1 Systematic assignment of a new, *N*, and old *O*, variety to plots in a field with a fertility gradient.

high at one end to low at the other. In each pair of plots in this field plan the new variety has been assigned to the plot with the higher fertility level. A difference between the new and the old varieties will reflect a fertility difference in favor of the new variety as well as a variety difference if it exists. In a good experimental design we want to assign the treatments to the experimental units in such a way that no treatment is consistently favored by being placed under the best conditions. To do this we use a process called *randomization*. By randomization we mean that treatments are assigned to the units in such a way that any unit is equally likely to receive any treatment. In some designs randomization is restricted in certain ways, but in no design is it completely eliminated. We will consider the randomization procedure as we look at the various designs.

There are at least two reasons for randomization in an experimental design:

1. To *eliminate bias*. Randomization ensures that no treatment is favored or discriminated against by systematic assignment to units in a design.
2. To *ensure independence* among the observations. This is necessary to provide valid significance tests and interval estimates.

1.4 BLOCKING

We now need to consider one additional feature of a good experimental design. This is the feature called *blocking*, or local control. The procedure here is to arrange the experimental material into groups, or blocks, of more or less uniform experimental units. Treatments are then assigned at random to the units within the blocks. In the analysis of data from a blocked design the variation among blocks is eliminated from the experimental error and, hence, precision is increased. As an example of blocking consider an experiment to be conducted with two batches of raw material. We could have one block of units selected from one batch and another block of units from the other batch. In this way, treatments would be compared within batches, and differences between batches would not enter the treatment comparisons. Further, such an arrangement would provide some information about the difference between batches of material. We will consider criteria for blocking and the consequences of blocking in further detail as we consider blocked designs.

There are a number of reasons for blocking in a designed experiment:

1. It can *increase the precision* of an experiment. Differences among blocks are removed from the experimental error in the analysis of the results.
2. *Treatments are compared under more nearly equal conditions* because comparisons are made within blocks of uniform units.
3. It can sometimes *increase the information* from an experiment. Blocks need not be placed at the same location nor run at the same time. By placing blocks at different locations, for example, a wider variety of conditions can be sampled with a given experiment.

In summary, a good experimental design takes into consideration the following features:

1. Replication
2. Randomization
3. Blocking

We will consider these features in connection with each of the designs we examine.

1.5   AN OVERVIEW

In broad terms the usual experiment takes the following form: we have a number of alternative treatments, one of which is applied to each experimental unit; an observation (or several observations) is then made on each unit. The object is to separate the treatment effects from the uncontrolled variation among units. Once the treatments, the experimental units, and the nature of the observations have been decided upon, the principal requirements are:

1. Experimental units receiving different treatments should differ in no systematic way from one another.
2. Experimental error should be suitably small, and this should be achieved with as few units as possible.
3. The conclusions should have a wide range of validity.
4. The experimental design should be the simplest one which will provide the desired precision.
5. A proper statistical analysis of the results should be possible without making any artificial assumptions.

We have stated a number of times that one of our goals is to increase accuracy in conducting an experiment, either to reduce the width of interval estimates or to decrease the size of differences required for significance. Accuracy increases as the standard error of a treatment mean, $s_{\bar{y}}$, decreases. Since $s_{\bar{y}} = \sqrt{s^2/r}$ there are a number of ways we can increase accuracy:

1. *Increase the size* of the experiment by increasing replication or by including more treatments. This can be self-defeating, however, if increasing the size necessitates the inclusion of more heterogeneous experimental units.

2. *Proper selection of treatments.* Factorial combinations of treatments (to be discussed later) include the feature of hidden replication for some comparisons.

3. *Refine the experimental technique.* This should result in a reduction of experimental error as estimated by s^2.

4. *Group the experimental units into homogeneous blocks.* To the extent that block differences are removed from experimental error s^2 is reduced.

5. *Measure a concomitant variable.* This will permit a covariance analysis, which may result in reduced experimental error.

When we begin to look at specific experimental designs we will see that the principal difference between them is the way in which the units are classified by the treatments they receive, but in some designs they are further classified into other groupings such as blocks, strips, etc. These groupings generally constitute restrictions on the random assignment of treatments to the units.

Once the experiment has been completed, the resulting data are usually subjected to a procedure called *analysis of variance.* This is a systematic procedure for partitioning the total variation among observations into components, each of which is associated with a source of variation. One of the sources is the variation among units treated alike. This is a measure of experimental error which provides the basis for interval estimates and significance tests. Other sources of variation are such things as treatments, blocks, strips, etc.

The variance or, more correctly, the mean square associated with each of the other sources of variation may be compared with the experimental error mean square. This comparison provides an F statistic for testing the significance of the difference among means for the particular variance source. In addition, the analysis of variance provides information from which standard errors of means and differences may be computed, and from which interval estimates may be constructed. We will outline and illustrate the analysis of variance for each of the experimental designs we consider.

1.6 DEFINITIONS

1. An *experiment* is a planned inquiry to discover new facts, or to confirm or deny the results of previous investigations.

2. A *treatment* is a procedure whose effect on the experimental material is to be measured.
3. A particular class of related treatments is often called a *factor*.
4. The states of a factor, i.e., the treatments within the class, are called the *levels* of the factor.
5. An *experimental unit* is the piece of experimental material to which one trial of a single treatment is applied.
6. A *sampling unit* is that fraction of the experimental unit on which the effect of the treatment is measured.
7. A group of homogeneous experimental units is called a *block*.
8. The quantity which is measured on the experimental material is often called the *yield*.
9. An *experimental design* is a set of rules by which the treatments to be used in an experiment are assigned to the experimental units.
10. *Experimental error* is the variation among experimental units which have been treated alike.
11. If treatments are assigned to a set of units in such a way that every unit is equally likely to receive any treatment, the assignment is said to be *random*.
12. When a treatment appears more than once in an experiment, the treatment is said to be *replicated*.

FURTHER READING*

Anderson, V. L., and R. A. McLean (1974), Ch. 3.
Cochran, W. G., and G. M. Cox (1957), Ch. 1 and 2.
Cox, D. R. (1958), Ch. 1–6.
Federer, W. T. (1955), Ch. 1 and 3.
Steel, R. G. D., and J. H. Torrie (1980), Ch. 6.

---

*Complete citations for the books listed for further reading are given in the Bibliography at the end of the book.

2
Completely Randomized Design

2.1 INTRODUCTION

The simplest, least restrictive design we can use may be developed as follows: Suppose we have p treatments and $n > p$ experimental units to be included in an experiment. We can assign the first treatment to r_1 units randomly selected from among the n, assign the second treatment to r_2 units randomly selected from the remaining $n - r_1$, and continue in this way until the pth treatment is randomly assigned to the final r_p units. The only restriction we make is that before treatments are assigned each unit must have an equal chance of being assigned to any treatment. The experimental design obtained in this way is called a *completely randomized design* (CRD).

2.1.1 Advantages

There are a number of advantages of a completely randomized design:

1. *Flexibility:* Any number of treatments and any number of replications may be used. Further, the number of replications need not be the same from one treatment to another, although comparisons are most precise when the treatments are equally replicated.
2. *Statistical analysis is simple* even with unequal replication, and it is not complicated by loss of data or missing observations.
3. The design provides *maximum degrees of freedom for error.*

2.1.2 Disadvantages

There is one principal disadvantage to the design:

1. Its *precision is low* if the experimental units are not uniform. Often the units can be grouped into blocks of homogeneous units with a resulting increase in precision.

7

2.1.3 Uses

Although other designs may have more precision, the completely randomized design has a number of *uses*:

1. It is most precise if the experimental material is uniform.
2. It is useful when a large fraction of the units may not respond or may be lost during the experiment.
3. It may be useful for experiments in which the total number of units is limited, because it provides maximum degrees of freedom for error.

2.2 RANDOMIZATION

In this design treatments are assigned to the experimental units completely at random. There are a number of ways to do this and we will consider two of them. Suppose we want to conduct an experiment with four treatments each replicated three times. This will require 12 units, which we number from 1 to 12 as in Fig. 2.1. We can now assign treatments to the units in one of the following ways.

2.2.1 By Lot

1. Obtain 12 identical slips of paper. Label three of them "Tmt. A," three of them "Tmt. B," etc.
2. Place the slips in a box (or hat) and mix thoroughly.
3. Pick a piece of paper at random. The treatment named on this piece is assigned to unit 1.
4. Without returning the first slip to the box, select another slip. The treatment named on this piece is assigned to unit 2.
5. Continue this way until all 12 slips of paper have been drawn.

1	2	3	4
5	6	7	8
9	10	11	12

FIG. 2.1 Assignment of numbers to experimental units.

2.2.2 Random Number Table

1. Select a starting point in a table of random numbers (Appendix Table A1) by closing your eyes and placing a finger on the table.
2. Moving up or down, right or left from the starting point, record the first 12 three-digit numbers in sequence. (Three-digit numbers are used because they are less likely to include ties.)
3. Rank the numbers from smallest to largest. These ranks correspond to experimental unit numbers.
4. Assign treatment A to the first three units in sequence, treatment B to the second three units in sequence, etc.

2.2.3 Example

To illustrate the use of the random number table, suppose the following had been obtained:

Sequence	Random number	Rank (unit number)	Treatment
1	448	4	A
2	699	7	A
3	340	3	A
4	733	9	B
5	580	6	B
6	852	12	B
7	514	5	C
8	723	8	C
9	152	2	C
10	744	10	D
11	828	11	D
12	041	1	D

The numbers are listed, in the sequence in which they are drawn, in column 2. They are then ranked in column 3. The ranks are unit numbers. Treatment A is assigned to the first three in sequence, treatment B to the second three in sequence, etc. The final experimental plan is given in Fig. 2.2.

2.3 DATA ANALYSIS

In the analysis of data from a completely randomized design there are a number of things we want to do:

1. Estimate the true mean yields for the p treatments. That is, estimate μ_i.

1	2	3	4
D	C	A	A
5	6	7	8
C	B	A	C
9	10	11	12
B	D	D	B

FIG. 2.2 Experimental plan with treatments assigned at random to experimental units in a completely randomized design.

2. Test the significance of the differences among the treatment means. For this purpose we want to test

$$H_0 : \mu_1 = \mu_2 = \cdots = \mu_p; \text{ all means are equal against}$$

$$H_a : \mu_1 \neq \mu_2 \neq \cdots \neq \mu_p; \text{ at least one mean differs from the others}$$

3. Obtain an estimate of the standard error of a treatment mean or of the difference between two treatment means for the purpose of computing interval estimates.

We will look, first, at the analysis for designs in which the treatments are equally replicated. We will then consider the modifications required for unequal replication.

2.4 EQUAL REPLICATION

Suppose we have p treatments each replicated r times for a total of rp observations. We begin by constructing a table, Table 2.1, of yields, treatment totals, and treatment means, in which

$y_{ij} = $ yield of the jth unit on the ith treatment

$y_{i.} = \sum_j y_{ij} = $ total of yields for the ith treatment

$y_{..} = \sum_i y_{.} = \sum_i \sum_j y_{ij} = $ grand total of all yields

$\bar{y}_i = y_{i.}/r = $ mean yield of the ith treatment

$\bar{\bar{y}} = y_{..}/rp = $ grand mean of all yields

At this point we note that the true mean yield of the ith treatment, μ_i, is estimated by the ith sample mean, $\bar{y}_i$.

TABLE 2.1 Data Table for CRD with Equal Numbers

	Treatment				
	1	2	$\cdots$	p	
	y_{11}	y_{21}	$\cdots$	y_{p1}	
	y_{12}	y_{22}	$\cdots$	y_{p2}	
	$\vdots$	$\vdots$		$\vdots$	
	y_{1r}	y_{2r}	$\cdots$	y_{pr}	Sum
Sum	$y_{1.}$	$y_{2.}$	$\cdots$	$y_{p.}$	$y_{..}$
Mean	$\bar{y}_1$	$\bar{y}_2$	$\cdots$	$\bar{y}_p$	$\bar{\bar{y}}$

TABLE 2.2 ANOVA for CRD

Source of variation	Degrees of freedom	Sum of squares	Mean square	F
Total	$rp - 1$	SSTot		
Treatments	$p - 1$	SST	MST	F_T
Error	$p(r - 1)$	SSE	MSE	

2.4.1 Analysis of Variance

We now set up an analysis of variance table with the format given in Table 2.2. To obtain the entries in this table we note that

1. In this design the total variation is partitioned into two components:
   a. Variation among treatment means (treatments).
   b. Variation among units within treatments (error).
2. The degrees of freedom are one less than the number of observations in each source of variation.
   a. There are rp observations in the experiment so that the total variability has $rp - 1$ degrees of freedom, d.f.
   b. There are p treatments, giving $p - 1$ d.f. for treatment.
   c. Within each treatment there are r observations so that each treatment contributes $r - 1$ d.f. to error, and since there are p treatments there are $p(r - 1)$ d.f. for error. Note, we can also obtain the error d.f. by subtracting the d.f. for treatment from the total d.f.:

$$rp - 1 - (p - 1) = p(r - 1)$$

3. The entries in the sum-of-squares column are obtained as follows:
   a. First compute a correction term C as

   $$C = y_{..}^2/rp \quad \text{then}$$

   b. $\text{SSTot} = \sum_i \sum_j y_{ij}^2 - C$

   c. $\text{SST} = (1/r) \sum_i y_{i.}^2 - C$

   d. $\text{SSE} = \text{SSTot} - \text{SST}$

4. The entries in the mean-square column are obtained by dividing the sums of squares by their associated degrees of freedom. Normally the mean square is not computed on the total line. We have
   a. $\text{MST} = \text{SST}/(p - 1)$
   b. $\text{MSE} = \text{SSE}/p(r - 1)$
5. Finally
   a. $F_T = \text{MST}/\text{MSE}$

Having completed the ANOVA table, how can we use it? First, $\text{MSE} = s^2$ is an unbiased estimate of σ^2, the random variation associated with the experimental units. Further,

$$F_T = \text{MST}/\text{MSE} \quad \text{with} \quad p - 1, p(r - 1) \text{ d.f.}$$

is a test statistic for $H_0 : \mu_1 = \mu_2 = \cdots = \mu_p$ against $H_a : \mu_1 \neq \mu_2 \neq \cdots \neq \mu_p$. If $F_T > F_{\alpha(p-1),p(r-1)}$ we say that the means are significantly different at the $100\alpha\%$ level. Note: $F_{\alpha(p-1),p(r-1)}$ is the one-tailed F at the α probability level with $p - 1$ and $p(r - 1)$ d.f.

2.4.2 Means and Standard Errors

For the completely randomized design an unbiased estimate of the ith treatment mean, $\mu_i = \mu + \tau_i$, is given by the sample mean, $\bar{y}_i = y_{i.}/r$, of the ith treatment.

If we want an interval estimate of μ_i we require the sample standard error of $\bar{y}_i$. This is found to be

$$s_{\bar{y}} = \sqrt{s^2/r} = \sqrt{\text{MSE}/r}$$

where $s_{\bar{y}} =$ the sample standard error of $\bar{y}_i$. Note that $s_{\bar{y}}$ is the same for all means if the treatments are equally replicated. Given the sample standard error the $(1 - \alpha)$ 100% confidence interval estimate, $L(\mu_i)$, of μ_i is given by

$$L(\mu_i) = \bar{y}_i \pm t_\alpha \sqrt{\text{MSE}/r}$$

where t_α is the two-tailed t at the α probability level with $p(r - 1)$ d.f.

Often we are not so much interested in estimating treatment means as we are in the difference between two treatment means. As might be expected, the difference $\mu_i - \mu_{i'}$ between the means of treatments i and i' is given by the difference between the sample means, $\bar{y}_i - \bar{y}_{i'}$. The sample standard error, $s_{\bar{d}} = s(\bar{y}_i - \bar{y}_{i'})$, of this difference is

$$s_{\bar{d}} = \sqrt{2s^2/r} = \sqrt{2MSE/r}$$

Analogous to the case with a single mean, the $(1 - \alpha)100\%$ confidence interval estimate, $L(\mu_i - \mu_{i'})$, of the difference between two means is given by

$$L(\mu_i - \mu_{i'}) = (\bar{y}_i - \bar{y}_{i'}) \pm t_\alpha\sqrt{2MSE/r}$$

Again, t_α is the two-tailed t at the α level with $p(r - 1)$ d.f.

2.4.3 Coefficient of Variation

An additional statistic is usually computed and presented along with the ANOVA table. This is the coefficient of variation, CV, which is computed as

$$CV = (\sqrt{MSE}/\bar{y})100$$

expressed as a percentage. The CV is a measure of relative variation. It has little meaning by itself. Since it is dimensionless, however, it can be used to compare the precision of one measure with another in a single experiment. It can also be used to compare the precision of one experiment with that of another.

2.4.4 Presentation of Results

The results of the analysis of data from a completely randomized design may be conveniently summarized in a table of treatment means along with their standard errors and, possibly, a statement about the significance of the differences. This is shown in Table 2.3.

Ordinarily, the ANOVA table is not presented since its use is to provide the experimental error variance, $s^2 = MSE$, and the F for the test of the significance of differences among treatment means. A numerical example of a completely randomized design and the analysis of the resulting data follows.

TABLE 2.3 Mean Yields

Treatment	1	2	$\cdots$	p	Standard error
Mean	$\bar{y}_1$	$\bar{y}_2$	$\cdots$	$\bar{y}_p$	$\sqrt{MSE/r}$

Note: Differences significant at the $100\alpha\%$ level.

Note that simple t tests can sometimes be used to answer questions which are based on the nature of the treatments.

2.4.5 Numerical Example: Completely Randomized Design

An anthropologist was interested in studying physical differences, if any, among the various races of people inhabiting Hawaii. As a part of her study she obtained a random sample of eight 5-year-old girls from each of three races: Caucasian, Japanese, and Chinese. She made a number of anthropometric measurements on each girl. She wanted to determine whether the Oriental races differ from the Caucasian, and whether the Oriental races differ from each other. The results of the head width measurements are given in Table 2.4.

The statistical analysis begins with the analysis of variance in Table 2.5. The sums of squares in Table 2.5 are computed as

1. Correction term, $C = y_{..}^2/rp = (332.00)^2/24 = 4{,}592.6667$

2. $\text{SSTot} = \sum_i \sum_j y_{ij}^2 - C = 4{,}604.9350 - 4{,}592.6667 = 12.2683$

TABLE 2.4 Head Width (cm)

	Caucasian	Japanese	Chinese	
	14.20	12.85	14.15	
	14.30	13.65	13.90	
	15.00	13.40	13.65	
	14.60	14.20	13.60	
	14.55	12.75	13.20	
	15.15	13.35	13.20	
	14.60	12.50	14.05	
	14.55	12.80	13.80	Total
Sum	116.95	105.50	109.55	332.00
Mean	14.619	13.188	13.694	13.833

TABLE 2.5 ANOVA of Head Width

Source	d.f.	SS	MS	F
Total	23	12.2683		
Race	2	8.4277	4.2139	23.04
Error	21	3.8406	.1829	

3. SST $= (1/r) \sum_i y_{i.}^2 - C = (1/8)(36,808.7550) - 4,592.6667$

$= 8.4277$

4. SSE $=$ SSTot $-$ SST $= 12.2683 - 8.4277 = 3.8406$

Further computations are

CV $= (\sqrt{s^2/\bar{y}})100 = (\sqrt{.1829}/13.83)100 = 3.1\%$

Standard error of a race mean, $s_{\bar{y}}$

$s_{\bar{y}} = \sqrt{s^2/r} = \sqrt{.1829/8} = .1512$

The anthropologist is interested in answers to the following questions:

1. Do head width means differ among the races?

$H_0: \mu_1 = \mu_2 = \mu_3$

$F =$ MST/MSE $= 4.2139/.1829 = 23.04$

$F_{.05(2,21)} = 3.47$

$F_{.01(2,21)} = 5.78$

Since $23.04 > 5.78$, the differences among means are highly significant ($\alpha = .01$).

2. Is there a difference between the Caucasian race and the Oriental races?

$H_0: \mu_1 = (1/2)(\mu_2 + \mu_3)$

$t = [\bar{y}_1 - (1/2)(\bar{y}_2 + \bar{y}_3)]/(s^2\{[1/n_1] + [1/(n_2 + n_3)]\})^{1/2}$

$= [14.619 - (1/2)(13.188 + 13.694)]/$

$(.1829\{[1/8] + [1/(8 + 8)]\})^{1/2}$

$= (1.1780)/.1852 = 6.361$

$t_{.01(21)} = \pm 2.831$. Since $2.831 < 6.361$ the difference between Caucasian and Oriental head width means is highly significant ($\alpha = .01$).

3. Do the Oriental races differ in head width?

$H_0: \mu_2 = \mu_3$

$t = (\bar{y}_2 - \bar{y}_3)/\sqrt{2s^2/r} = (13.188 - 13.694)/\sqrt{(2)(.1829)/8}$

$= -.5060/\sqrt{.0457} = -2.366$

$t_{.05(21)} = \pm 2.080 \qquad t_{.01(21)} = \pm 2.831$

Since $-2.831 < -2.366 < -2.080$ the difference between Japanese and Chinese is significant at the 5% level but not at the 1% level.

Report of statistical analysis.   The mean head width of 5-year-old girls of three different races is summarized in the following table:

Race	Caucasian	Japanese	Chinese	Standard error
Mean width (cm)	14.619	13.188	13.694	.151

Differences among the races were highly significant. Head width of Caucasian girls was about 1.18 cm larger than the average for Oriental girls. Head width of Chinese girls was about 0.50 cm larger than that of Japanese girls, a small, but statistically significant, difference.

2.5   UNEQUAL REPLICATION

If the treatments are not equally replicated, either by design or because units are lost during the experiment, only minor modifications are necessary in the analysis. Suppose that there are r_1 observations on treatment 1, r_2 observations on

TABLE 2.6   Data Table for CRD with Unequal Replication

	Treatment				
	1	2	$\cdots$	p	
	y_{11}	y_{21}	$\cdots$	y_{p1}	
	y_{12}	y_{22}	$\cdots$	y_{p2}	
	$\vdots$	$\vdots$		$\vdots$	
	y_{1r_1}	y_{2r_2}	$\cdots$	y_{pr_p}	Sum
Total	$y_{1.}$	$y_{2.}$	$\cdots$	$y_{p.}$	$y_{..}$
r_i	r_1	r_2	$\cdots$	r_p	$r_.$
Mean	$\bar{y}_1$	$\bar{y}_2$	$\cdots$	$\bar{y}_p$	$\bar{\bar{y}}$

Note: $y_{i.} = \sum\limits_{j=1}^{r_i} y_{ij}$

$y_{..} = \sum\limits_{i=1}^{p} y_{i.}$

$r_. = \sum\limits_{i=1}^{p} r_i$

$\bar{y}_i = y_{i.}/r_i$

$\bar{\bar{y}} = y_{..}/r_.$

treatment 2, . . . , r_p observations on treatment p. The data will take the form shown in Table 2.6.

2.5.1 Analysis of Variance

The analysis of variance table, Table 2.7 takes essentially the same form as for equal replication of treatments. The sums of squares in Table 2.7 are computed as follows:

1. $C = y_{..}^2/r.$

2. $\text{SSTot} = \sum_i \sum_j y_{ij}^2 - C$

3. $\text{SST} = \sum_i (y_{i.}^2/r_i) - C$

4. $\text{SSE} = \text{SSTot} - \text{SST}$

As before, the mean squares are computed by dividing the sums of squares by their associated degrees of freedom.

2.5.2 Standard Errors, Interval Estimates, and Significance Tests

1. *Significance test:* As before, we test

$$H_0{:}\mu_1 = \mu_2 = \cdots = \mu_p$$

against

$$H_a{:}\mu_1 \neq \mu_2 \neq \cdots \neq \mu_p$$

using as a test statistic

$$F_T = \text{MST/MSE} \qquad \text{with} \qquad p - 1, r. - p \quad \text{d.f.}$$

TABLE 2.7 ANOVA of CRD with Unequal Replication

Source of variation	Degrees of freedom	Sum of squares	Mean square	F
Total	$r. - 1$	SSTot		
Treatment	$p - 1$	SST	MST	F_T
Error	$r. - p$	SSE	MSE	

2. *Treatment means:* μ_i, the true mean of the ith treatment, is estimated by

$$\bar{y}_i = y_{i.}/r_i$$

    the sample mean for the ith treatment. Because of unequal replication a separate standard error, $s_{\bar{y}_i}$, must be computed for each treatment mean. We have

$$s_{\bar{y}_i} = \sqrt{s^2/r_i} = \sqrt{\text{MSE}/r_i}$$

    Interval estimates are also computed separately for each treatment:

$$L(\mu_i) = \bar{y}_i \pm t_\alpha \sqrt{\text{MSE}/r_i}$$

    where t_α has $r. - p$ d.f.

3. *Treatment differences:* The difference between two treatment means, $\mu_i - \mu_{i'}$, is estimated by the difference between the sample means, $\bar{y}_i - \bar{y}_{i'}$. Again, the sample standard error is computed separately for each difference. We have

$$s_{(\bar{y}_i - \bar{y}_{i'})} = \left[\left(\frac{1}{r_i} + \frac{1}{r_{i'}} \right) \text{MSE} \right]^{1/2}$$

    The interval estimate of a difference is

$$L(\mu_i - \mu_{i'}) = (\bar{y}_i - \bar{y}_{i'}) \pm t_\alpha \left[\left(\frac{1}{r_i} + \frac{1}{r_{i'}} \right) \text{MSE} \right]^{1/2}$$

    and t_α has $r. - p$ d.f.

2.5.3 Numerical Example: Unequal Replication

An experiment was conducted to compare the yields of five lentil varieties under rainfed conditions in northern Syria. The experiment was planted in a completely randomized design with each variety replicated four times. During the growing season, however, sheep broke through the fence and heavily grazed four plots along the edge of the experiment before they were detected. The yields, in kilograms per hectare, from the remaining plots are given in Table 2.8.

The analysis of variance for the data is given in Table 2.9. The sums of squares in Table 2.9 are computed as:

TABLE 2.8 Yields (kg/ha) of Five Lentil Varieties

	Variety					
	1	2	3	4	5	
	740	545	325	740	605	
	430	440	290	630	505	
	760	390		870	430	
	640				540	Sum
Σ	2,570	1,375	615	2,240	2,080	8,880
r_i	4	3	2	3	4	16
$\bar{y}$	642.5	458.3	307.5	746.7	520.0	555.0

TABLE 2.9 ANOVA of Lentil Yields

Source of variation	Degrees of freedom	Sum of squares	Mean square	F
Total	15	422,700		
Variety	4	296,279	74,070	6.44**
Error	11	126,421	11,493	

**Significant at the 1% level.
$CV = (\sqrt{11,493}/555.0)100 = 19.3\%$.

1. Correction term

$$C = \left(\sum_i \sum_j y_{ij} \right)^2 / \left(\sum_i r_i \right)$$

$$= (8,880)^2/16 = 4,928,400$$

2. $SSTot = \sum_i \sum_j y_{ij}^2 - C$

$$= 740^2 + 430^2 + \cdots + 540^2 - 4,928,400$$

$$= 5,351,100 - 4,928,400 = 422,700$$

3. $SST = \sum_i (y_{i.}^2/r_i) - C$

$$= (2570^2/4) + (1375^2/3) + \cdots + (2080^2/4) - 4,928,400$$

$$= 5,224,679 - 4,928,400 = 296,279$$

4. $SSE = SSTot - SST = 422,700 - 296,279 = 126,421$

Mean squares are computed by dividing the sums of squares by their associated degrees of freedom.

$$F_T = MST/MSE = 74{,}070/11{,}493 = 6.44$$

$$F_{.01(4,11)} = 5.67$$

Since $6.44 > 5.67$ we conclude that the differences among means are significant at the 1% level.

Standard errors must be computed separately for each variety because of unequal replication. In general,

$$s_{\bar{y}} = \sqrt{MSE/r_i}$$

1. For means (Var. 1, 5) with $r_i = 4$

$$s_{\bar{y}} = \sqrt{11{,}493/4} = 53.6$$

2. For means (Var. 2, 4) with $r_i = 3$

$$s_{\bar{y}} = \sqrt{11{,}493/3} = 61.9$$

3. For means (Var. 3) with $r_i = 2$

$$s_{\bar{y}} = \sqrt{11{,}493/2} = 75.8$$

Report of statistical analysis. Analysis of the results of a trial conducted to compare yields among lentil varieties indicates that there is a highly significant difference in yield among the five varieties. The results are summarized in Table 2.10.

Because some of the original plots were accidentally destroyed before harvest the variability in this trial may have been somewhat high. The coefficient of variation was 19.3% compared to the usual range of 10–15% for CVs in other trials of this type.

2.6 FIXED AND RANDOM EFFECTS MODELS

For the completely randomized design the mathematical model which describes an observation, and which provides the basis for the data analysis, depends on the assumptions we make about the treatments we include in the experiment. With this design there are two types of models:

TABLE 2.10 Mean Yields (kg/ha) of Five Lentil Varieties Grown Under Rainfed Conditions in Northern Syria

Variety	1	2	3	4	5
Mean yield	642.5	458.3	307.5	546.7	520.0
Standard error	53.6	61.9	75.8	61.9	53.6

1. Fixed effects models
2. Random effects models

The inferences we can make with the data from a completely randomized design depend on which of the two models is in effect.

2.6.1 Description and Assumptions

Up to this point we have assumed that the treatments used in an experiment include the only ones of possible interest. The goal is to estimate the treatment means and the mean differences among the treatments. If the experiment were to be repeated the same treatments would be included. In such a situation the model is called a *fixed effects model*. It is written

$$y_{ij} = \mu + \tau_i + \epsilon_{ij}$$

in which

y_{ij} = *j*th observation on the *i*th treatment

μ = overall mean of all observations

τ_i = added effect of the *i*th treatment measured as a deviation from μ

ϵ_{ij} = a random error associated with the *j*th observation on the *i*th treatment

The assumptions underlying this model are:

1. The τ_i are fixed, and since they are measured as deviations from μ, $\Sigma_i \tau_i = 0$.
2. The ϵ_{ij} are a random sample from a population which is normally distributed, has a mean of zero, and has a common variance σ^2. This assumption is often symbolized $\epsilon_{ij} \sim N(0, \sigma^2)$.

In other situations the treatments in a particular experiment may be a random sample from a large population of similar treatments. The goal here is to estimate the variation among the treatment means. We are not interested in the means themselves. If the experiment were to be repeated a different sample of treatments would be included. In this situation the model is called a *random effects model*. This model is written

$$y_{ij} = \mu + \tau_i + \epsilon_{ij}$$

where

y_{ij} = the *j*th observation on the *i*th treatment

μ = overall mean of all observations

τ_i = a random effect associated with the *i*th treatment

ϵ_{ij} = a random error associated with the jth observation on the ith treatment

The assumptions underlying this model are:

1. The τ_i are a random sample of τ's from a population which is normally distributed, has a mean of zero, and has a common variance, σ_τ^2. This assumption is symbolized

    $\tau_i \sim N(0, \sigma_\tau^2)$

2. The ϵ_{ij} are random errors, $\epsilon_{ij} \sim N(0, \sigma^2)$.

A few examples might point up the difference between the two models:

1. *Fixed:* A scientist develops three new fungicides. His interest is in these fungicides only.
   Random: A scientist is interested in the way a fungicide works. He selects, at random, three fungicides from a group of similar fungicides to study the action.
2. *Fixed:* Measure the rate of production of five particular machines.
   Random: Choose five machines to represent machines as a class.
3. *Fixed:* Conduct an experiment to obtain information about four specific soil types.
   Random: Select, at random, four soil types to represent all soil types.

Note:

Random effects models are more common in sample surveys. In designed experiments the grouping categories are usually random effects.

Fixed effects models as regards treatment effects are the rule in designed experiments. Unless otherwise specified, we will assume that the treatments in a designed experiment are fixed.

Calculations for the ANOVA are the same under both models. However, the inferences differ from one model to the other. We can see this by considering the expected values of the mean squares, E(MS), under the two models. (Note:

TABLE 2.11 Expected Values of the Mean Squares

			E(MS)	
Source	d.f.	MS	Fixed	Random
Treatment	$p - 1$	MST	$\sigma^2 + \dfrac{r \Sigma_i \tau_i^2}{p - 1}$	$\sigma^2 + r\sigma_\tau^2$
Error	$p(r - 1)$	MSE	σ^2	σ^2

the expected value of a variable is the mean of all possible values of the variable in a population whose members are observations of the variable.) Suppose we have p treatments with r replications per treatment. The expectations under the two models would be as shown in Table 2.11.

2.6.2 Significance Tests

Under the *fixed effects* model we see that MSE estimates σ^2, the variance of ϵ_{ij}, while MST estimates σ^2 plus a function of the sum of squares of the fixed treatment effects. The ratio $F = \text{MST/MSE}$ estimates

$$\left[\sigma^2 + \left(r \sum_i \tau_i^2 \right)/(p - 1) \right]/\sigma^2$$

This ratio follows the F distribution only if $\Sigma_i \tau_i^2 = 0$. Hence, with the fixed effects model F is a test statistic for $H_0{:}\tau_1 = \tau_2 = \cdots = \tau_p = 0$ against $H_a{:}\tau_1 \neq \tau_2 \neq \cdots \neq \tau_p \neq 0$. That is, under the fixed effects model F tests the hypothesis that all means are equal against the alternative that at least one mean differs from the others.

For the *random effects* model MSE is also an estimate of σ^2, the variance of ϵ_{ij}. The treatment mean square, MST, estimates σ^2 plus $r\sigma_\tau^2$, the variance of the population of τ's. The ratio $F = \text{MST/MSE}$ is an estimate of

$$(\sigma^2 + r\sigma_\tau^2)/\sigma^2$$

which follows the F distributon only if $\sigma_\tau^2 = 0$. Hence, under this model F is a test statistic for $H_0{:}\sigma_\tau^2 = 0$ against $H_a{:}\sigma_\tau^2 > 0$. That is, F tests the hypothesis that there is no variation among the treatment means against the alternative that the means vary.

For either model the critical region for F is the α region of the upper tail. The numerator cannot be smaller than the denominator, except by chance, when H_0 is true.

2.6.3 Estimation

Under the fixed effects model our goal is to estimate $\mu_i = \mu + \tau_i$, the true mean of the ith treatment. We have seen that

$$\bar{y}_i = y_i./r_i$$

is an unbiased point estimate of μ_i.

Under the random effects model our goal is to test $H_0{:}\sigma_\tau^2 = 0$ against

$H_a:\sigma_\tau^2 > 0$. However, we might want to estimate the variance components: σ^2 and σ_τ^2. If we consider the expected values of the mean squares we see that

$$\sigma^2 \doteq s^2 = \text{MSE}$$
$$\sigma_\tau^2 \doteq s_t^2 = (\text{MST} - \text{MSE})/r$$

(Note: " $\doteq$ " means "is estimated by.")

2.7 SUBSAMPLING (NESTED DATA, HIERARCHICAL CLASSIFICATION)

In some experimental situations it is inconvenient to measure yield on the entire experimental unit. In this case one or more sampling units (subsamples) are selected at random from each experimental unit. Yield is then measured on each of the subsamples. We can illustrate the idea with a couple of examples:

1. A treatment might consist of a supplement to an individual's (experimental unit) diet, while yield might be the level of a blood constituent. We could not measure the level in all of the blood. Instead, we take two or more blood samples and measure the constituent in each sample.
2. An agronomist, studying the response of a perennial forage to added fertilizer, applied fertilizer treatments to large plots of the forage. At harvest the mass of forage on the entire plot was too great to handle, so he selected three small areas on each plot and measured yield on these.

Whenever the experiment involves subsampling there are two sources of variability in the experimental error:

1. Variation among sampling units within experimental units
2. Variation among experimental units on the same treatment

These two sources of variation must be included both in the mathematical model underlying the data and in the data analysis.

2.7.1 Model and Assumptions

To describe an observation on a subsample we use the model

$$y_{ijk} = \mu + \tau_i + \epsilon_{ij} + \delta_{ijk}$$

where

y_{ijk} = yield of the kth sampling unit from the jth experimental unit on the ith treatment

μ = true overall mean of all observations

τ_i = an effect characteristic of the ith treatment

ϵ_{ij} = random variation of the jth experimental unit on the ith treatment

δ_{ijk} = random variation of the kth sampling unit from the jth experimental unit on the ith treatment.

When subsampling is used in a completely randomized design ϵ and δ are invariably random elements. We assume that

$$\epsilon_{ij} \sim N(0, \sigma_\epsilon^2)$$

and

$$\delta_{ijk} \sim N(0, \sigma^2)$$

As was the case when we did not subsample, the τ_i may be either fixed or random effects. If the τ_i are fixed then they are measured as deviations from μ, and we assume that

$$\sum_i \tau_i = 0$$

In this case our interest is in estimating the true mean of the ith treatment,

$$\mu_i = \mu + \tau_i$$

If the τ_i are random then we assume that they are a random sample from a population of τ's such that

$$\tau_i \sim N(0, \sigma_\tau^2)$$

Here our interest is in testing $H_0: \sigma_\tau^2 = 0$ and in estimating σ_τ^2, the variance of the population of τ's.

2.7.2 Analysis of Variance

The analysis proceeds in much the same way as if sampling had not been done. The ANOVA table, however, contains one additional source of variation: variation among sampling units within experimental units. This source is often called *sampling error*.

Suppose we have a completely randomized design for p treatments each replicated r times. Suppose, also, that we take n samples from each experimental unit. We can define the following totals and means. Let:

y_{ijk} = the yield of the kth sampling unit in the jth experimental unit on the ith treatment

$y_{ij.} = \sum_k y_{ijk}$ = total for jth unit on ith treatment

$$y_{i..} = \sum_j y_{ij.} = \text{total for the } i\text{th treatment}$$

$$y_{...} = \sum_i y_{i..} = \text{grand total}$$

$$\bar{y}_{i..} = y_{i..}/rn = \text{mean of the } i\text{th treatment}$$

$$\bar{\bar{y}} = y_{...}/prn = \text{grand mean}$$

The sample data may be tabulated as in Table 2.12. The format for the analysis of variance table is given in Table 2.13.

TABLE 2.12   Data Table for CRD with Subsampling

Unit	Sample	Treatment 1	2	$\cdots$	p	
1	1	y_{111}	y_{211}	$\cdots$	y_{p11}	
	2	y_{112}	y_{212}	$\cdots$	y_{p12}	
	$\vdots$	$\vdots$	$\vdots$		$\vdots$	
	n	y_{11n}	y_{21n}	$\cdots$	y_{p1n}	
Sum		$y_{11.}$	$y_{21.}$		$y_{p1.}$	
2	1	y_{121}	y_{221}	$\cdots$	y_{p21}	
	2	y_{122}	y_{222}	$\cdots$	y_{p22}	
	$\vdots$	$\vdots$	$\vdots$		$\vdots$	
	n	y_{12n}	y_{22n}	$\cdots$	y_{p2n}	
Sum		$y_{12.}$	$y_{22.}$	$\cdots$	$y_{p2.}$	
	$\vdots$	$\vdots$	$\vdots$		$\vdots$	
r	1	y_{1r1}	y_{2r1}	$\cdots$	y_{pr1}	
	2	y_{1r2}	y_{2r2}	$\cdots$	y_{pr2}	
	$\vdots$	$\vdots$	$\vdots$		$\vdots$	
	n	y_{1rn}	y_{2rn}	$\cdots$	y_{prn}	
Sum		$y_{1r.}$	$y_{2r.}$	$\cdots$	$y_{pr.}$	Grand sum
Grand sum		$y_{1..}$	$y_{2..}$	$\cdots$	$y_{p..}$	$y_{...}$
Mean		$\bar{y}_{1..}$	$\bar{y}_{2..}$	$\cdots$	$\bar{y}_{p..}$	$\bar{\bar{y}}$

TABLE 2.13 ANOVA for CRD with Subsampling

Source	d.f.	SS	MS	F
Total	$prn - 1$	SSTot		
Treatment	$p - 1$	SST	MST	F_T
Experimental error	$p(r - 1)$	SSE	MSE	F_E
Sampling error	$pr(n - 1)$	SSS	MSS	

The sums of squares for Table 2.13 are computed in much the same way as before:

1. Correction term, C

   $$C = y^2_{...}/prn$$

2. $$\text{SSTot} = \sum_i \sum_j \sum_k y^2_{ijk} - C$$

3. $$\text{SST} = (1/rn) \sum_i y^2_{i..} - C$$

4. $$\text{SSE} = (1/n) \sum_i \sum_j y^2_{ij.} - C - \text{SST}$$

5. $$\text{SSS} = \text{SSTot} - \text{SST} - \text{SSE}$$

The mean squares are computed by dividing the sums of squares by their associated degrees of freedom. The F's are

$$F_T = \text{MST/MSE} \qquad p - 1, p(r - 1) \quad \text{d.f.}$$
$$F_E = \text{MSE/MSS} \qquad p(r - 1), pr(n - 1) \quad \text{d.f.}$$

2.7.3 Significance Tests

To see what hypotheses are tested by the F values we consider the expected values of the mean squares under the different assumptions about the τ_i. The expectations are shown in Table 2.14.

Under both models F_E estimates the ratio

$$(\sigma^2 + n\sigma_\epsilon^2)/\sigma^2$$

Hence, under either model F_E is a test statistic for $H_0:\sigma_\epsilon^2 = 0$ against $H_a:\sigma_\epsilon^2 > 0$. That is, F_E tests the significance of the variation among units treated alike.

With the random treatment effects model F_T estimates

TABLE 2.14 Expected Mean Squares

Source	MS	E(MS) τ_i fixed	τ_i random
Treatment	MST	$\sigma^2 + n\sigma_\epsilon^2 + (rn\Sigma_i\tau_i^2)/(p - 1)$	$\sigma^2 + n\sigma_\epsilon^2 + rn\sigma_\tau^2$
Experimental error	MSE	$\sigma^2 + n\sigma_\epsilon^2$	$\sigma^2 + n\sigma_\epsilon^2$
Sampling error	MSS	σ^2	σ^2

$$(\sigma^2 + n\sigma_\epsilon^2 + rn\sigma_\tau^2)/(\sigma^2 + n\sigma_\epsilon^2)$$

This ratio follows the F distribution only if $\sigma_\tau^2 = 0$. Hence, with random treatment effects F_T is a test statistic for $H_0:\sigma_\tau^2 = 0$ against $H_a:\sigma_\tau^2 > 0$, a test of the significance of the variation among treatment effects.

For the fixed treatment effects model F_T estimates

$$[\sigma^2 + n\sigma_\epsilon^2 + (rn \sum_i \tau_i^2)/(p - 1)]/(\sigma^2 + n\sigma_\epsilon^2)$$

This ratio follows the F distribution only if $\Sigma_i\tau_i^2 = 0$. For the fixed effects model, then, F_T is a test statistic for $H_0:\tau_1 = \tau_2 = \cdots = \tau_p = 0$ against $H_a:\tau_1 \neq \tau_2 \neq \cdots \neq \tau_p \neq 0$. Hence, F_T tests the significance of differences among the treatment means.

2.7.4 Estimation

Under both models the variation among sampling units, σ^2, is estimated by

$$\sigma^2 \doteq s^2 = MSS$$

The additional variation among experimental units treated alike, σ_ϵ^2, is estimated as

$$\sigma_\epsilon^2 \doteq s_e^2 = (MSE - MSS)/n$$

If the treatment effects are assumed to be random we might want to estimate σ_τ^2, the variation among the treatment effects. This estimate is obtained as

$$\sigma_\tau^2 \doteq s_t^2 = (MST - MSE)/rn$$

Under the fixed treatment effects model our interest is in the treatment means or in differences among the treatment means. With subsampling, as was the case when subsamples were not taken, the true mean, μ_i, of the ith treatment is estimated by

$$\mu_i = \mu + \tau_i \doteq \bar{y}_{i..} = y_{i..}/rn$$

The variance, $\sigma_{\bar{y}i..}^2$, of this estimated mean involves both the variation among experimental units and the variation among sampling units. We have

$$\sigma^2_{\bar{y}_{i..}} = (\sigma^2 + n\sigma^2_\epsilon)/rn = (\sigma^2/rn) + (\sigma^2_\epsilon/r)$$

The estimated variance, $V(\bar{y}_{i..})$, of a sample mean is obtained from the entries in the ANOVA as

$$V(\bar{y}_{i..}) = (s^2 + ns^2_e)/rn = MSE/rn$$

Assuming, as we do, that both ϵ and δ are normally distributed, the $(1 - \alpha)100\%$ confidence interval estimate of μ_i is given by

$$L(\mu_i) = \bar{y}_{i..} \pm t_\alpha\sqrt{V(\bar{y}_{i..})}$$
$$= \bar{y}_{i..} \pm t_\alpha\sqrt{MSE/rn}$$

in which t_α is the tabular t at the α probability level with $p(r - 1)$ d.f.

If we are interested in the difference, $\mu_i - \mu_{i'}$, between two means our estimate is

$$\mu_i - \mu_{i'} \doteq \bar{y}_{i..} - \bar{y}_{i'..}$$

This estimate has sample variance

$$V(\bar{y}_{i..} - \bar{y}_{i'..}) = 2MSE/rn$$

The $(1 - \alpha)$ 100% confidence interval estimate of the mean difference between two treatments is obtained as

$$L(\mu_i - \mu_{i'}) = (\bar{y}_{i..} - \bar{y}_{i'..}) \pm t_\alpha\sqrt{2MSE/rn}$$

where, again, t_α has $p(r - 1)$ d.f.

2.7.5   Sampling Plans

We can use the data from the ANOVA to see how the variance of a treatment mean might be expected to change under different sampling schemes. We have

$$V(\bar{y}_{i..}) = (s^2 + ns^2_e)/rn = (s^2/rn) + (s^2_e/r)$$

From the mean squares in the ANOVA we can obtain sample values for s^2 and s^2_e as

$$s^2 = MSS$$

and

$$s^2_e = (MSE - MSS)/n$$

These values may then be substituted into the equation for $V(\bar{y}_{i..})$ along with any chosen values for r and n to estimate the variance under these values of r and n. To illustrate, suppose we have run an experiment with $p = 6$ treatments replicated $r = 5$ times with $n = 3$ subsamples per unit. Suppose, further, we have obtained the ANOVA given in Table 2.15.

From the ANOVA we compute

$$s^2 = MSS = 2$$

$$s_e^2 = (MSE - MSS)/n = (14 - 2)/3 = 4$$

Under the present scheme $r = 5$ and $n = 3$, and

$$V(\bar{y}_{i..}) = MSE/rn = 14/(5)(3) = .9333$$

Now, suppose we propose a plan with $r = 3$ replications and $n = 5$ samples per unit. With this plan we would expect the variance of a treatment mean to be on the order of

$$V(\bar{y}_{i..}) = \frac{s^2 + ns_e^2}{rn} = \frac{2 + (5)(4)}{(3)(5)} = \frac{22}{15} = 1.4667$$

Alternatively, suppose $r = 4$ and $n = 4$. Then the expected variance would be

$$V(\bar{y}_{i..}) = \frac{2 + (4)(4)}{(4)(4)} = \frac{18}{16} = 1.1250$$

Note that r is a divisor of both components of $V(\bar{y}_{i..})$ while n is a divisor of only one component. Hence, for a fixed total number, $m = rn$, of observations on a treatment the variance is minimized by setting $n = 1$ and $r = m$. If costs must be considered, however, this will not necessarily produce the most economical sampling scheme.

The accompanying numerical example illustrates the computations involved in a completely randomized design with subsampling. As usual, a suggested "report of statistical analysis" is included. These are presented to emphasize the idea that statistical analyses are conducted solely as an aid to the interpretation of research data. In this regard they serve as an analytical tool, just as does a chemical procedure. Once they have served this purpose, statistical analyses, per se, have no place in the report of the research.

TABLE 2.15 ANOVA

Source	d.f.	MS	E(MS)
Treatment	5	134	$\sigma^2 + 3\sigma_\epsilon^2 + 15\Sigma_i\tau_i^2/5$
Experimental error	24	14	$\sigma^2 + 3\sigma_\epsilon^2$
Sampling error	60	2	σ^2

2.7.6 Numerical Example: Completely Randomized Design with Subsampling

A chemist wanted to measure the ability of three chemicals to retard the spread of fire when used to treat plywood panels. He obtained 12 panels and sprayed 4 of the panels with each of the three chemicals. He then cut two small pieces from each panel and measured the time required for each to be completely consumed in a standard flame. He obtained the results in Table 2.16.

TABLE 2.16 Time (min) for Complete Burn

Panel	Sample	Chemical			
		A	B	C	
1	1	10.3	4.4	3.1	
	2	9.8	4.7	3.3	
Σ		20.1	9.1	6.4	
2	1	5.8	2.7	6.5	
	2	5.4	1.6	5.4	
Σ		11.2	4.3	11.9	
3	1	8.7	4.6	5.1	
	2	10.0	4.0	7.5	
Σ		18.7	8.6	12.6	
4	1	8.9	5.6	5.6	
	2	9.4	3.4	4.2	
Σ		18.3	9.0	9.8	Total
ΣΣ		68.3	31.0	40.7	140.0
Mean		8.54	3.88	5.09	5.83

TABLE 2.17 ANOVA of Burn Time

Source	d.f.	SS	MS	F
Total	23	146.0733		
Chemical	2	93.6308	46.8154	9.68**
Experimental error	9	43.5322	4.8369	6.51**
Sampling error	12	8.9103	.7425	

**Significant at the 1% level.

The analysis of variance of this (nested) set of data is given in Table 2.17. Note: For this data set $p = 3$, $r = 4$, $n = 2$. The sums of squares in Table 2.17 are computed as:

1. Correction term, $C = y_{...}^2/prn = (140.00)^2/24$

    $$= 816.6667$$

2. $\text{SSTot} = \sum_i \sum_j \sum_k y_{ijk}^2 - C = 962.7400 - 816.6667$

    $$= 146.0733$$

3. $\text{SST} = (1/rn) \sum_i y_{i..}^2 - C = [1/(4 \times 2)](7282.38) - 816.6667$

    $$= 93.6308$$

4. $\text{SSE} = (1/n) \sum_i \sum_j y_{ij.}^2 - C - \text{SST}$

    $$= (1/2)(1907.66) - 816.6667 - 93.6308 = 43.5325$$

5. $\text{SSS} = \text{SSTot} - \text{SST} - \text{SSE} = 146.0733 - 93.6308 - 43.5322$
    $$= 8.9102$$

To test the significance of the differences among chemicals, $H_0: \mu_1 = \mu_2 = \mu_3$ against $H_a: \mu_1 \neq \mu_2 \neq \mu_3 \neq$, use

$$F_T = \text{MST/MSE} = 46.8154/4.8369 = 9.68$$

Since $F_{.01(2,9)} = 8.02$ and $9.68 > 8.02$ the differences among chemical means are significant at the 1% level.

To test whether there is significant variation among panels, $H_0: \sigma_\epsilon^2 = 0$ against $H_a: \sigma_\epsilon^2 > 0$, use

$$F_E = \text{MSE/MSS} = 4.8369/.7425 = 6.51$$

Since $F_{.01(9,12)} = 4.39$ and $6.51 > 4.39$ the variation among panels treated alike is significant at the 1% level.

The standard error of a treatment mean, $s_{\bar{y}}$, is

$$s_{\bar{y}} = \sqrt{\text{MSE}/rn} = \sqrt{4.8369/(4 \times 2)} = .7776$$

The mean squares in this analysis have the following expectations, E(MS):

Source	d.f.	E(MS)
Chemical	2	$\sigma^2 + 2\sigma_\epsilon^2 + 8\Sigma\tau_i^2/2$
Experimental error	9	$\sigma^2 + 2\sigma_\epsilon^2$
Sampling error	12	σ^2

TABLE 2.18 Time (min) Required for Complete Burn of Plywood Samples Treated with Three Chemicals

Chemical	A	B	C	Standard error
Mean burn time	8.54	3.88	5.09	0.78

Hence, the estimates of the variance components are:

1. Variation among samples within panels, σ^2

$$\sigma^2 \doteq s^2 = MSS = .7425$$

2. Variation among panels within treatments, σ_ϵ^2

$$\sigma_\epsilon^2 \doteq s_e^2 = \frac{MSE - MSS}{n} = \frac{4.8369 - .7425}{2} = 2.0472$$

3. Variance of a sample mean, $s_{\bar{y}}^2$

$$s_{\bar{y}}^2 = \frac{MSE}{rn} = \frac{s^2 + ns_e^2}{rn} = \frac{s^2}{rn} + \frac{s_e^2}{r} = \frac{.7425}{rn} + \frac{2.0472}{r}$$

Report of statistical analysis. There were highly significant differences among the three chemicals in their ability to retard the spread of fire. The mean time required for a standard flame to completely consume a sample of treated plywood is given in Table 2.18. It is apparent that chemical A is the most effective and chemical B is the least effective fire retardant.

The variation in burn time among samples within panels is relatively small. The variation among panels treated alike, however, is significantly (1% level) greater than zero. To increase precision (reduce standard error) in future experiments, more panels should be tested for each chemical. Increasing the number of panels from four to eight and reducing the number of samples per panel from two to one, for example, would be expected to reduce the standard error from .78 to .59.

FURTHER READING*

Anderson, V. L., and R. A. McLean (1974), Ch. 4.
Cochran, W. A., and G. M. Cox (1957), Ch. 3 and 4.
Federer, W. T. (1955), Ch. 4.
Snedecor, G. W., and W. G. Cochran (1980), Ch. 12 and 13.
Steel, R. G. D., and J. H. Torrie (1980), Ch. 7.

---

*See Bibliography for complete citation.

3
Randomized Block Design

3.1 BLOCKING

Recall that a completely randomized design is characterized by assigning the treatments completely at random to the experimental units. This is always a valid technique although it is seldom a good one. If we know something about the material we are using, it is often possible to group the material into groups, or blocks, of homogeneous units, and then to compare the several treatments on the units within the blocks. In this way differences among blocks can be eliminated from experimental error, with a resulting increase in precision.

There are a number of criteria on which the units may be grouped into blocks. In field experiments, for example, neighboring plots tend to behave more nearly alike than do plots at some distance from each other. The blocks, then, should be composed of compact groups of adjacent plots. In animal experiments the animals of one breed or sex tend to behave more nearly alike than do animals of different breeds or sexes. Here breed or sex could serve as a blocking factor. In other cases time of treatment application might act to differentiate one block from another.

In industrial experiments all of the units in one block might be selected from one batch of raw material while other batches of raw material might serve as sources of experimental units for other blocks. Other blocking factors might include time, proximity, genetic background, etc. In general, any grouping of units into blocks is valid so long as it is done before applying the treatments. Any property of the units which can be determined before the experiment begins can be used as a basis for blocking. Blocking will be effective, however, only if the variance among units within blocks is smaller than the variance over the whole set of units. This leads to a number of general statements as regards blocking:

1. Precision usually decreases as the size, i.e., number of units per block, of the block increases, hence block size should be kept as small as possible.
2. In field experiments with no obvious blocking criteria blocks which are nearly square have been found to be more effective than blocks of other shapes.
3. If gradients, such as slope, fertility, or time trend, exist in an experimental situation, then the blocks should be long and narrow and should be placed perpendicular to the gradient.
4. The key to success in blocking is to minimize the variance among units within blocks while maximizing the variance among blocks.
5. In some disciplines, particularly agriculture, the terms "block" and "replication" are used interchangeably. In the strict sense this is not valid because replication applies only to repetitions of the treatments in an experiment, while blocking implies only grouping of the units whether or not the blocks carry a full set of treatments.

Once the units have been grouped into blocks it is best to handle each block as a unit. Treatments should be applied in one block at a time. Field operations should be completed in one block before moving to another. Similarly, if observations are to be made by more than one individual each should be assigned to a different block.

On the other hand, it is not necessary that blocks be of the same shape. Sometimes it is necessary to use irregular blocks in order to obtain sufficient homogeneous units. Similarly, it is not necessary that all blocks be conducted at the same location or at the same time. Sometimes placing the blocks at different locations will provide additional information on treatment response under varied conditions.

3.2 RANDOMIZED BLOCK DESIGN

The most basic experimental design which includes the blocking feature is the *randomized block design* (RBD), probably the most often used experimental design. To construct a randomized block design for p treatments each replicated r times we require rp experimental units. We first group the units into r blocks of p units each in such a way that units within blocks are as nearly alike as possible. We then assign, at random, the treatments to the units within the blocks, subject to the restriction that each treatment occurs once, and only once, in each block.

3.2.1   Advantages

There are a number of advantages of a randomized block design:

1. Blocking can increase precision by removing one source of variation from experimental error.
2. Any number of blocks and any number of treatments can be used so long as each treatment is replicated the same number of times in each block.
3. Statistical analysis is relatively simple.

3.2.2   Disadvantages

There are a few disadvantages to this design:

1. Missing data can cause some difficulty in the analysis.
2. Assignment of treatments by mistake to units in the wrong block can lead to problems in the analysis.
3. The design is less efficient than others in the presence of more than one source of variation.
4. The efficiency of the design decreases as the number of treatments and, hence, block size increases.
5. If the experimental units are homogeneous the completely randomized design is more efficient.

3.2.3   Uses

The randomized block design has a number of uses:

1. It can remove one source of variation from the experimental error and thus increase precision.
2. It provides unbiased estimates of the means for the blocking categories, providing additional information from the experiment.
3. It provides satisfactory precision in most cases without the use of a more complex design.

3.2.4   Randomization

In the randomized block design the experimental units are first grouped into blocks. Treatments are then randomly assigned to the units within the blocks. A separate randomization is used in each block. To illustrate the procedure, suppose we want to run an experiment with five treatments replicated four times in a field with a fertility gradient. We proceed as follows:

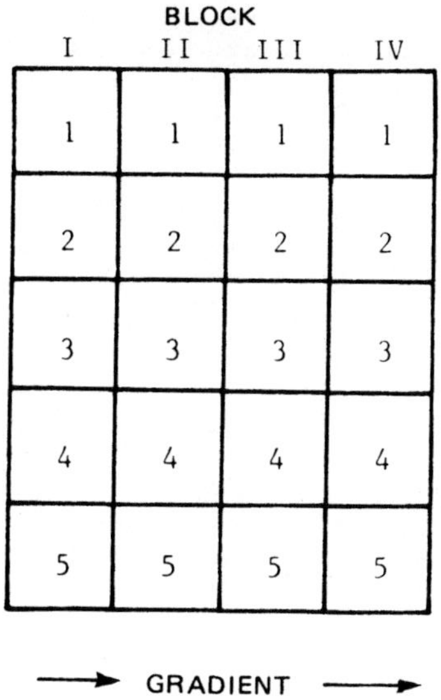

BLOCK

Fig. 3.1 Assignment of numbers to units blocked to remove effects of a gradient.

1. Form four blocks of five plots each perpendicular to the gradient. Number the plots from 1 to 5 within each block as shown in Fig. 3.1.
2. Use a table of random numbers (Appendix, Table A1), or some other procedure, to assign the treatments to the units in the first block. To illustrate for block I:

Sequence	Random number	Rank (plot)	Treatment
1	293	2	A
2	078	1	B
3	721	5	C
4	569	3	D
5	612	4	E

3. Repeat step 2 for the remaining three blocks:

Block II

1	962	4	A
2	036	1	B
3	844	3	C
4	963	5	D
5	097	2	E

Sequence	Random number	Rank (plot)	Treatment
Block III			
1	675	3	A
2	936	5	B
3	709	4	C
4	591	1	D
5	665	2	E
Block IV			
1	230	1	A
2	981	5	B
3	687	4	C
4	604	3	D
5	454	2	E

The final field plan is given in Fig. 3.2.

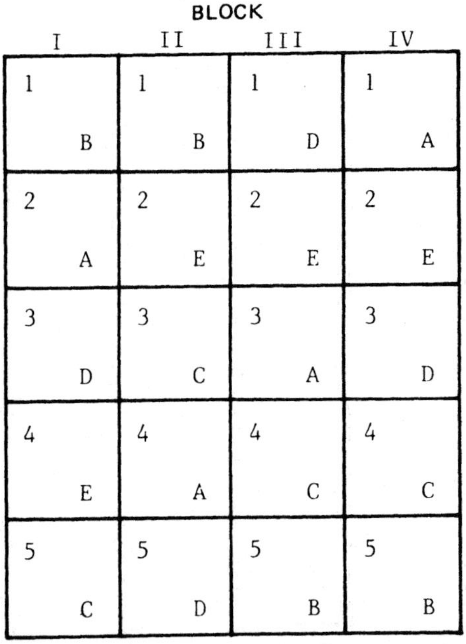

Fig. 3.2 Final experimental plan with treatments randomly assigned to units within blocks in a randomized block design.

3.2.5 Analysis of Variance

Suppose we have p treatments replicated r times in a randomized block design with r blocks of p units each. Let

y_{ij} = yield of the jth treatment in the ith block

TABLE 3.1 Data Table for Randomized Block Design

Block	Treatment 1	2	$\cdots$	p	Sum	Mean
1	y_{11}	y_{12}	$\cdots$	y_{1p}	$y_{1.}$	$\bar{y}_{1.}$
2	y_{21}	y_{22}	$\cdots$	y_{2p}	$y_{2.}$	$\bar{y}_{2.}$
$\vdots$	$\vdots$	$\vdots$		$\vdots$	$\vdots$	$\vdots$
r	y_{r1}	y_{r2}	$\cdots$	y_{rp}	$y_{r.}$	$\bar{y}_{r.}$
Sum	$y_{.1}$	$y_{.2}$	$\cdots$	$y_{.p}$	$y_{..}$	
Mean	$\bar{y}_{.1}$	$\bar{y}_{.2}$	$\cdots$	$\bar{y}_{.p}$	$\bar{\bar{y}}$	

We begin the analysis by constructing a table, Table 3.1, of yields, totals, and means. In Table 3.1 the symbols are defined as:

Totals

$$y_{i.} = \sum_j y_{ij} \quad , \qquad \bar{y}_{i.} = y_{i.}/p \qquad \text{Block}$$

$$y_{.j} = \sum_i y_{ij} \quad , \qquad \bar{y}_{.j} = y_{.j}/r \qquad \text{Treatment}$$

$$y_{..} = \sum_i y_{i.} = \sum_j y_{.j} \quad , \qquad \bar{\bar{y}} = y_{..}/rp \qquad \text{Total}$$

Means

The format for the analysis of variance table is the same as that for the completely randomized design except that a line is added for variation among blocks. This is shown in Table 3.2.

The computations in Table 3.2 proceed in much the same way as for the completely randomized design:

1. Sums of squares:

   a. First compute a correction term, $C = y_{..}^2/rp$; then

TABLE 3.2 ANOVA for RBD

Source	d.f.	SS	MS	F
Total	$rp - 1$	SSTot		
Block	$r - 1$	SSR	MSR	F_R
Treatment	$p - 1$	SST	MST	F_T
Error	$(r - 1)(p - 1)$	SSE	MSE	

b. $\text{SSTot} = \sum_i \sum_j y_{ij}^2 - C$

c. $\text{SSR} = (1/p) \sum_i y_{i.}^2 - C$

d. $\text{SST} = (1/r) \sum_j y_{.j}^2 - C$

e. $\text{SSE} = \text{SSTot} - \text{SSR} - \text{SST}$

2. Mean Squares:
    a. $\text{MSR} = \text{SSR}/(r - 1)$
    b. $\text{MST} = \text{SST}/(p - 1)$
    c. $\text{MSE} = \text{SSE}/(r - 1)(p - 1)$
3. F ratios:
    a. $F_R = \text{MSR}/\text{MSE}$; $(r - 1)$, $(r - 1)(p - 1)$ d.f.
    b. $F_T = \text{MST}/\text{MSE}$; $(p - 1)$, $(r - 1)(p - 1)$ d.f.
4. Coefficient of variation:

$$\%\text{CV} = \left(\frac{\sqrt{\text{MSE}}}{\bar{\bar{y}}} \right) 100$$

3.2.6 Significance Tests

The hypothesis tested by F_T depends on whether the treatments are fixed or random effects. In the usual designed experiment the treatments are fixed. In this case F_T tests $H_0{:}\tau_1 = \tau_2 = \cdots = \tau_p = 0$ against $H_a{:}\tau_1 \neq \tau_2 \neq \cdots \neq \tau_p \neq 0$. This is a test of the significance of the differences among the treatment means. On the other hand, if the τ_i are random, then F_T tests $H_0{:}\sigma_\tau^2 = 0$ against $H_a{:}\sigma_\tau^2 > 0$.

In the usual situation we use blocks as an error control device. That is, we group the units in order to remove interblock variance from experimental error. Note that while treatments are replicated the blocking classifications are not. Hence we have no true error for testing blocks, and F_R provides, at best, only an approximate test. If F_R exceeds F_α we can conclude that blocking has been effective in reducing error, but we cannot put a true significance level on our conclusion. If F_R does not exceed F_α we conclude that we have gained nothing by blocking. In effect, we have decreased efficiency because we lose $r - 1$ d.f. from error without a compensating reduction in our error mean square. In addition, since we deliberately try to maximize the interblock variance by making block means as different as possible, it seems pointless to test the hypothesis that they are equal.

3.2.7 Estimates and Variances

In the randomized block design, as in the completely randomized design, the unbiased estimate of the true treatment mean is

$$\mu_j = \mu + \tau_j \doteq \bar{y}_{.j} = y_{.j}/r$$

That is, the true treatment mean is estimated by the sample mean. The sample standard error, $s_{\bar{y}}$, of this estimate is

$$s_{\bar{y}} = \sqrt{MSE/r}$$

Since, in this design, all of the treatments are equally replicated a single standard error applies to all treatment means. Given the sample standard error the $(1 - \alpha)$ 100% confidence interval estimate, $L(\mu_j)$ of the jth treatment mean is obtained as

$$L(\mu_j) = \bar{y}_{.j} \pm t_\alpha \sqrt{MSE/r}$$

where t_α is the two-tailed t at the α probability level with $(r - 1)(p - 1)$ d.f.

Normally, in a designed experiment we will be concerned about the difference between treatment means rather than about the means themselves. The mean difference, $\mu_j - \mu_{j'}$, between treatments j and j' is estimated as

$$\mu_j - \mu_{j'} \doteq \bar{y}_{.j} - \bar{y}_{.j'}$$

the difference between the sample means. The sample standard error of the estimate difference is

$$s_{(\bar{y}_{.j} - \bar{y}_{.j'})} = \sqrt{2MSE/r}$$

Finally, the $(1 - \alpha)100\%$ confidence interval estimate, $L(\mu_j - \mu_{j'})$, of a mean difference is

$$L(\mu_j - \mu_{j'}) = (\bar{y}_{.j} - \bar{y}_{.j'}) \pm t_\alpha \sqrt{2MSE/r}$$

where, again, t_α has $(r - 1)(p - 1)$ d.f.

3.2.8 Presentation of Results

The results of the analysis of data from a randomized block design may be concisely summarized in a table of treatment means along with the standard error and a statement about the significance of the differences. This is shown in Table 3.3.

With this design it is also possible to present a table of means for the blocking factor. Because there is no estimate of error for these means, however, no standard error or statement of significance is appropriate. This table has the format of Table 3.4.

TABLE 3.3 Treatment Mean Yields

Treatment	1	2	$\cdots$	p	Standard error
Mean	$\bar{y}_{.1}$	$\bar{y}_{.2}$	$\cdots$	$\bar{y}_{.p}$	$\sqrt{MSE/r}$

Note: Differences significant at the $100\alpha\%$ level.

TABLE 3.4 Block Mean Yields

Block	1	2	$\cdots$	r
Mean	$\bar{y}_{1.}$	$\bar{y}_{2.}$	$\cdots$	$\bar{y}_{r.}$

3.3 MODEL AND ASSUMPTIONS

The linear model which describes an observation from a randomized block design, and on which the analysis is based, is

$$y_{ij} = \mu + \beta_i + \tau_j + \epsilon_{ij}$$

in which

y_{ij} = yield of the jth treatment in the ith block

μ = overall mean yield of all observations

β_i = added effect characteristic of the ith block

τ_j = added effect of the jth treatment

ϵ_{ij} = random error associated with the unit on the jth treatment in the ith block

We make the following assumptions:

1. Blocks are almost always used for error control, hence β_i are random effects; $\beta_i \sim N(0, \sigma_\beta^2)$.
2. In designed experiments the treatments are usually fixed. The τ_j are measured as a deviation from μ and $\Sigma_j \tau_j = 0$.
3. Occasionally the τ_j are random, in which case it is assumed that $\tau_j \sim N(0, \sigma_\tau^2)$.
4. ϵ_{ij} is always random, hence $\epsilon_{ij} \sim N(0, \sigma^2)$.

Note also that

1. $\mu_i = \mu + \beta_i$ = true mean for units in the ith block
2. $\mu_j = \mu + \tau_j$ = true mean for units on the jth treatment

3.4 MISSING VALUES

The analysis of variance algorithms for analyzing data from a randomized block, or higher order, design are correct only if the treatments are equally replicated in all of the blocks and if all of the blocks contain the same number of experimental units. If loss of data causes a great deal of imbalance among treatments or blocks, the only theoretically feasible way to handle the remaining data involves a least squares analysis with multiple regression techniques on a computer. There are some remedial measures which can be taken, however, if only a few observations are missing.

If only one value is missing the procedure is to calculate a substitute value, $\hat{y}_{ij}$, using the formula

$$\hat{y}_{ij} = \frac{ry_{i.} + py_{.j} - y_{..}}{(r - 1)(p - 1)}$$

in which

$y_{i.}$ = the sum of the remaining values in the block with the missing observation

$y_{.j}$ = the sum of the remaining observations on the treatment with the missing value

$y_{..}$ = grand total of all available observations

The value $\hat{y}_{ij}$ is substituted and the ANOVA is carried out as usual. Because of the substitution, however, one degree of freedom is lost from both the total and the error lines in the ANOVA. It should also be noted that the substitute value adds no information to the experiment. Its only purpose is to provide a valid analysis of variance. It should not be used in the computation of means.

When two or three values, say a, b, c, are missing, we first guess a value for b and c and use the formula to find an approximation for a. With this value for a and a guessed value for c we use the formula to solve for b. Then with these two values we use the equation to find an approximation for c. This cycle is repeated until the new values do not differ appreciably from previous ones. Then the ANOVA is computed with these values. One degree of freedom for each substituted value must be subtracted from the total and error lines in the ANOVA.

When all of the missing values are in the same block or treatment the simplest solution is to act as if the block or treatment had not been included in the experiment. The ANOVA is run with the entire block or treatment omitted.

3.5 RELATIVE EFFICIENCY

Sometimes it is of interest to see how much precision has been gained by using a randomized block design in place of the simpler, less restrictive completely randomized design. We can get some information about whether blocking has been effective by using the approximate F test with $F_R = $ MSR/MSE as a test statistic. A more informative procedure is to compute the relative efficiency of the randomized block design compared to the completely randomized design had it been used for the same experiment. The relative efficiency, RE, is computed as

$$RE = \frac{(r - 1)MSR + r(p - 1)MSE}{(rp - 1)MSE}$$

If blocking has been effective in increasing precision, then RE will be greater than one. The quantity $(RE - 1)100$ measures the percentage increase in precision due to blocking.

The accompanying numerical example illustrates the computations involved in the analysis of data from a randomized block design. Also illustrated is the computation of relative efficiency and the analysis when an observation is missing. As is customary, an example of the type of report which might be developed from the statistical analysis is also included.

3.6 NUMERICAL EXAMPLE: RANDOMIZED BLOCK DESIGN

An agronomist wanted to compare the effect of five different sources of nitrogen on the dry matter yield of barley used as a forage crop. The five sources were:

1.	$(NH_4)_2SO_4$	4.	$Ca(NO_3)_2$
2.	NH_4NO_3	5.	$NaNO_3$
3.	$CO(NH_2)_2$	6.	Control (no N)

He also decided to use a control with no nitrogen. Because he wanted the results to apply over a range of conditions, he decided to conduct the experiment on four types of soil. For his experimental design he chose a randomized block design with soil type as the blocking factor. He located six plots on each of the four soil types, then randomly assigned the treatments to the plots within types. At maturity he clipped each plot and measured the dry matter of forage produced. He obtained the yields (kilograms per plot) shown in Table 3.5.

The analysis begins with the analysis of variance in Table 3.6. Note: In this experiment $r = 4$, $p = 6$. The sums of squares in Table 3.6 are computed as:

TABLE 3.5 Barley Yields Under Six Nitrogen Treatments

Treatment	Soil type				Total	Mean
	I	II	III	IV		
1	32.1	35.6	41.9	35.4	145.0	36.25
2	30.1	31.5	37.1	30.8	129.5	32.38
3	25.4	27.4	33.8	31.1	117.7	29.42
4	24.1	33.0	35.6	31.4	124.1	31.02
5	26.1	31.0	33.8	31.9	122.8	30.70
6	23.2	24.8	26.7	26.7	101.4	25.35
Total	161.0	183.3	208.9	187.3	740.5	
Mean	26.83	30.55	34.82	31.22		30.85

TABLE 3.6 ANOVA of Forage Yields

Source	d.f.	SS	MS	F
Total	23	492.36		
Soils (blocks)	3	192.55	64.18	
N source	5	255.28	51.06	17.19**
Error	15	44.53	2.97	

**Significant at the 1% level.

$$CV = \left(\frac{\sqrt{2.97}}{30.85}\right) 100 = 5.6\%.$$

1. $C = y_{..}^2/rp = (740.5)^2/(4 \times 6) = 22,847.51$

2. $SSTot = \sum_i \sum_j y_{ij}^2 - C = 23,339.87 - 22,847.51 = 492.36$

3. $SSR = (1/p)\sum_i y_{i.}^2 - C = (1/6)(138,240.39) - 22,847.51 = 192.55$

4. $SST = (1/r)\sum_j y_{.j}^2 - C = (1/4)(92,411.15) - 22,847.51 = 255.28$

5. $SSE = SSTot - SSR - SST = 492.36 - 192.55 - 255.28 = 44.53$

1. Standard error of a treatment mean, $s_{\bar{y}_i}$

    $$s_{\bar{y}_i} = \sqrt{MSE/r} = \sqrt{(2.97)/4} = 0.86$$

TABLE 3.7 Yield of Forage with One Observation Missing

Treatment	Soil type				Total
	I	II	III	IV	
1	32.1	35.6	41.9	35.4	145.0
2	30.1	31.5	X	30.8	92.4
3	25.4	27.4	33.8	31.1	117.7
4	24.1	33.0	35.6	31.4	124.1
5	26.1	31.0	33.8	31.9	122.8
6	23.2	24.8	26.7	26.7	101.4
Total	161.0	183.3	171.8	187.3	703.4

2. Difference between two treatment means, $s_{(\bar{y}_i - \bar{y}_{i'})}$

$$s_{(\bar{y}_i - \bar{y}_{i'})} = \sqrt{2MSE/r} = \sqrt{(2)(2.97)/4} = 1.22$$

$$RE = \frac{(r-1)MSR + r(p-1)MSE}{(rp-1)(MSE)} = \frac{(3)(64.18) + (4)(5)(2.97)}{(23)(2.97)}$$
$$= 3.69$$

We can use this example to illustrate the computations involved when an observation is missing from a randomized block design. Suppose that the plot treated with NH_4NO_3 on soil type III was missing. The data in Table 3.7 would remain.

We use the missing value formula to obtain a substitute value for $\hat{y}_{32}$.

$$\hat{y}_{ij} = \frac{ry_{i.} + py_{.j} - y_{..}}{(r-1)(p-1)} = \frac{(4)(171.8) + (6)(92.4) - 703.4}{(4-1)(6-1)} = 35.9 = \hat{y}_{32}$$

We substitute this for the missing value and obtain the ANOVA (Table 3.8).

Report of statistical analysis. Analysis of the yield data from an experiment to determine the effect of different sources of nitrogen on the forage dry matter

TABLE 3.8 ANOVA of Yields with One Missing Value

Source	d.f.	SS	MS	F
Total	22	478.75		
Block	3	183.22	61.07	
Treatment	5	251.93	50.38	16.18
Error	14	43.60	3.11	

TABLE 3.9 Mean Yield (kg/plot) of Dry Matter from Barley Treated with Different Nitrogen Sources

	Source of N					
	$(NH_4)_2SO_4$	NH_4NO_3	$CO(NH_2)_2$	$Ca(NO_3)_2$	$NaNO_3$	None
Mean Yield	36.25	32.38	29.42	31.02	30.70	25.35
Standard error $= 0.86$						

produced by barley indicates highly significant differences among sources. Ammonium sulfate, $(NH_4)_2SO_4$, produced the highest mean yield, 36.25 kg/plot, while $CO(NH_2)_2$ produced the lowest, 29.42 kg/plot, among the nitrogen treatments. When no nitrogen was added the mean yield was only 25.35 kg/plot. These results are summarized in Table 3.9.

Blocking was effective in increasing the precision of this experiment. The relative efficiency of the randomized block design compared to a completely randomized design in the same area was 3.69. Blocking by soil type increased the efficiency by 269%. If soil type had been ignored we would have required $(3.69)(4) = 14.8$ or about 15 replications with a completely randomized design to attain the same precision as provided by 4 replications of the randomized block design.

FURTHER READING*

Anderson, V. L., and R. A. McLean (1974), Ch. 5.
Cochran, W. G., and G. M. Cox (1957), Ch. 4.
Federer, W. T. (1955), Ch. 5.
Snedecor, G. W., and W. G. Cochran (1980), Ch. 14.
Steel, R. G. D., and J. H. Torrie (1980), Ch. 9.

---

*See Bibliography for complete citation.

4

More Restrictive Designs

4.1 INTRODUCTION

In the randomized block design we control one source of variation by grouping our experimental units into blocks of units which are more or less homogeneous with respect to that variance source. Differences among blocks are then removed from experimental error. The requirement that all treatments must appear equally often in every block constitutes one restriction in the assignment of treatments to the experimental units. In some cases we might like to control two or more sources of variation. We cannot do this with a randomized block design unless we use one source as a blocking factor and the other sources as concomitant variables in a covariance analysis.

4.2 LATIN SQUARE

There are experimental designs, however, which allow us to control two or more sources under proper grouping of the experimental units. One design which controls two sources of variation is called a *Latin square*. To construct a Latin square design for p treatments we require p^2 experimental units. These units are first classified into p groups, of p units each, based on one of the sources of variation. This is commonly called the *row* classification. The units are then classified into p groups, of p units each, based on the second source of variation. This is commonly called the *column* classification. Treatments are then assigned in such a way that each treatment occurs once, and only once, in each row and column.

The basic pattern of a Latin square can be seen by considering a design, shown in Fig. 4.1, for $p = 4$ treatments, A, B, C, and D, in a 4×4 square. Note that in Fig. 4.1 each row and column form a complete block, and that each

	COLUMN			
ROW	1	2	3	4
1	A	B	C	D
2	B	C	D	A
3	C	D	A	B
4	D	A	B	C

Fig. 4.1 Basic design for a 4 × 4 Latin square.

treatment occurs once, and only once, in each row and column. The basic Latin square for any number, p, of treatments is easy to construct:

1. Designate the treatments by the first p letters of the alphabet.
2. Assign the letters in alphabetic order, beginning with A, to units in the first row.
3. In the second and succeeding rows, shift the letters one unit to the left in each succeeding row.

4.2.1 Advantages

There is one principal advantage of the Latin square, and this constitutes the primary reason for its *use*:

1. It allows the experimenter to control two sources of variation.

4.2.2 Disadvantages

The Latin square has a number of disadvantages:

1. It requires p^2 units to study p treatments. This limits the number of treatments, from a practical standpoint, to 10 or fewer.
2. As p increases the experimental error per unit is likely to increase.
3. If p is small the d.f. for error becomes very small.
4. The analysis becomes very complicated if there are missing data or if treatments are misassigned.

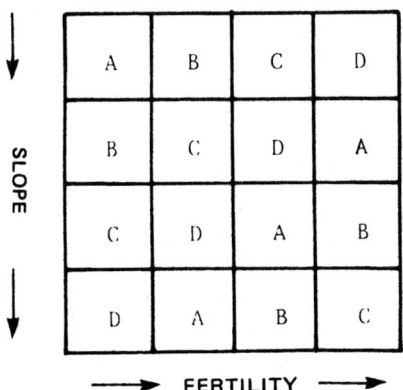

Fig. 4.2 Latin square in a square field plan to remove the effects of a slope and a fertility gradient.

4.2.3 Field Layout

In field experiments the Latin square could be used to remove the effect of two gradients, say slope and fertility, at right angles to each other. In this case the best field plan would be a square with plots themselves which are nearly square, as shown in Fig. 4.2.

Although textbook plans seem to indicate that a square field plan is required for a Latin square design, this is not necessarily so. Consider the construction of a design, shown in Fig. 4.3, for an experiment with a time factor. Suppose there is a time trend affecting the units, but suppose that the experiment involves the application of treatments at succeeding time periods. We can use the Latin square to advantage in this situation by using time periods as the "row" grouping and order within time periods as the "column" grouping. This would remove the effect of the time trend more effectively than would time periods alone.

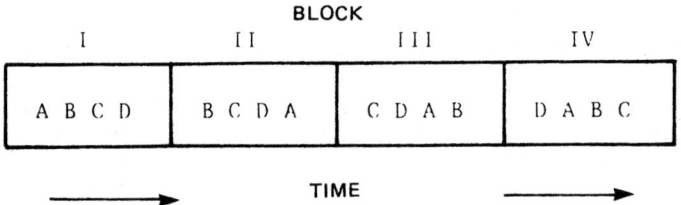

Fig. 4.3 Latin square arranged as a strip to remove a time trend. Blocks serve as rows and order within block serves as columns.

4.2.4 Randomization

To develop a randomized field plan we begin with a basic Latin square of the size required. Then, using a table of random numbers or some other means, we randomize the order of rows. The work plan is then constructed by randomizing the order of columns. For example, suppose we want a work plan for a Latin square with five treatments:

1. Start with the basic plan:

    ```
    A  B  C  D  E
    B  C  D  E  A
    C  D  E  A  B
    D  E  A  B  C
    E  A  B  C  D
    ```

2. Obtain a random order for rows:

Random number	Rank (order)	Row
214	2	1
130	1	2
710	5	3
331	3	4
457	4	5

    This gives the following row randomization:

    ```
    B  C  D  E  A
    A  B  C  D  E
    E  A  B  C  D
    C  D  E  A  B
    D  E  A  B  C
    ```

3. Use the same procedure to randomize the order for columns:

Random number	Rank (order)	Column
772	5	1
476	4	2
425	2	3
463	3	4
257	1	5

    The final work plan after column randomization is:

    ```
    A  E  C  D  B
    E  D  B  C  A
    D  C  A  B  E
    B  A  D  E  C
    C  B  E  A  D
    ```

Note that after randomization of both rows and columns the plan still maintains the property that each treatment occurs once, and only once, in each row and column.

4.2.5 Analysis of Variance

The analysis of data from a Latin square design for p treatments in a $p \times p$ square proceeds in much the same way as previously described analyses. Let

$y_{ij(k)}$ = yield of the experimental unit in the ith row and jth column, subjected to the kth treatment

We first set up a table, Table 4.1, of observations and totals. In Table 4.1 the symbols are defined as

$$y_{i.} = \sum_j y_{ij} = \text{sum of observations in the } i\text{th row}$$

$$y_{.j} = \sum_i y_{ij} = \text{sum of observations in the } j\text{th column}$$

$$y_{..} = \sum_i y_{i.} = \sum_j y_{.j} = \text{grand total of all observations}$$

We also require a table, Table 4.2, of treatment totals and means where

T_k = sum of observations on the kth treatment

$\bar{y}_k = (T_k)/p$ = sample mean for the kth treatment

The format for the analysis of variance is given in Table 4.3.

TABLE 4.1   Data Table for Latin Square

Row	Column				Sum
	1	2	$\cdots$	p	
1	y_{11}	y_{12}	$\cdots$	y_{1p}	$y_{1.}$
2	y_{21}	y_{22}	$\cdots$	y_{2p}	$y_{2.}$
$\vdots$	$\vdots$	$\vdots$		$\vdots$	$\vdots$
p	y_{p1}	y_{p2}	$\cdots$	y_{pp}	$y_{p.}$
Sum	$y_{.1}$	$y_{.2}$	$\cdots$	$y_{.p}$	$y_{..}$

TABLE 4.2 Table of Treatment Totals and
Means for Latin Square

	Treatment				
	1	2	$\cdots$	p	Sum
Sum	T_1	T_2	$\cdots$	T_p	$y_{..}$
Mean	$\bar{y}_1$	$\bar{y}_2$	$\cdots$	$\bar{y}_p$	$\bar{y}$

TABLE 4.3 ANOVA for Latin Square Design

Source	d.f.	SS	MS	F
Total	$p^2 - 1$	SSTot		
Rows	$p - 1$	SSR	MSR	F_R
Columns	$p - 1$	SSC	MSC	F_C
Treatment	$p - 1$	SST	MST	F_T
Error	$(p - 1)(p - 2)$	SSE	MSE	

The computations for the entries in the sum-of-squares column in Table 4.3 are

1. First compute a correction term, C, as

$$C = y_{..}^2/p^2$$

then

2. $\text{SSTot} = \sum_i \sum_j y_{ij}^2 - C$

3. $\text{SSR} = (1/p) \sum_i y_{i.}^2 - C$

4. $\text{SSC} = (1/p) \sum_j y_{.j}^2 - C$

5. $\text{SST} = (1/p) \sum_k T_k^2 - C$

6. $\text{SSE} = \text{SSTot} - \text{SSR} - \text{SSC} - \text{SST}$

Mean squares are, as before, obtained by dividing the sums of squares by their associated degrees of freedom. The F ratios are computed by dividing the appropriate mean squares by MSE. Finally, $s^2 = \text{MSE}$ is an unbiased estimate of σ^2.

4.2.6   Significance Tests

Before looking at the significance tests we should first consider the model and the assumptions underlying it. The model which describes an observation from a Latin square design is

$$y_{ij(k)} = \mu + \rho_i + \gamma_j + \tau_{(k)} + \epsilon_{ij}$$

in which

> $y_{ij(k)}$ = yield of the unit in the ith row, jth column subjected to the kth treatment. Note: The subscript for treatments is enclosed in parentheses because an observation is uniquely identified by its row-column position.

> μ = overall mean yield.

> ρ_i = an added effect common to the ith row.

> γ_j = an added effect common to the jth column.

> $\tau_{(k)}$ = an added effect of the ith treatment.

> ϵ_{ij} = random error, $\epsilon_{ij} \sim N(0,\sigma^2)$

In the Latin square rows and columns are invariably used for error control. Hence, these are random effects and we make the following assumptions:

$$\rho_i \sim N(0,\sigma_\rho^2)$$

$$\gamma_j \sim N(0,\sigma_\gamma^2)$$

Treatment effects may be fixed (the usual case) or random. For fixed treatments we assume that the $\tau_{(k)}$ are measured as deviations from μ, and that

$$\sum_k \tau_{(k)} = 0$$

If the treatments are random we assume that

$$\tau_{(k)} \sim N(0,\sigma_\tau^2)$$

Under the usual assumptions (ρ_i, γ_j random, $\tau_{(k)}$ fixed), the mean squares in the Latin square have the expectations shown in Table 4.4.

On considering the expected mean squares it is clear that $F_T = \text{MST/MSE}$, with $p - 1$, $(p - 1)(p - 2)$ d.f., is a test statistic for

$$H_0{:}\tau_1 = \tau_2 = \cdots = \tau_p = 0$$

against

$$H_a{:}\tau_1 \neq \tau_2 \neq \cdots \neq \tau_p \neq 0$$

TABLE 4.4 Expected Values of the Mean Squares

Source	MS	E(MS)
Row	MSR	$\sigma^2 + p\sigma_\rho^2$
Column	MSC	$\sigma^2 + p\sigma_\gamma^2$
Treatment	MST	$\sigma^2 + p\Sigma_k\tau_k^2/(p - 1)$
Error	MSE	σ^2

Analogous to the situation in the randomized block design, treatments are replicated in the Latin square but row and column classes are not. Thus, $F_R = $ MSR/MSE and $F_C = $ MSC/MSE provide only approximate significance test statistics. They can be used to give an indication of whether the row classification, column classification, or both, were effective in reducing the experimental error.

4.2.7 Estimates and Standard Errors

The point estimate of a treatment mean, μ_k, is

$$\mu_k \doteq \bar{y}_k = T_k/p$$

The sample variance, $V(\bar{y}_k)$, of this estimate is found to be

$$V(\bar{y}_k) = s^2/p = \text{MSE}/p$$

Hence, the sample standard error, $s_{\bar{y}_k}$, of a treatment mean is

$$s_{\bar{y}_k} = \sqrt{\text{MSE}/p}$$

Under the normality assumption about ϵ_{ij} the $(1 - \alpha)100\%$ confidence interval estimate, $L(\mu_k)$, of a treatment mean is computed as

$$L(\mu_k) = \bar{y}_k \pm t_\alpha\sqrt{\text{MSE}/p}$$

where t_α has $(p - 1)(p - 2)$ degrees of freedom.

The difference between two treatment means, $\mu_k - \mu_{k'}$, is estimated by the difference between the two sample means

$$\mu_k - \mu_{k'} \doteq \bar{y}_k - \bar{y}_{k'}$$

This estimate has a sample variance of

$$V(\bar{y}_k - \bar{y}_{k'}) = 2\text{MSE}/p$$

and a sample standard error, $s_{(\bar{y}_k - \bar{y}_{k'})}$, of

$$s_{(\bar{y}_k - \bar{y}_{k'})} = \sqrt{2\text{MSE}/p}$$

The $(1 - \alpha)100\%$ confidence interval estimate, $L(\mu_k - \mu_{k'})$ of the difference is

$$L(\mu_k - \mu_{k'}) = (\bar{y}_k - \bar{y}_{k'}) \pm t_\alpha\sqrt{2MSE/p}$$

Again, t_α has $(p - 1)(p - 2)$ d.f.

4.3 MISSING OBSERVATIONS

As was the case with the randomized block design, the ANOVA algorithms for the Latin square apply only if all of the data are available. With only one missing value we can compute a substitute value, $\hat{y}_{ij(k)}$, using the equation

$$\hat{y}_{ij(k)} = \frac{p(y_{i.} + y_{.j} + T_{(k)}) - 2y_{..}}{(p - 1)(p - 2)}$$

where

$y_{i.}$, $y_{.j}$, and $T_{(k)}$ = sums of the remaining observations in row, column, and treatment with the missing value

$y_{..}$ = grand total of the available observations

If several values are missing, an iterative procedure, similar to that used with the randomized block design, is applied to obtain substitute values.

4.4 RELATIVE EFFICIENCY

The effectiveness of either the row or the column grouping may be estimated by computing the relative efficiency with either rows or columns as blocks in a randomized block design. The relative efficiency, RE, with rows omitted and columns as blocks is

$$RE = \frac{MSR + (p - 1)MSE}{pMSE}$$

With columns omitted and rows as blocks we have

$$RE = \frac{MSC + (p - 1)MSE}{pMSE}$$

To go one step farther, we can compare the relative efficiency of the Latin square to a completely randomized design with

$$RE = \frac{MSR + MSC + (p - 1)MSE}{(p + 1)MSE}$$

4.5 NUMERICAL EXAMPLE: LATIN SQUARE DESIGN

A ceramics engineer wanted to test the strength of high-tension insulators made from four new clay mixtures, *A, B, C, D*, and a control, *E*. He made five insulators from each mixture. He suspected that there was a temperature gradient from front to back and from top to bottom in his oven. He decided to use a Latin square design with shelves (top to bottom) as rows and positions on the shelves (front to back) as columns. The insulators were placed in the oven in the Latin square arrangement. After firing, the strength of each insulator was measured. The experimental layout and strength measurements were as shown in Fig. 4.4.

Required summary tables are given as Table 4.5 and Table 4.6. The analysis of variance is presented in Table 4.7.

Note: In this example $p = 5$. The sums of squares in Table 4.7 are computed as:

FRONT   BACK

A 33.8	B 33.7	D 30.4	C 32.7	E 24.4
D 37.0	E 28.8	B 33.5	A 34.6	C 33.4
C 35.8	D 35.6	A 36.9	E 26.7	B 35.1
E 33.2	A 37.1	C 37.4	B 38.1	D 34.1
B 34.8	C 39.1	E 32.7	D 37.4	A 36.4

TOP (first row) ... BOTTOM (last row)

Fig. 4.4 Latin square experimental plan and strength measurements on insulators.

TABLE 4.5 Row × Column Summary Table of Insulator Strength (lbs/in³)

Row (top to bottom)	Column (front to back)					Total
	1	2	3	4	5	
1	33.8	33.7	30.4	32.7	24.4	155.0
2	37.0	28.8	33.5	34.6	33.4	167.3
3	35.8	35.6	36.9	26.7	35.1	170.1
4	33.2	37.1	37.4	38.1	34.1	179.9
5	34.8	39.1	32.7	37.4	36.4	180.4
Total	174.6	174.3	170.9	169.5	163.4	852.7

TABLE 4.6 Treatment Totals and Means of Insulator Strength

Treatment	A	B	C	D	E	Total
Total	178.8	175.2	178.4	174.5	145.8	852.7
Mean	35.76	35.04	35.68	34.90	29.16	34.11

TABLE 4.7 ANOVA of Insulator Strength

Source	d.f.	SS	MS	F
Total	24	296.66		
Top to bottom (row)	4	87.40	21.85	7.12**
Front to back (column)	4	16.56	4.14	1.35
Mixture	4	155.89	38.97	12.69**
Error	12	36.81	3.07	

**Significant at the 1% level.

Note: $CV = (\sqrt{MSE}/\bar{y})100 = (\sqrt{3.07}/34.11)100 = 5.1\%$.

1. $C = y_{..}^2/p^2 = (852.7)^2/(5)^2 = 29{,}083.89$

2. $\text{SSTot} = \sum_i \sum_j y_{ij(k)}^2 - C = 29{,}380.55 - 29{,}083.89 = 296.66$

3. $\text{SSR} = (1/p)\sum_i y_{i.}^2 - C = (1/5)(145{,}856.47) - 29{,}083.89 = 87.40$

4. $\text{SSC} = (1/p)\sum_j y_{.j}^2 - C = (1/5)(145{,}502.27) - 29{,}083.89 = 16.56$

5. $\text{SST} = (1/p)\sum_k T_k^2 - C = (1/5)(146{,}198.93) - 29{,}083.89 = 155.89$

6. SSE = SSTot − SSR − SSC − SST

$$= 296.66 - 87.40 - 16.56 - 155.89 = 36.81$$

Standard errors:

a. Treatment mean, $s_{\bar{y}}$

$$s_{\bar{y}} = \sqrt{\text{MSE}/p} = \sqrt{3.07/5} = 0.78$$

b. Difference between two means, $s_{(\bar{y}_k - \bar{y}_{k'})}$

$$s_{(\bar{y}_k - \bar{y}_{k'})} = \sqrt{2\text{MSE}/p} = \sqrt{(2)(3.07)/5} = 1.11$$

Report of statistical analysis. Analysis of data from an experiment designed to measure the effect of different clay mixtures on the strength of high-tension insulators indicated that differences among mixtures are highly significant. On considering the mean strength, presented in Table 4.8, it is seen that there is little difference among the new mixtures. However, the new mixtures appeared to be uniformly stronger than the control.

There appeared to be a significant increase in strength from top to bottom of the oven. Position from front to back, however, appeared to have little effect. Future experiments in this oven might use a randomized block design with shelves as blocks.

4.6 RESTRICTIONS

From a practical standpoint there are some restrictions on the number of treatments which can be tested in a Latin square.

1. If p is small the d.f. for error is very small. For example

$$p = 2, \qquad \text{d.f. } E = 0$$
$$p = 3, \qquad \text{d.f. } E = 2$$

2. If p is large the required number of experimental units, p^2, is very large.
3. The practical range for p in a single square is $4 \leq p \leq 10$.

TABLE 4.8 Strength of High-Tension Insulators Made from Different Clay Mixtures

	Mixture					Standard error
	A	B	C	D	Control	
Mean	35.76	35.04	35.68	34.90	29.16	0.78

TABLE 4.9 ANOVA for Replicated Latin Squares

Source	d.f.
Total	$np^2 - 1$
Squares	$n - 1$
Rows in squares	$n(p - 1)$
Columns in squares	$n(p - 1)$
Treatments	$p - 1$
Error	$n(p - 1)(p - 2) + (n - 1)(p - 1)$

If p is small ($p = 3$ or 4) the degrees of freedom for error may be increased by using more than one square. With n squares the ANOVA format is given by Table 4.9.

We will not go through the general details of the computations involved. They are a straightforward extension of the computations for a single square. We will, however, present a numerical example.

4.7 NUMERICAL EXAMPLE : MULTIPLE LATIN SQUARES

An animal nutritionist was interested in the effect of three feed additives, *A, B, C,* on several measures of production and nutrition in dairy cattle. He was primarily concerned with fat-corrected milk production, but he also wanted to look at energy intake and digestibility. For this reason the animals had to be individually fed in digestion stalls, of which only three were available. In designing the experiment the nutritionist wanted to control a number of possible sources of variation:

1. Animal to animal
2. Time trends within animals
3. Variation from trial to trial

Recognizing the several sources of variation, the limitations of the apparatus, and the need to attain reasonable precision, the nutritionist decided to run a series of Latin squares as his experimental design. Since three additives were to be tested, and since only three digestion stalls were available, he decided to use 3 × 3 Latin squares with rows representing feeding periods and columns representing cows assigned to the individual digestion stalls. To obtain sufficient degrees of freedom for error and to obtain sufficient replication for reasonable precision, he decided to run four squares using three different animals in each square.

The experimental plan and the yield of 4% fat-corrected milk (FCM) (pounds) during the last 5 days of each 7-day feeding period are as given in Table 4.10.

TABLE 4.10 Experimental Plan and 4% FCM Yields

		Animal		
Square	Period	11	12	13
I	11	C 115	A 139	B 127
	12	A 138	B 209	C 224
	13	B 125	C 186	A 172
		21	22	23
II	21	C 176	B 163	A 135
	22	A 186	C 201	B 175
	23	B 146	A 101	C 134
		31	32	33
III	31	C 186	B 194	A 166
	32	A 130	C 180	B 152
	33	B 123	A 97	C 137
		41	42	43
IV	41	A 128	C 154	B 150
	42	C 137	B 129	A 106
	43	B 164	A 138	C 168

TABLE 4.11 Period Totals

Period	Square				
	I	II	III	IV	
1	381	474	546	432	
2	571	562	462	372	
3	483	381	357	470	Sum
Sum	1435	1417	1365	1274	5491

TABLE 4.12   Cow Totals

Cow	Square				
	I	II	III	IV	
1	378	508	439	429	
2	534	465	471	421	
3	523	444	455	424	Sum
Sum	1435	1417	1365	1274	5491

TABLE 4.13   Additive Totals and Means

	Additive			
	A	B	C	Sum
Sum	1636	1857	1998	5491
Mean	136.33	154.75	166.50	152.53

TABLE 4.14   ANOVA of 4% FCM Data

Source	d.f.	SS	MS	F
Total	35	33,624.97		
Squares	3	1,738.30	579.43	6.25
Period in square	8	19,094.67	2,386.83	25.74
Cow in square	8	5,944.67	743.08	8.01
Treatment	2	5,549.05	2,774.52	29.92**
Error	14	1,298.28	92.73	

**Significant at the 1% level.

The required summary tables are presented as Tables 4.11, 4.12, and 4.13. The analysis of variance of the 4% FCM data is given in Table 4.14.

In this example: n = number of squares = 4, p = number of cows per square = number of periods per square = number of treatments = 3.

The sums of squares for Table 4.14 are computed as:

1. $C = y^2_{...}/np^2 = (5,491)^2/(4)(9) = 837,530.03$

2. $\text{SSTot} = \sum_i \sum_j \sum_k y^2_{ijk} - C = 871,155 - 837,530.03 = 33,624.97$

3. $\text{SS(sq)} = (1/p^2) \sum y^2_{i..} - C = (1/9)(7,553,415) - 837,530.03$

   $= 1,738.30$

4. $\text{SS(per/sq)} = (1/p) \sum_i \sum_j y_{ij.}^2 - C - \text{SS(sq)}$

$$= (1/3)(2,575,089) - 837,530.03 - 1,738.30$$

$$= 19,094.67$$

5. $\text{SS(cow/sq)} = (1/p) \sum_i \sum_k y_{i.k}^2 - C - \text{SS(sq)}$

$$= (1/3)(2,535,639) - 837,530.03 - 1,738.30$$

$$= 5,944.67$$

6. $\text{SST} = (1/np) \sum_l T_l^2 - C$

$$= (1/12)(10,116,949) - 837,530.03 = 5,549.05$$

7. $\text{SSE} = \text{SSTot} - \text{SS(sq)} - \text{SS(per/sq)} - \text{SS(cow/sq)} - \text{SST}$

$$= 33,624.97 - 1,738.30 - 19,094.67 - 5,944.67 - 5,549.05$$

$$= 1,298.28$$

$\text{CV} = (\sqrt{92.73}/152.53)100 = 6.3\%$

Standard error of a treatment mean $= \sqrt{92.73/12} = 2.78$

Standard error of a difference between two treatment means $= \sqrt{(2)(92.73)/12} = 3.93$

Report of statistical analysis.   A trial was conducted to determine the effect of three feed additives on a number of measures of nutritive status of dairy cattle. This report is concerned with the statistical analysis of fat-corrected milk yields.

There were highly significant differences in mean 5-day yields among the three additives. Consideration of Table 4.15 indicates that the highest yield was obtained with additive *C* while the least was produced under additive *A*.

Controlling on differences among squares, differences among animals within squares, and differences among feeding periods within squares contributed to a reduction in experimental error and a consequent increase in precision. Removal of period differences was most effective in this regard, while removal of differences among squares was least effective. It is suggested that a similar design be used in future work of this type.

TABLE 4.15 Mean 5-Day Yields (pounds) of 4% Fat-Corrected Milk for Three Feed Additives

Additive	A	B	C	Standard error
Mean yield	136.33	154.75	166.50	2.78

4.8 FURTHER RESTRICTION: THE GRAECO-LATIN SQUARE

Using the Latin square permits us to remove two sources of variation from experimental error by the use of the row and column classifications for the experimental units. We can go one step farther and control on a third source of variation by using a design called a *Graeco-Latin square*.

In a Graeco-Latin square the experimental units are grouped in three different ways. As is the case with a Latin square, a design for p treatments requires p^2 experimental units. These are grouped by rows and columns and by an additional classification usually designated by Greek letters. The assignment of Greek letter groupings is restricted so that each Greek letter occurs once, and only once, in each row and column. Thus, the Greek letters form a Latin square with respect to rows and columns. Treatments, designated by Latin letters, are now assigned to the experimental units in such a way that each treatment occurs once, and only once, in each row and column and with each Greek letter. The treatments thus form a different Latin square on the rows and columns, one which is independent of the square formed by the Greek letters. An example of a Graeco-Latin square for $p = 4$ is presented in Fig. 4.5.

4.8.1 Advantages

There is one principal advantage to a Graeco-Latin square and this describes its primary *use*:

1. It permits control of variability for three sources of variation.

	COLUMN			
	1	2	3	4
ROW 1	C γ	A β	D α	B δ
2	D δ	B α	C β	A γ
3	A α	C δ	B γ	D β
4	B β	D γ	A δ	C α

Fig. 4.5 Basic 4 × 4 Graeco-Latin square.

4.8.2 Disadvantages

The Graeco-Latin square has a number of disadvantages:

1. The required number of experimental units increases rapidly as the number of treatments is increased.
2. The number of degrees of freedom for error is small if the number of treatments is small.
3. Loss of data greatly complicates the statistical analysis.
4. Balance on three groupings is seldom possible.

4.8.3 Restrictions

Graeco-Latin square designs have been constructed for all numbers of treatment from 3 to 12 except for 10. Theoretically, these designs are possible for any p provided that p is a prime number or the power of a prime number. From a practical standpoint, however, the limits are $5 \le p \le 12$:

1. For small p the error d.f. are too small.

$$p = 3, \qquad \text{d.f. } E = 0$$
$$p = 4, \qquad \text{d.f. } E = 3$$

2. For large p the required number of experimental units is too large.

$$p = 12, \qquad \text{units} = 144$$

Because of its limitations, the Graeco-Latin square has seldom been used in research.

4.8.4 Analysis

We will not go into the details of the analysis of results from a Graeco-Latin square. The computations are a straightforward extension of those involved in the analysis of data from a Latin square.

The format of the ANOVA for a Graeco-Latin square for p treatments is given in Table 4.16.

4.8.5 Numerical Example: Graeco-Latin Square Design

A food processor wanted to determine the effect of package design on the sale of one of his products. He had five designs to be tested: A, B, C, D, and E. There were a number of sources of variation, however, whose possible effect he wanted to remove from consideration. These included: (1) day of the week, (2) differences among stores, and (3) effect of shelf height. He decided to conduct a trial using a Graeco-Latin square design with five weekdays corresponding

TABLE 4.16 ANOVA of
Graeco-Latin Square

Source	d.f.
Total	$p^2 - 1$
Rows	$p - 1$
Columns	$p - 1$
Greek letters	$p - 1$
Treatments	$p - 1$
Error	$(p - 1)(p - 3)$

to the row classification, five different stores assigned to the column classification, and five shelf heights corresponding to the Greek letter classification.

The experimental plan and the sales (dollars) for the trial were as shown in Fig. 4.6. The required summary tables are combined in Table 4.17. The analysis of variance is given in Table 4.18.

Report of statistical analysis. A pilot study was conducted to determine the effect of package design on the acceptance, as measured by sales, of a new

STORE

DAY	I	II	III	IV	V
MON.	E α 238	C δ 228	B γ 158	D ϵ 188	A β 74
TUES.	D δ 149	B β 220	A α 92	C γ 169	E ϵ 282
WED.	B ϵ 222	E γ 295	D β 104	A δ 54	C α 213
THUR.	C β 187	A ϵ 66	E δ 242	B α 122	D γ 90
FRI.	A γ 65	D α 118	C ϵ 279	E β 278	B δ 176

Fig. 4.6 Experimental plan and sales (dollars) for package design trial. (Greek letters designate shelf height; Latin letters designate package design.)

TABLE 4.17 Summary Tables for Analysis of Package Design Data

Day (row)	Mo	Tu	We	Th	Fr	Sum
Sum	886	912	888	707	916	4309
Store (column)	I	II	III	IV	V	Sum
Sum	861	927	875	811	835	4309
Shelf (Greek)	α	β	γ	δ	ε	Sum
Sum	783	863	777	849	1037	4309
Design (treatment)	A	B	C	D	E	Sum
Sum	351	898	1076	649	1335	4309
Mean	70.2	179.6	215.2	129.8	267.0	172.4

TABLE 4.18 ANOVA of Package Design Sales Data

Source	d.f.	SS	MS	F
Total	24	139,395.76		
Day	4	6,138.56	1,534.64	1.66
Store	4	1,544.96	386.24	.42
Shelf height	4	8,852.16	2,213.04	2.39
Design	4	115,462.15	28,865.54	31.21**
Error	8	7,397.92	924.74	

**Significant at the 1% level.

Note: CV $= (\sqrt{924.74}/172.4)100 = 17.6\%$.

Standard error of a treatment mean $= \sqrt{924.74/5} = 13.60$.

Standard error of a difference between two treatment means $= \sqrt{(2)(924.74)/5} = 19.23$.

TABLE 4.19 Mean Sales ($) of Product in Packages of Five Different Designs

Design	A	B	C	D	E	Standard error
Mean sales	70.2	179.6	215.2	129.8	267.0	13.60

food product. Analysis of the resulting data indicates that there were highly significant differences in mean sales among the five package designs tested. Mean sales are given in Table 4.19.

Clearly, the design with the highest acceptance, as measured by sales, was design *E*, while design *A* was least acceptable.

The analysis indicates further that there was no real effect of day of the week, store, or shelf position within the store on the sale of the new food product.

4.9 SUMMARY OF BASIC EXPERIMENTAL DESIGNS

For each design assume that there are *p* treatments with *r* replications per treatment.

4.9.1 Completely Randomized Design (CRD)

1. Treatments assigned completely at random to *r* out of *rp* experimental units.
2. Field layout for $p = 4$, $r = 4$ is shown in Fig. 4.7.
3. Analysis of variance of a CRD is displayed in Table 4.20
4. Uses:
   a. When experimental units are homogeneous.

C	B	A	A
A	B	D	D
D	B	C	B
C	C	A	D

Fig. 4.7 Experimental plan for a completely randomized design.

TABLE 4.20 ANOVA of a CRD

Source	d.f.
Total	$rp - 1$
Treatment	$p - 1$
Error	$p(r - 1)$

    b. When number of units is small.

    c. When loss of units during trial is expected.

5. Advantages:

    a. Easy to lay out.

    b. Simple analysis, even with unequal numbers.

    c. Maximum d.f. for error.

6. Disadvantage:

    a. Inefficient if experimental units are heterogeneous.

4.9.2 Randomized Block Design (RBD)

1. Experimental units grouped into r blocks of p units each. Treatments assigned at random to units within blocks. Each treatment occurs once, and only once, in each block.

2. Field layout for $p = 4$, $r = 4$ is presented in Fig. 4.8.

3. Analysis of variance format for a RBD is given in Table 4.21.

BLOCK

I	II	III	IV
C	B	A	C
A	C	C	A
B	D	D	D
D	A	B	B

Fig. 4.8 Experimental plan for a randomized block design.

TABLE 4.21 ANOVA of a RBD

Source	d.f.
Total	$rp - 1$
Blocks	$r - 1$
Treatments	$p - 1$
Error	$(r - 1)(p - 1)$

4. Uses:
   a. When units are heterogeneous, but can be grouped into blocks of homogeneous units.
   b. To control one source of variability.
5. Advantages:
   a. Fairly easy to lay out.
   b. Analysis is usually not complicated.
   c. Proper blocking increases efficiency.
6. Disadvantages:
   a. Missing data cause difficulty in the analysis.
   b. Less efficient than other designs with more than one source of controllable variation.

4.9.3 Latin Square Design

1. Experimental units are grouped into p rows of p homogeneous units each. They are then grouped into p columns of p homogeneous units each. Treatments are assigned so that each treatment occurs once, and only once, in each row and column.
2. Field layout for $p = 4$, $r = p = 4$, $c = p = 4$. Latin square design is displayed in Fig. 4.9.
3. Analysis of variance format for a Latin square is presented in Table 4.22.
4. Use:
   a. To remove two sources of variability.
5. Advantage:
   a. Increases efficiency in the presence of two-way variability.

| | COLUMN | | | |
ROW	1	2	3	4
1	B	C	A	D
2	A	B	D	C
3	D	A	C	B
4	C	D	B	A

Fig. 4.9 Experimental plan for a Latin square design.

TABLE 4.22 ANOVA of a Latin Square

Source	d.f.
Total	$p^2 - 1$
Rows	$p - 1$
Columns	$p - 1$
Treatments	$p - 1$
Error	$(p - 1)(p - 2)$

6. Disadvantages:
    a. Too few d.f. for error when p is small.
    b. Requires large number p^2 of units when p is large.
    c. Complicated analysis with missing data or misassigned treatments.

FURTHER READING*

Anderson, V. L., and R. A. McLean (1974), Ch. 8.
Cochran, W. G., and G. M. Cox (1957), Ch. 4.
Federer, W. T. (1955), Ch. 6.
Kempthorne, O. (1973), Ch. 10.

---

*See Bibliography for complete citation.

5

Separation of Means

5.1 INTRODUCTION

In considering the analysis of experimental data thus far we have restricted ourselves to rather elementary types of hypotheses:

1. Simple hypotheses, each with one degree of freedom. These include
   a. $H_0: \mu = \mu_0$, where μ is a true mean and μ_0 is a hypothetical mean. This is the hypothesis that the true mean has some previously specified numerical value. The test statistic for this hypothesis is either $t = (\bar{y} - \mu_0)/\sqrt{V(\bar{y})}$ or else $F = (\bar{y} - \mu_0)^2/V(\bar{y})$, in which $\bar{y}$ is the sample mean and $V(\bar{y})$ is the sample variance of $\bar{y}$.
   b. $H_0: \mu_1 = \mu_2$ or, equivalently, $H_0: \mu_1 - \mu_2 = 0$, where μ_1 and μ_2 are the true means of treatments 1 and 2, respectively. This is the hypothesis that two treatment means are equal. The test statistic for this hypothesis is either $t = (\bar{y}_1 - \bar{y}_2)/\sqrt{V(\bar{y}_1 - \bar{y}_2)}$ or $F = (\bar{y}_1 - \bar{y}_2)^2/V(\bar{y}_1 - \bar{y}_2)$, where $\bar{y}_1$ and $\bar{y}_2$ are the sample means and $V(\bar{y}_1 - \bar{y}_2)$ is the sample variance of the difference between the means.
2. Multiple hypotheses of the form $H_0: \mu_1 = \mu_2 = \cdots = \mu_p$. This is the hypothesis that all of p true means are the same. This hypothesis is tested, usually within the framework of the analysis of variance, with the test statistic $F = \text{MST/MSE}$, where MST is the treatment mean square and MSE is the error mean square from the analysis of variance.

Usually, however, we want to make a more detailed examination of the data than is permitted by these elementary hypotheses. Depending on the objectives of the experiment and the nature of the treatments, we might want to answer one or more of the following questions:

1. Can we pick a winner, or winners, from among a set of unstructured treatments?
2. What about comparing each mean with every other mean? Is there any evidence of real differences?
3. Do the data conform to a previously specified linear function of the true means?

There are a number of procedures for examining these questions, and we will look at some of them.

5.2 SELECTED PAIRWISE COMPARISONS

The basic idea behind procedures for this type of problem can be developed in the following way:

Suppose we want to compare the means of two randomly chosen treatments. That is, suppose we want to test $H_0:\mu_1 = \mu_2$ against $H_a:\mu_1 \neq \mu_2$. We can use as a test statistic

$$t = \frac{\bar{y}_1 - \bar{y}_2}{\sqrt{V(\bar{y}_1 - \bar{y}_2)}}$$

If $|t| > t_\alpha$, where $|t|$ is the sample value of t with sign ignored and t_α is the two-tailed t at the α probability level, we reject H_0. If H_0 is rejected the conclusion is that μ_1 is significantly different from μ_2 at the $100\alpha\%$ significance level. Hence, if

$$t_\alpha < |t| = |\bar{y}_1 - \bar{y}_2|/\sqrt{V(\bar{y}_1 - \bar{y}_2)}$$

or, equivalently, if

$$t_\alpha\sqrt{V(\bar{y}_1 - \bar{y}_2)} < |\bar{y}_1 - \bar{y}_2|$$

then μ_1 is significantly different from μ_2. Any difference, $\bar{y}_1 - \bar{y}_2$, between sample means which exceeds $t_\alpha\sqrt{V(\bar{y}_1 - \bar{y}_2)}$ would indicate a significant difference between μ_1 and μ_2. Further, $t_\alpha\sqrt{V(\bar{y}_1 - \bar{y}_2)}$ is the smallest difference for which significance would be declared. Based on this argument we can define a "least significant difference," LSD, as

$$LSD = t_\alpha\sqrt{V(\bar{y}_1 - \bar{y}_2)}$$

In particular, suppose we want the LSD for two means: $\bar{y}_1$ with n_1 observations and $\bar{y}_2$ with n_2 observations. We would compute

$$LSD = t_\alpha\sqrt{(1/n_1 + 1/n_2)s^2}$$

where s^2 = MSE is the pooled variance of the two samples. Note that if $n_1 = n_2 = r$ then

$$\text{LSD} = t_\alpha\sqrt{2s^2/r} = t_\alpha\sqrt{2\text{MSE}/r}$$

Whether the treatments are equally replicated or not, the d.f. for t are the d.f. for s^2.

The LSD is in disfavor among some statisticians because it can easily be misused. There are a couple of situations for which the LSD is a valid test statistic and for which the significance level, α, holds for each comparison:

1. Making comparisons planned in advance of seeing the data, for example, comparing a new treatment with a control.
2. Comparing of adjacent ranked means. This is, essentially, a procedure for picking the winner.

There are a number of situations in which the use of the LSD is not legitimate from a theoretical standpoint:

1. Making comparisons suggested by the data, for example, comparing the largest mean with the smallest.
2. Making all possible paired comparisons among means.

We illustrate the legitimate use of the LSD in Sections 5.2.1 and 5.2.2.

5.2.1 Comparisons Planned in Advance

An experiment was conducted to compare each of five new insecticides against a standard insecticide used as a control. Three samples of each insecticide were sprayed into closed containers each holding 100 flies. After a 5-min exposure the number of dead flies was counted and the results are shown in Table 5.1.

TABLE 5.1 Results of an Insecticide Experiment

Insecticide	Mean no. of flies killed	Difference new − control
Control	10	
1	30	20
2	15	5
3	12	2
4	25	15
5	35	25

Note: MSE(12 d.f.) = 54.

$$\text{LSD} = t_\alpha\sqrt{\frac{2\text{MSE}}{r}} = \frac{2.179}{3.055}\sqrt{\frac{(2)(54)}{3}} = \frac{13.07\ (5\%)}{18.33\ (1\%)}.$$

The differences between the control and insecticides 1 and 5 are highly significant (α = .01), while insecticide 4 is only significantly (α = .05) different from the control. Insecticides 2 and 3 do not differ significantly from the control.

5.2.2 Pick the Winner

A plant breeder wanted to measure the resistance of six varieties of wheat to a particular race of stem rust. He planted five seeds of each variety in each of four pots of well-mixed soil. He then randomly assigned the 24 pots to locations on a greenhouse bench (thus, the experimental unit is a pot of five seedlings replicated four times in a completely randomized design). After the plants emerged he uniformly inoculated the experiment with a culture of the stem rust. At maturity he measured the yield (grams per pot) of wheat in each pot. The results he obtained (arranged in order of decreasing mean yield) are shown in Table 5.2.

If the differences are compared with the LSDs we see that the two highest yielding varieties, 6 and 4, are not significantly different, but that there is a significant (α = .05) gap between varieties 4 and 5. Another significant (α = .05) gap exists between varieties 2 and 1, while the difference between variety 3 and the others is highly significant (α = .01).

5.3 MULTIPLE COMPARISONS (DATA SNOOPING)

When using the LSD to test independent comparisons planned in advance or to look for significant gaps among a set of ranked means, the maximum number of legitimate comparisons is limited to the number of treatment degrees of

TABLE 5.2   Results of a Stem Rust Experiment

Variety	Rank (i)	Mean yield ($\bar{y}_i$) g/pot	Difference $\bar{y}_{(i-1)} - \bar{y}_i$
6	1	95.3	
4	2	94.0	1.3
5	3	75.0	19.0*
2	4	69.0	6.0
1	5	50.3	18.7*
3	6	24.0	26.3**

*Significant at the 5% level.
**Significant at the 1% level.
Note: MSE(18 d.f.) = 120.0

$$\text{LSD} = t_\alpha \sqrt{\frac{2\text{MSE}}{r}} = \frac{2.101}{2.878} \sqrt{\frac{(2)(120.0)}{4}} = \frac{16.27 \ (5\%)}{22.29 \ (1\%)}.$$

freedom. In some cases, however, a researcher may want to look at comparisons among all possible pairs of means. For a set of k treatments this would involve as many as $k(k - 1)/2$ comparisons, which is greater than the $k - 1$ independent comparisons that are possible using the LSD. We require a procedure that will allow us to draw a conclusion about which differences, if any, are real, and to assign a measure of confidence in our conclusions.

Whenever we compare two means by any statistical procedure there are two kinds of error we can make in the conclusion we reach. A type I error is the error of concluding that two means are different when, in fact, they are not. The rate at which this kind of error will be made (probability of a type I error) is designated α, and is called the significance level for the comparison. The second kind of erroneous conclusion, a type II error, is the error of concluding that two means are not different when, in fact, they are different. The probability of making this kind of error is designated β. The probability that the comparison procedure will detect a real difference when it exists is called the power of the procedure, and is equal to $1 - \beta$.

If we use the ordinary LSD as a test statistic against which to make all possible pairwise comparisons the probability of a type I error may no longer be at the chosen significance level. Cochran and Cox (1957), for example, indicate that if the LSD is used to compare the highest and lowest sample means when, in fact, there are no real differences among the true means, then the sample difference will exceed the 5% LSD about 13% of the time with 3 means and as much as 90% of the time with 20 means. To overcome this difficulty a number of test statistics have been proposed against which to judge all possible pairwise comparisons, of which the highest against the lowest is the most extreme case. We will examine four of these procedures in detail:

1. Fisher's protected LSD (FPLSD) (Fisher, 1966)
2. Tukey's honestly significant difference (HSD) (Tukey, 1951)
3. Student-Newman-Keuls test (SNK) (Steel and Torrie, 1980)
4. Waller and Duncan's Bayes LSD (BLSD) (Waller and Duncan, 1969; Duncan, 1975)

The first three of these procedures are the more traditional ones. With these the experimenter chooses a significance level, α, to protect himself against a type I error. He then allocates his error rate by his choice of procedure. With the FPLSD the error rate is a comparisonwise error rate. It measures the proportion of all comparisons expected to be declared significant when, in fact, they are not. A criticism of this type of procedure is that the probability of falsely declaring at least one difference to be significant is high and increases as the number of means being compared is increased. The FPLSD is the most liberal of the three procedures since it requires the smallest difference for significance.

Procedures like the HSD involve an experimentwise error rate. Here the error rate measures the expected proportion of experiments in which one or more differences per experiment will be falsely declared significant. A criticism of this type of procedure is that its power to detect real differences decreases as the number of means increases. The HSD is the most conservative of the three procedures because it requires the largest difference for significance.

With the SNK procedure the error rate allocation varies with the degree of separation between the means being compared. For adjacent means the error rate is comparisonwise, and the difference required for significance is the same as the LSD. For means with the greatest separation, highest versus lowest, the error rate is experimentwise, and the difference required for significance is the HSD. For other comparisons the difference required for significance is somewhere between these extremes.

The Waller-Duncan BLSD takes into consideration the homogeneity, as measured by the F ratio in the ANOVA, of the set of means under consideration. If the F ratio is low, indicating a set of homogeneous means, the BLSD has the conservative character of an experimentwise error approach. On the other hand for larger F ratios ($F \geq 4$), indicating heterogeneous means, the BLSD has the characteristics of a comparisonwise error rate. Unlike the other procedures, the significance level for the comparison is not based on the probability of making a type I error. Rather the factor called a "minimum average risk t" is chosen based on a "k ratio," which is the ratio of the seriousness of the type I to that of the type II error. These k ratios can be chosen so that the Bayes t values approximate the α level t values commonly used in other multiple comparison procedures. The advantage of the BLSD, in contrast to experimentwise error rate approaches, is that its power does not decrease as the number of means increases.

5.3.1 Fisher's Protected LSD (FPLSD)

The FPLSD uses the ordinary LSD as a standard against which pairwise comparisons are judged. The added provision is that it is not used unless the F test for treatments in the analysis of variance is judged to be significant. To use FPLSD the procedure is:

1. Conduct the usual ANOVA on the data.
2. Make the F test as

    $$F = MST/MSE$$

3. If $F \leq F_\alpha$ do not continue.
4. If $F > F_\alpha$ find t_α in the usual two-tailed t table and compute FPLSD $= t_\alpha\sqrt{2MSE/r}$
5. For any pair of means, $\bar{y}_i$, $\bar{y}_j$, if $|\bar{y}_i - \bar{y}_j| >$ FPLSD the difference is judged to be significant.

5.3.2 Tukey's Honestly Significant Difference (HSD)

The HSD is computed in a manner similar to the LSD except that the standard error of a mean replaces the standard error of a difference, and the studentized range, Q_α, is used in place of a t_α. For the HSD the procedure is:

1. From the table of percentage points of the studentized range (Appendix Table A4) select a value of Q_α, which depends on n = number of means and υ = error degrees of freedom.
2. Compute

    $$HSD = Q_\alpha\sqrt{MSE/r}$$

3. For any pair of means, $\bar{y}_i$, $\bar{y}_j$, if $|\bar{y}_i - \bar{y}_j| > HSD$ the difference is judged to be significant.

5.3.3 Student-Newman-Keuls Test (SNK)

The SNK is a test in which the difference needed for significance varies with the degree of separation between the means being compared. For example, means separated by one mean require a smaller difference for significance than do means separated by two means. If there are p means in the set there will be $p - 1$ required differences, one for each of the $p - 1$ possible separations. The SNK test procedure is:

1. Rank the p means from high, $\bar{y}_h$, to low, $\bar{y}_l$.
2. Compute $p - 1$ significant differences, SNK_i, using the equation for the HSD $p - 1$ times as

    $$SNK_i = Q_{\alpha i}\sqrt{MSE/r} \qquad i = 2, 3, \ldots, p$$

    using $Q_{\alpha i}$ for υ d.f. and, in turn, $n = 2, 3, \ldots, p$ means.
3. If $|\bar{y}_h - \bar{y}_l| < SNK_p$ no differences are significant, and the test is finished.
4. If $|\bar{y}_h - \bar{y}_l| > SNK_p$ this difference is significant, and we proceed by comparing $\bar{y}_h - \bar{y}_{l+1}$ and $\bar{y}_{h-1} - \bar{y}_l$ against SNK_{p-1}.
5. Continue in this way until all significant pairwise comparisons have been found.

5.3.4 Waller-Duncan Bayes LSD (BLSD)

The BLSD is computed in a way analogous to the FPLSD. We find the standard error of a difference between means and multiply it by an appropriate minimum average risk t, say t_B, to obtain the BLSD. The problem is to select a risk ratio comparable to the significance levels of the other procedure. For this purpose Duncan (1975) has shown that a k ratio of 100 is equivalent to an α of .05, while a k ratio of 500 is equivalent to an α of .01. The procedure is as follows:

1. Do the usual ANOVA on the data.
2. Compute the F ratio of MST/MSE with (to correspond to the table symbols) q and f degrees of freedom.
3. Choose an appropriate k ratio (or error weight ratio). Use $k = 100$ for an approximately 5% significance test or $k = 500$ for an approximately 1% significance test.
4. Obtain the minimum average risk t, t_B from Appendix Table A5. The value for t_B is dependent on k, F, q, and f.
5. Compute BLSD as

$$\text{BLSD} = t_B \sqrt{2\text{MSE}/r}$$

6. For any pair of means, $\bar{y}_i$, $\bar{y}_j$, if $|\bar{y}_i - \bar{y}_j| > \text{BLSD}$ the difference is judged to be a real difference.

5.3.5 Example

To illustrate the use of these multiple comparison procedures we will take the data from the stem rust experiment used previously to illustrate the use of the LSD to find significant gaps. The data are from a completely randomized design with six varieties replicated four times. The necessary data are given in Table 5.3.

For each procedure we will use a 5% significance level ($\alpha = .05, k = 100$) and illustrate two methods of presenting the results.

FPLSD

1. $F = \text{MST/MSE} = 2976.44/120.00 = 24.80$

$F_{.05(5,18)} = 2.77$, $24.80 > 2.77$, F is significant

TABLE 5.3 Stem Rust Data and Their Analysis of Variance

Variety	1	2	3	4	5	6
Rank	5	4	6	2	3	1
Mean	50.3	69.0	24.0	94.0	75.0	95.3

ANOVA

Source	d.f.	MS	F
Variety	5	2976.44	24.80
Error	18	120.00	

TABLE 5.4 Grouping of Ranked Means with FPLSD

Variety	6	4	5	2	1	3
Mean	95.3	94.0	75.0	69.0	50.3	24.0*

*Means underlined by the same line are not significantly (α = .05) different.

2. $FPLSD = t_{.05(18)}\sqrt{2MSE/r}$

   $= 2.101\sqrt{(2)(120.00)/4} = 16.27$

3. Results
   a. In Table 5.4 the ranked means are grouped using FPLSD. Note: Computations for Table 5.4 are:

   $95.3 - 16.27 = 79.03$
   $94.0 - 16.27 = 77.73$
   $75.0 - 16.27 = 58.73$
   $69.0 - 16.27 = 52.73$
   $50.3 - 16.27 = 34.03$

   b. Differences between means are tested using FPLSD in Table 5.5.

HSD

1. $n = 6, \upsilon = 18, Q_{.05(6,18)} = 4.49$

2. $HSD = Q_\alpha\sqrt{MSE/r} = 4.49\sqrt{120.00/4} = 24.59$

3. Results
   a. In Table 5.6 the ranked means are grouped using HSD.
   b. Differences between means are tested using HSD in Table 5.7.

TABLE 5.5 Differences Between Means Compared with FPLSD

Mean		24.0	50.3	69.0	75.0	94.0
	Rank	6	5	4	3	2
95.3	1	71.3*	45.0*	26.3*	20.3*	1.3
94.0	2	70.0*	43.7*	25.0*	19.0*	
75.0	3	51.0*	24.7*	6.0		
69.0	4	45.0*	18.7*			
50.3	5	26.3*				

*Significant at the 5% level.

TABLE 5.6 Grouping of Ranked Means With HSD

Variety	6	4	5	2	1	3
Mean	95.3	94.0	75.0	69.0	50.3	24.0

*Means underlined by same line are not significantly (α = .05) different.

TABLE 5.7 Differences Between Means Compared with HSD

Mean		24.0	50.3	69.0	75.0	94.0
	Rank	6	5	4	3	2
95.3	1	71.3*	45.0*	26.3*	20.3	1.3
94.0	2	70.0*	43.7*	25.0*	19.0	
75.0	3	51.0*	24.7*	6.0		
69.0	4	45.0*	18.7			
50.3	5	26.3*				

*Significant at the 5% level.

TABLE 5.8 Grouping of Ranked Means with SNK

Variety	6	4	5	2	1	3
Mean	95.3	94.0	75.0	69.0	50.3	24.0*

*Means underlined by same line are not significantly (α = .05) different.

SNK

1. $n = 6, 5, \ldots , 2; \upsilon = 18$

2. $\text{SNK} = Q_{\alpha(n,\upsilon)}\sqrt{\text{MSE}/r}$

   $= Q_{.05(n,18)}\sqrt{120.00/4}$

n	2	3	4	5	6
$Q_{.05(n,18)}$	2.97	3.61	4.00	4.28	4.49
SNK	16.27	19.77	21.91	23.44	24.59

3. Results
   a. In Table 5.8 the ranked means are grouped using SNK.
   b. Differences between means are tested using SNK in Table 5.9.

TABLE 5.9 Differences Between Means Compared with SNK

Mean		24.0	50.3	69.0	75.0	94.0
	SNK	24.59	23.44	21.91	19.77	16.27
95.3	24.59	71.3*	45.0*	26.3*	20.3*	1.3
94.0	23.44	70.0*	43.7*	25.0*	19.0*	
75.0	21.91	51.0*	24.7*	6.0		
69.0	19.77	45.0*	18.7*			
50.3	16.27	26.3*				

*Significant at the 5% level.

BLSD

1. Error weight ratio corresponding to $\alpha = .05$ is $k = 100$.

   $F = \text{MST/MSE} = 24.80$, with $q = 5$ and $f = 18$ degrees of freedom.

   $t_B = t_{(k,F,q,n)} = t_{(100,25,5,18)} = 1.93$

2. $\text{BLSD} = t_B\sqrt{2\text{MSE}/r} = 1.93\sqrt{(2)(120.00)/4}$

   $= 14.95$

3. Results
   a. In Table 5.10 the ranked means are grouped using BLSD.
   b. Differences between means are tested using BLSD in Table 5.11.

Note: It is not possible to prepare tables of t_B for every possible value of F. If the computed F from the ANOVA of a set of data does not correspond to one of the F values in the table there are two alternatives:

TABLE 5.10 Grouping of Ranked Means with BLSD.

Variety	6	4	5	2	1	3
Mean	95.3	94.0	75.0	69.0	50.3	24.0*

*Means underlined by the same line are not significantly (k = 100) different.

TABLE 5.11 Differences Between Means Compared with BLSD

Mean		24.0	50.3	69.0	75.0	94.0
	Rank	6	5	4	3	2
95.3	1	71.8*	45.0*	26.3*	20.3*	1.3
94.0	2	70.0*	43.7*	25.0*	19.0*	
75.0	3	51.0*	24.7*	6.0		
69.0	4	45.0*	18.7*			
50.3	5	26.3*				

*Significant at k = 100.

1. For most practical purposes it is sufficient to use the t_B from the table for which F is closest to the ANOVA F. For a conservative test the t_B from the table for which the F is smaller than the ANOVA F may be used.

2. For a more precise value of t_B an interpolated value may be computed. This interpolation may be done using either $a = 1/\sqrt{F}$ or $b = \sqrt{F/(F - 1)}$ as weights. Note that values of a and b are given for the F values in each section of the t_B table (Appendix Table A5). This interpolation process is illustrated below:

Example: Suppose we want to find t_B in a situation for which k = 100, q = 6, f = 12, and we have a sample ANOVA F of 3.38.

1. In Table A5 we have

$$F = 3.00, \quad a = .577, \quad b = 1.225, \quad t_B = 2.49$$
$$F = 4.00, \quad a = .500, \quad b = 1.155, \quad t_B = 2.37$$

2. For the sample data

$$a = 1/\sqrt{F} = 1/\sqrt{3.38} = .544$$
$$b = \sqrt{F/(F - 1)} = \sqrt{3.38/2.38} = 1.192$$

3. To interpolate using a's as weights we have

$$t_B = 2.37 + (2.49 - 2.37) \left(\frac{.544 - .500}{.577 - .500} \right) = 2.44$$

4. To interpolate using b's as weights we have

$$t_B = 2.49 - (2.49 - 2.37) \left(\frac{1.225 - 1.192}{1.225 - 1.155} \right) = 2.43$$

5. Whether to use a or b in the interpolation depends on the values of F, q, and f. The choice is indicated as follows (Duncan, 1975):

F	f	$q \le 100$	$q > 100$	F	f	$q \le 20$	$q > 20$
≤ 2.4	≤ 60	a	a	> 2.4	≤ 20	a	b
	> 60	a	b		> 20	b	b

Comparison of Methods

The differences required for significance under the four procedures may be summarized as follows:

Procedure	Difference for significance
FPLSD	16.27
HSD	24.59
SNK	16.27–24.59
BLSD	14.95

5.4   COMMENTS ON MULTIPLE COMPARISON PROCEDURES

A few comments should be made about multiple comparison procedures and their use:

1. Because of the challenge they present, and because of the broad spectrum of philosophies embraced by different statisticians, a number of multiple comparison procedures have been developed. For the most part, these have been described and compared by Miller (1966).

2. In a computer simulation study involving 88,000 differences, Carmer and Swanson (1973) evaluated a number of multiple comparison procedures. They compared the LSD, FPLSD, HSD, SNK, and BLSD, among others, from the standpoint of protection against a type I error and, more important, their power to detect real differences when they exist. Their conclusion was that there

is little difference between the FPLSD and the BLSD procedures. None of the other methods approach these two.

3. From the intuitive point of view the BLSD is very appealing. It is a single value which, as it should be, is large when the sample F indicates that the means are homogeneous, and small when the means appear to be heterogeneous. Further, its power to detect real differences does not depend on the number of means being compared. Its principal drawback is that tables of t_B are not as readily available as are ordinary t tables.

4. The preference of this author is the BLSD. However, because it is familiar to many scientists, and because the necessary tables are widely available, the FPLSD is also quite acceptable.

5. A number of authors (Carmer and Walker, 1982; Chew, 1976; Little, 1978; Petersen, 1977) have pointed out that multiple comparison procedures are frequently used in situations for which they were never intended. Their primary use is to make comparisons among means for unstructured treatments, of which a typical example would be a variety trial. For experiments involving factorial sets of treatments (Chapter 6) or graded levels of quantitative factors there is almost always a more meaningful statistical procedure. In particular, consideration should be given to the construction and testing of contrasts which can be formed in advance of seeing the data, and which can address the specific objectives of the experiment. Contrasts are considered in detail beginning in Section 5.5.

5.5 CONTRASTS

We are now ready to consider the question of determining whether the data conform to a previously specified linear function of the true means. We can introduce the concepts involved by considering an experiment designed to determine the wearing quality of a new paint. Suppose that the paint was tested under four conditions:

1. Hard wood, dry climate (μ_1)
2. Hard wood, wet climate (μ_2)
3. Soft wood, dry climate (μ_3)
4. Soft wood, wet climate (μ_4)

There are a number of questions we might like the experimental results to answer:

1. Is the average life on hard wood the same as the average life on soft wood?
2. Is the average life in a dry climate the same as the average life in a wet climate?

3. Does the difference in paint life between wet and dry climates depend on whether the wood is hard or soft?

We can think that these questions were generated by considering the nature of the treatments. Alternatively, we can think that the treatments were selected with the goal of providing answers to the questions. In any case, the questions are meaningful and can be formulated before the experiment is run.

The questions can be answered by comparing means under the following hypotheses, numbered to correspond to the questions:

1. $H_0:(1/2)(\mu_1 + \mu_2) = (1/2)(\mu_3 + \mu_4)$

   or $H_0:(1/2)(\mu_1 + \mu_2) - (1/2)(\mu_3 + \mu_4) = 0$

2. $H_0:(1/2)(\mu_1 + \mu_3) = (1/2)(\mu_2 + \mu_4)$

   or $H_0:(1/2)(\mu_1 + \mu_3) - (1/2)(\mu_2 + \mu_4) = 0$

3. $H_0:(1/2)(\mu_1 - \mu_2) = (1/2)(\mu_3 - \mu_4)$

   or $H_0:(1/2)(\mu_1 - \mu_2) - (1/2)(\mu_3 - \mu_4) = 0$

A systematic way of forming and testing hypotheses such as these is provided by the concept of contrasts among means. These contrasts can be considered to be particular linear functions of the means. Suppose that

$$L = k_1\mu_1 + k_2\mu_2 + \cdots + k_p\mu_p$$

is a linear function of a set of p means. The function, L, is said to be a *contrast* of means if

$$\sum_{i=1}^{p} k_i = 0$$

provided that at least one $k_i \neq 0$. For example

$$L = (1/2)(\mu_1 + \mu_2) - (1/2)(\mu_3 + \mu_4)$$
$$= (1/2)\mu_1 + (1/2)\mu_2 - (1/2)\mu_3 - (1/2)\mu_4$$

is a contrast because $1/2 + 1/2 - 1/2 - 1/2 = 0$.

We use contrasts in hypothesis testing by framing the null hypothesis as a contrast and then testing the hypothesis that the contrast is zero against the alternative that it is not. That is, we test

$$H_0:L = 0$$

or $\quad H_0:k_1\mu_1 + k_2\mu_2 + \cdots + k_p\mu_p = 0$

against

$$H_a:L \neq 0$$

or $H_a:k_1\mu_1 + k_2\mu_2 + \cdots + k_p\mu_p \neq 0$

To test the hypothesis we need a sample estimate of L and the sample variance of the estimate. The sample estimate, l, is obtained by substituting the sample mean, $\bar{y}_i$, for the population means in the contrast under test. We have

$$L \doteq l = k_1\bar{y}_1 + k_2\bar{y}_2 + \cdots + k_p\bar{y}_p$$

Under the condition that the samples are independent, which is the case with balanced experimental designs, the variance, σ_l^2, of l is given by

$$\sigma_l^2 = \left(\frac{k_1^2}{r_1} + \frac{k_2^2}{r_2} + \cdots + \frac{k_p^2}{r_p} \right) \sigma^2$$

where r_i is the number of observations from which the ith mean is computed. The sample estimate, $V(l)$, of σ_l^2 is obtained by substituting $s^2 = $ MSE for σ^2 in σ_l^2. We have

$$V(l) = \left(\frac{k_1^2}{r_1} + \frac{k_2^2}{r_2} + \cdots + \frac{k_p^2}{r_p} \right) \text{MSE}$$

Under the usual assumption that the experimental errors are normally distributed, a test statistic for testing $H_0:L = 0$ against $H_a:L \neq 0$ is

$$t = l/\sqrt{V(l)}$$

which has degrees of freedom equal to the degrees of freedom for MSE.

If we reject $H_0:L = 0$ we might want to obtain an interval estimate, $L(L)$, of the contrast. Since we have a sample estimate, l, of the contrast and the sample variance, $V(l)$, of the estimate we can use the t distribution to obtain the interval estimate, $L(L)$, as

$$L(L) = l \pm t_\alpha \sqrt{V(l)}$$

where, again, t_α has the d.f. of MSE.

5.5.1 Numerical Example: Contrasts

Suppose we have conducted a gasoline mileage test and have obtained the results shown in Table 5.12.

Suppose we want to compare the true mean performance of standards with that of compacts. We could set up the contrast

$$L = (1/2)(\mu_1 + \mu_2) - (1/3)(\mu_3 + \mu_4 + \mu_5)$$

and test the hypothesis $H_0:L = 0$ against $H_a:L \neq 0$. Although $H_0:L = 0$ is testable in its present form, it is easier to do some simplification first:

TABLE 5.12   Results of Gasoline Mileage Test

Type model	Standard		Compact		
	F	*C*	*F*	*C*	*P*
No. of cars	3	2	4	1	2
Mean M.P.G.	13	14	20	19	19

Note: $s^2 = $ MSE $= 2.00$, d.f. $= 7$.

$$H_0: (1/2)(\mu_1 + \mu_2) - (1/3)(\mu_3 + \mu_4 + \mu_5) = 0$$

or $\quad H_0: ((1/2)\mu_1 + (1/2)\mu_2 - (1/3)\mu_3 - (1/3)\mu_4 - (1/3)\mu_5) = 0$

or $\quad H_0: (1/6)(3\mu_1 + 3\mu_2 - 2\mu_3 - 2\mu_4 - 2\mu_5) = 0$

or $\quad H_0: 3\mu_1 + 3\mu_2 - 2\mu_3 - 2\mu_4 - 2\mu_5 = 0$

These forms are equivalent, but the last form is easiest to use because it has been cleared of fractions.

The sample estimate of the contrast is

$$l = 3\bar{y}_1 + 3\bar{y}_2 - 2\bar{y}_3 - 2\bar{y}_4 - 2\bar{y}_5$$
$$= 3(13) + 3(14) - 2(20) - 2(19) - 2(19) = -35.0$$

The sample variance is

$$V(l) = \left[\frac{3^2}{3} + \frac{3^2}{2} + \frac{(-2)^2}{4} + \frac{(-2)^2}{1} + \frac{(-2)^2}{2} \right] 2.00 = 29.00$$

To test $H_0: 3\mu_1 + 3\mu_2 - 2\mu_3 - 2\mu_4 - 2\mu_5 = 0$ the test statistic is

$$t = \frac{-35}{\sqrt{29}} = -6.50$$

In the *t* table we find that $t_{.01(7)} = 3.499$. Since $|-6.50| > 3.499$ we reject H_0 and conclude that gas consumption by standards is significantly ($\alpha = .01$) different from that of compacts.

To illustrate the fact that the results of the test are the same for equivalent forms of the contrast consider the original form

$$H_0: (1/2)\mu_1 + (1/2)\mu_2 - (1/3)\mu_3 - (1/3)\mu_4 - (1/3)\mu_5 = L = 0$$
$$l = (1/2)\bar{y}_1 + (1/2)\bar{y}_2 - (1/3)\bar{y}_3 - (1/3)\bar{y}_4 - (1/3)\bar{y}_5$$
$$= (1/2)(13) + (1/2)(14) + (-1/3)(20) + (-1/3)(19) + (-1/3)(19)$$
$$= -5.833$$

$$V(l) = \left[\frac{(1/2)^2}{3} + \frac{(1/2)^2}{2} + \frac{(-1/3)^2}{4} + \frac{(-1/3)^2}{1} + \frac{(-1/3)^2}{2} \right] 2.00$$

$$= .8056$$

$$t = \frac{l}{\sqrt{V(l)}} = \frac{-5.8333}{\sqrt{.8056}} = -6.50$$

5.6 ORTHOGONAL CONTRASTS

It is possible to form more than one meaningful contrast with a given set of treatments. In fact, with p treatments it is possible to construct $p - 1$ contrasts which are statistically independent of each other. Now, for the $p - 1$ contrasts to be independent they must be mutually orthogonal. Hence, we require the following definition:

Two contrasts

$$L_1 = k_{11}\mu_1 + k_{12}\mu_2 + \cdots + k_{1p}\mu_p$$

and

$$L_2 = k_{21}\mu_1 + k_{22}\mu_2 + \cdots + k_{2p}\mu_p$$

are orthogonal if, and only if,

$$\sum_{j=1}^{p} k_{1j}k_{2j} = 0$$

With p treatments it is always possible to find $p - 1$ orthogonal contrasts even though they may not all be meaningful in the context of the experimental objectives. Further, more than one set of mutually orthogonal contrasts can be found among p treatment means, but each set can contain no more than $p - 1$ mutually orthogonal contrasts.

5.6.1 Examples of Orthogonal Contrasts

Consider the following set of treatments designed to test the effect of gasoline additives:

(1). No Additive   P. P added
   T. T added   TP. Both T and P added

We can construct a couple of different sets of mutually orthogonal contrasts among these four treatments. Each set will contain three contrasts.

TABLE 5.13 A Set of Orthogonal Contrasts Among
Gasoline Additives

Contrast	Treatment			
	(1)	T	P	TP
T vs. P	0	$+1$	-1	0
Alone vs. together	0	-1	-1	$+2$
None vs. some	-3	$+1$	$+1$	$+1$

It is convenient to display the contrasts in tabular form with treatment designators as column headings, contrast descriptions as row designators, and coefficients of the contrasts in the body of the table.

One set of orthogonal contrasts is shown in Table 5.13. A different set of orthogonal contrasts for the same set of treatments is given in Table 5.14.

Because of their usefulness in the examination of experimental data, it might be worthwhile to look at another example of a set of orthogonal contrasts. Consider an experiment designed to compare a number of washday products. Suppose that the following treatments were used:

S. Soap G. Granular detergent
L. Liquid detergent F. Detergent flakes

A meaningful set of orthogonal contrasts might be as presented in Table 5.15.

TABLE 5.14 Different Set of Orthogonal Contrasts
Among Gasoline Additives

Contrast	Treatment			
	(1)	T	P	TP
T	-1	$+1$	-1	$+1$
P	-1	-1	$+1$	$+1$
$T \times P$	$+1$	-1	-1	$+1$

Note: "T" compares all treatments containing T with all treatments without T.

"P" compares all treatments with P against those without.

"$T \times P$" compares the difference between T and no T in the absence and presence of P.

TABLE 5.15 Orthogonal Contrasts Among Washday Products

	Treatment			
Contrast	S	L	G	F
Soap vs. detergent	-3	$+1$	$+1$	$+1$
Liquid vs. solid	0	-2	$+1$	$+1$
Granule vs. flake	0	0	-1	$+1$

5.7 CONTRASTS OF TOTALS

We have considered in some detail the formation and testing of contrasts among means. In many instances it would be convenient if we could construct and test contrasts within the context of the analysis of variance. That is, we should like to be able to partition the treatment sum of squares into a number of components, each associated with a contrast of interest. For this purpose it is convenient to work with contrasts of totals instead of contrasts of means, and we can develop a set of rules which apply when using totals.

For convenience we make the following assumptions, which apply in the usual designed experiment:

1. Treatments are equally replicated.
2. p = number of treatments.
3. r = number of replications per treatment.
4. T_j = sum of the yields on the jth treatment.
5. $SST = (1/r) \sum_j T_j^2 - (\sum_j T_j)^2/rp$ = treatment SS.

Under these assumptions the following rules, which are a simplification of the rules given in Section 5.8, apply:

1. $L = k_1 T_1 + k_2 T_2 + \cdots + k_p T_p$ is a contrast of totals if

$$\sum_{j=1}^{p} k_j = 0$$

2. The sample variance, $V(L)$, of a contrast of totals is

$$V(L) = (r \sum_j k_j^2)s^2 = (r \sum_j k_j^2)MSE$$

3. If L_1 and L_2 are contrasts of totals, and if $\sum_j k_{1j}k_{2j} = 0$ then L_1 and L_2 are orthogonal.

4. If L_1 is a contrast of totals then

$$S_1^2 = (\sum_j k_{1j}T_j)^2/(r \sum_j k_{1j}^2) = L_1^2/(r \sum_j k_{1j}^2)$$

is a single-d.f. component of the treatment sum of squares, SST.

5. If L_1 and L_2 are orthogonal then

$$S_2^2 = L_2^2/(r \sum k_{2j}^2)$$

is a component of SST $- S_1^2$.

6. If $L_1, L_2, \ldots, L_{p-1}$ are mutually orthogonal contrasts of totals then

$$S_1^2 + S_2^2 + \cdots + S_{p-1}^2 = SST$$

7. If $T_1, T_2, \ldots, T_q$ are treatment totals for a subset of $q < p$ treatments then

$$S_{(q-1)}^2 = (1/r) \sum_{j=1}^{q} T_j^2 - (\sum_{j=1}^{q} T_j)^2/rq$$

is a component of SST with $q - 1$ d.f.

By the preceding rules it is possible, but not always easy, to partition the treatment SS into $p - 1$ components, each representing one of the $p - 1$ degrees of freedom for treatment. For a given set of p treatments we can find no more than $p - 1$ single-d.f. sums of squares, in a given partition, which sum to SST. The partition is not unique, however, since there are a number of sets of mutually orthogonal contrasts which define an orthogonal partition. However, for any partition no more than $p - 1$ independent S^2's may be found.

If we construct orthogonal contrasts of totals we can compute a single-degree-of-freedom sum of squares S_i^2, associated with each contrast. We can then use

$$F = S_i^2/MSE$$

with 1, d.f.E degrees of freedom as a test statistic for testing

$$H_0:L_i = k_{i1}\mu_1 + k_{i2}\mu_2 + \cdots + k_{ip}\mu_p = 0$$

against

$$H_a:L_i = k_{i1}\mu_1 + k_{i2}\mu_2 + \cdots + k_{ip}\mu_p \neq 0$$

5.7.1 Numerical Example: Partitioning SST

Suppose the data from the gasoline additive experiment, assuming that each treatment was replicated three times, are as given in Table 5.16.

We can look at two of the possible sets of three orthogonal contrasts. These are given in Tables 5.17 and 5.18.

TABLE 5.16 Data from Gasoline Additive Experiment

		Treatment		
	(1)	*T*	*P*	*TP*
Total (3 rep.) M.P.G.	36	48	42	51

Note: SST = 44.25.

TABLE 5.17 Set of Orthogonal Contrasts of Totals for Additive Experiment

Contrast	Treatment Sum	(1) 36	*T* 48	*P* 42	*TP* 51	*L*	S^2
T vs. *P*		0	+1	−1	0	6	6.00
Alone vs. together		0	−1	−1	+2	12	8.00
Some vs. none		−3	+1	+1	+1	33	30.25
						Σ	44.25

Note: $L_i = \sum\limits_{i=1}^{4} k_{ij}T_j,\ S_i^2 = L_i^2/(r \sum\limits_{i=1}^{4} k_{ij}^2).$

TABLE 5.18 A Different Set of Orthogonal Contrasts for Additive Experiment

Contrast	Treatment Sum	(1) 36	*T* 48	*P* 42	*TP* 51	*L*	S^2
T		−1	+1	−1	+1	21	36.75
P		−1	−1	+1	+1	9	6.75
T × *P*		+1	−1	−1	+1	−3	.75
						Σ	44.25

Note that the individual sums of squares are different in each of the partitions even though both sets contain three mutually orthogonal contrasts. On the other hand, the individual sums of squares add up to the treatment sum of squares in each case.

The rules for constructing and using contrasts of both means and totals are summarized in Section 5.8. The accompanying numerical example in Section 5.9 illustrates the use of contrasts of totals to help explain the results of a fertilizer experiment.

5.8 SUMMARY OF RULES FOR CONTRASTS

5.8.1 Means

1. A linear function, L, of a set of means

$$L = k_1\mu_1 + k_2\mu_2 + \cdots + k_p\mu_p$$

   is a contrast if, and only if, $\sum_{j=1}^{p} k_j = 0$ provided that at least one $k_j \neq 0$.

2. If L is a contrast then the sample estimate of L, l, is given by

$$l = k_1\bar{y}_1 + k_2\bar{y}_2 + \cdots + k_p\bar{y}_p$$

   where $\bar{y}_j$ is the sample estimate of μ_j.

3. If the p samples are independent, and if r_j is the number of observations in the jth sample, then the variance of l, σ_l^2, is given by

$$\sigma_l^2 = \left(\frac{k_1^2}{r_1} + \frac{k_2^2}{r_2} + \cdots + \frac{k_p^2}{r_p} \right) \sigma^2$$

   and the sample variance of l, $V(l)$, is obtained by substituting s^2 for σ^2 in σ_l^2.

4. Two contrasts:

$$L_1 = k_{11}\mu_1 + k_{12}\mu_2 \cdots + k_{1p}\mu_p$$

   and

$$L_2 = k_{21}\mu_1 + k_{22}\mu_2 + \cdots + k_{2p}\mu_p$$

   are *orthogonal* (statistically independent) if, and only if,

$$\sum_{j=1}^{p} k_{1j}k_{2j} = 0$$

5.8.2 Totals

Suppose that $T_j = \sum_{i=1}^{r_j} y_{ij} = y_{.j}$ is the sum of the r_j observations on the jth treatment.

1. A linear function, L, of a set of totals, where

$$L = k_1T_1 + k_2T_2 + \cdots + k_pT_p$$

   is a contrast of totals if, and only if,

$$r_1k_1 + r_2k_2 + \cdots + r_pk_p = 0$$

In particular, if $r_1 = r_2 = \cdots = r_p = r$, then L is a contrast if

$$\sum_{j=1}^{p} k_j = 0$$

2. If L is a contrast of totals and if the samples are independent then the variance of L, σ_L^2, is given by

$$\sigma_L^2 = (r_1 k_1^2 + r_2 k_2^2 + \cdots + r_p k_p^2)\sigma^2$$

and the sample variance of L, $V(L)$, is obtained by substituting s^2 for σ^2 in σ_L^2. In particular, if $r_1 = r_2 = \cdots = r_p = r$, the sample variance of L is:

$$V(L) = r(k_1^2 + k_2^2 + \cdots + k_p^2)s^2 = (r \sum_{j=1}^{p} k_j^2)\text{MSE}$$

3. Two contrasts of totals

$$L_1 = k_{11}T_1 + k_{12}T_2 + \cdots + k_{1p}T_p$$

and

$$L_2 = k_{21}T_1 + k_{22}T_2 + \cdots + k_{2p}T_p$$

are orthogonal if $r_1 k_{11} k_{21} + r_2 k_{12} k_{22} + \cdots + r_p k_{1p} k_{2p} = 0$. In particular, if $r_1 = r_2 = \cdots = r_p = r$, then L_1 and L_2 are orthogonal if

$$\sum_{j=1}^{p} k_{1j} k_{2j} = 0$$

4. If L_1 is a contrast of totals then

$$S_1^2 = L_1^2/D_1 = (\sum_{j=1}^{p} k_{1j}T_j)^2/(r_1 k_{11}^2 + r_2 k_{12}^2 + \cdots + r_p k_{1p}^2)$$

Note: $D_i = \sum_{j=1}^{p} r_j k_{ij}^2$

is single-degree-of-freedom component of the treatment sum of squares. In particular, if $r_1 = r_2 = \cdots = r_p = r$, then

$$S_1^2 = (\sum_{j=1}^{p} k_{1j}T_j)^2/r \sum_{j=1}^{p} k_{1j}^2$$

is a single-d.f. component of SST.

5. If L_1 and L_2 are orthogonal, and if SST is the treatment sum of squares, then $S_2^2 = L_2^2/D_2$ is a component of $\text{SST} - S_1^2$.

6. IF $L_1, L_2, \ldots, L_{p-1}$ are $p - 1$ mutually orthogonal contrasts of totals, then

$$S_1^2 + S_2^2 + \cdots + S_{p-1}^2 = \text{SST}$$

7. If $T_1, T_2, \ldots, T_q$ are totals for a subset of $q < p$ treatments, then

$$S_{(q-1)}^2 = \frac{T_1^2}{r_1} + \frac{T_2^2}{r_2} + \cdots + \frac{T_q^2}{r_q} - \frac{(T_1 + T_2 + \cdots + T_q)^2}{(r_1 + r_2 + \cdots + r_q)}$$

is a component of SST with $q - 1$ degrees of freedom. In particular, if $r_1 = r_2 = \cdots = r_q = r$,

$$S_{(q-1)}^2 = \left(\frac{1}{r}\right)\left(\sum_{j=1}^{q} T_j^2\right) - \left(\frac{1}{rq}\right)\left(\sum_{j=1}^{q} T_j\right)^2$$

is a component of SST with $q - 1$ d.f.

5.8.3 Hypothesis Tests

1. $F = L_i^2/D_i s^2 = S_i^2/\text{MSE}$, with 1 and d.f.E degrees of freedom, is a test statistic for $H_0:L_i = k_{i1}\mu_1 + k_{i2}\mu_2 + \cdots + k_{ip}\mu_p = 0$ against $H_a:L_i \neq 0$.
2. $F = [S_{(q-1)}^2/(q - 1)]/\text{MSE}$, with $q - 1$ and d.f.E degrees of freedom, is a test statistic for $H_0:\mu_1 = \mu_2 = \cdots = \mu_q$ against $H_a:\mu_1 \neq \mu_2 \neq \cdots \neq \mu_q$.

5.9 NUMERICAL EXAMPLE: DESIGN AND ANALYSIS

Suppose we are interested in measuring the response of grass to added nitrogen. Suppose we know that, if it will respond at all, 100 kg/ha is enough. The question to be answered is, "When should the nitrogen be applied?" To answer this question we choose four treatments: (1) check—no nitrogen, (2) fall—100 kg applied in October, (3) spring—100 kg applied in March, (4) split—50 kg in October and 50 kg in March.

The field in which we will conduct the trial has three soil types. We will use a randomized block design with soil type as a blocking factor. We select four plots of each soil type and randomly assign the four treatments to plots within each type. Our aim is to compare the treatments under as nearly uniform conditions as we can, and to remove soil type differences from experimental error. The field plan of the experiment is shown in Fig. 5.1.

Now, assume we have run the experiment, measured the yield of grass (kilograms per plot), and obtained the data shown in Table 5.19.

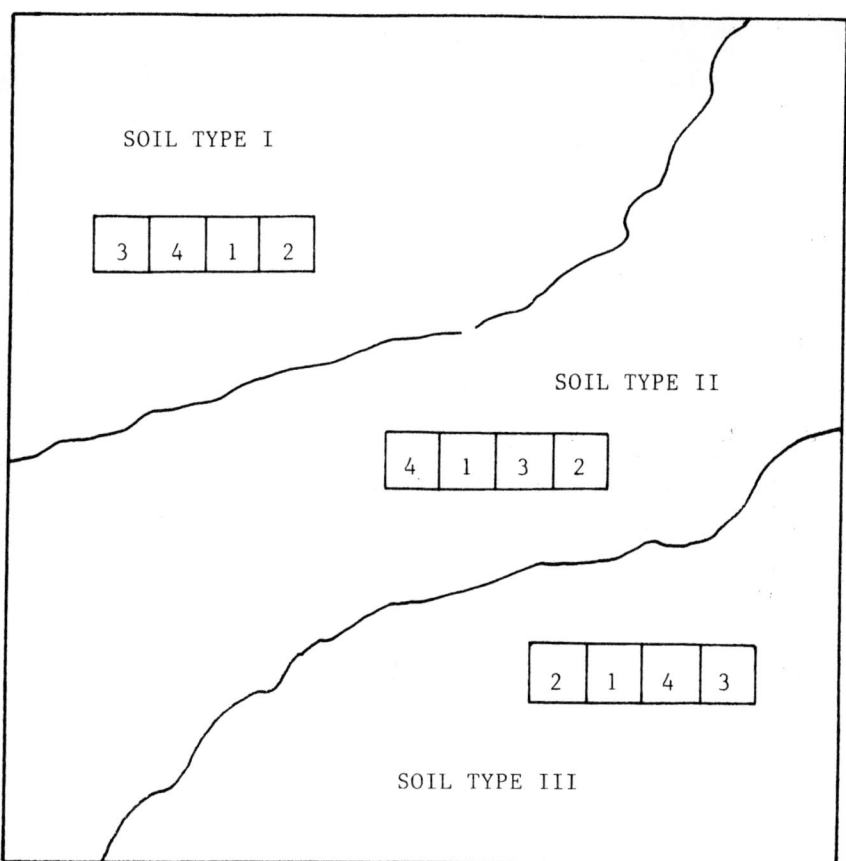

Fig. 5.1 Field plan for a nitrogen fertilizer trial.

The preliminary analysis is presented in Table 5.20. With this analysis we conclude that blocking on soil type resulted in a substantial reduction of experimental error. We also reject the hypothesis of equal treatment effects. We should examine the data in more detail.

We begin with the question, "If we ignore the check is there a difference among nitrogen treatment effects?" That is, can we reject $H_0:\tau_2 = \tau_3 = \tau_4$ in favor of $H_a:\tau_2 \neq \tau_3 \neq \tau_4$? We compute

$$S_N^2 = [(1/3)(37.0^2 + 38.4^2 + 34.2^2)] - [(37.0 + 38.4 + 34.2)^2/3 \times 3]$$

$$= 3.0489, \quad \text{with} \quad 3 - 1 = 2 \text{ d.f.}$$

$$\text{MS}_N = 3.0489/2 = 1.5244$$

$$F_N = 1.5244/.0892 = 17.09, \quad \text{with} \quad 2, 6 \text{ d.f.}$$

TABLE 5.19 Data Table for Nitrogen Fertilizer Trial

Treatment	Type I	Type II	Type III	Sum	Mean
Check	9.9	12.3	11.4	33.6	11.20
Fall	11.4	12.9	12.7	37.0	12.33
Spring	12.1	13.4	12.9	38.4	12.80
Split	10.1	12.2	11.9	34.2	11.40
Sum	43.5	50.8	48.9	143.2	
Mean	10.88	12.70	12.22		11.93

TABLE 5.20 Preliminary ANOVA of Yield of Grass

Source	d.f.	SS	MS	F
Total	11	12.9067		
Soil type	2	7.1717	3.5858	40.21**
Treatment	3	5.2000	1.7333	19.44*
Error	6	.5350	.0892	

**Significant at the 1% level.

Since $F_{.01(3,6)} = 9.78$ we reject H_0 and conclude that the difference among nitrogen treatments is highly significant.

Because there is an apparent difference among effects we might consider constructing a set of $4 - 1 = 3$ mutually orthogonal contrasts among the treatment totals. One set is defined by the set of coefficients given in Table 5.21.

We can display the preceeding computations and the pertinent hypothesis tests in a final ANOVA table, Table 5.22.

Report of statistical analysis. Statistical analysis of yields from an experiment designed to evaluate the timing of nitrogen fertilizer application on grass leads to the following conclusions:

1. Differences among treatment means were highly significant. These means are given in Table 5.23.
2. There was no real difference between fall and spring application. However, a single application in either fall or spring resulted in a significantly higher yield than did an application split between fall and spring.
3. On the average the application of 100 kg/ha of nitrogen resulted in increased yield. This application was probably effective only if it was applied as a single application since there was little increase apparent from a split application.

TABLE 5.21 Orthogonal Contrasts Among Nitrogen Treatments

	Treatment						
Contrast	Check 33.6	Fall 37.0	Spring 38.4	Split 34.2	L	D^*	S^2
Fall vs. spring.	0	−1	+1	0	1.4	6	.3267
Single vs. split	0	+1	+1	−2	7.0	18	2.7222
N vs. no N	−3	+1	+1	+1	8.8	36	2.1511

$^*D = r \sum k_j^2.$

Note: 1. The S^2 for the contrasts which do not involve the check add up to the previously computed SS for N treatments.

$$.3267 + 2.7222 = 3.0489$$

2. The three individual S^2's for the orthogonal contrasts, each with 1 d.f., sum to the treatment SS.

$$.3267 + 2.7222 + 2.1511 = 5.2000$$

TABLE 5.22 Final ANOVA of Grass Yields in Nitrogen Trial

Source	d.f.	SS	MS	F
Total	11	12.9067		
Soil type	2	7.1717	3.5858	40.20**
Treatment	3	5.2000	1.7333	19.43**
Fall vs. spring	1	.3267	.3267	3.66
Single vs. split	1	2.7222	2.7222	30.51**
N vs. no N	1	2.1511	2.1511	24.12**
Error	6	.5350	.0892	

**Significant at the 1% level.

Note: Standard error, $s_{\bar{y}} = \sqrt{MSE/r} = \sqrt{.0892/3} = .1724.$
$CV = 100(\sqrt{MSE}/\bar{y}) = 100(\sqrt{.0892}/11.93) = 2.5\%.$

TABLE 5.23 Mean Yield (kg/plot) of Grass

Treatment	Check	Fall	Spring	Split	Standard error
Mean	11.20	12.33	12.80	11.40	0.17

4. Blocking on soil type resulted in a significant increase in precision in this experiment.

5.10   REGRESSION COMPONENTS OF THE TREATMENT SS

Frequently the treatments in an experiment consist of graded levels of a quantitative variable. For example, we might be interested in the response of a chemical reaction to changes in operating temperature. In this case treatments would be different temperatures and response would be measured as yield of product. In this type of situation it often makes sense to consider the possibility of developing the equation for a curve which describes the relation between yield response and the level of the quantitative treatment variable. In this sense we want to consider the application of regression procedures within the analysis of variance framework.

To illustrate, suppose we have an experiment in which the treatments are four different levels of temperature. Suppose that each temperature is assigned to five randomly selected experimental units (making a completely randomized design). In the usual ANOVA framework we would use $F = \text{MST/MSE}$ to test $H_0: \mu_1 = \mu_2 = \mu_3 = \mu_4$ against $H_a: \mu_1 \neq \mu_2 \neq \mu_3 \neq \mu_4$.

We can get more insight into the nature of the response, however, if we look at temperature as an independent variable, x, and response to temperature as a dependent variable, y. We could then fit a model

$$y = \beta_0 + \beta_1 x + \epsilon$$

and test

$$H_0: \beta_1 = 0$$

to determine whether or not y is linearly related to x.

The analysis of variance table takes the form ($p = 4$, $r = 5$, completely randomized design) shown in Table 5.24.

The basic computations in Table 5.24 proceed in the usual way:

TABLE 5.24   ANOVA for an Experiment with Quantitative Treatments

Source	d.f.	SS	MS	F
Total	19	SSTot		
Treatment	3	SST	MST	F_T
Linear regression	1	SSR	MSR	F_R
Lack of fit	2	SSL	MSL	F_L
Error	16	SSE	MSE	

$$\text{SSTot} = \sum_i \sum_j y_{ij}^2 - \left(\sum_i \sum_j y_{ij}\right)^2 / rp$$

$$\text{SST} = (1/r) \sum_i y_{i.}^2 - \left(\sum_i \sum_j y_{ij}\right)^2 / rp$$

$$\text{SSE} = \text{SSTot} - \text{SST}$$

5.10.1 Partitioning SST

We now want to partition SST into two parts:

1. SSR = sum of squares for the linear regression of y on x
2. SSL = sum of squares for failure of the linear regression model to describe the relationship between x and y (lack of fit)

If we can find SSR then SSL = 'SST − SSR. Note that SSE in this situation corresponds to the sum of squares for pure error in the regression context.

We can compute SSR in one of two ways:

1. Run an ordinary linear regression analysis on the data. (This, in most cases, is the hard way.)
2. Find a set of coefficients which define a linear regression contrast among the treatment totals and compute S^2 for this contrast.

To use the second approach we can use the deviations of the treatment levels from the treatment level mean as coefficients (k's) in a contrast of totals. That is, if x_i is the level of the ith treatment and $\bar{x}$ is the mean of the treatment levels then

$$L_R = (x_1 - \bar{x})T_1 + (x_2 - \bar{x})T_2 + (x_3 - \bar{x})T_3 + (x_4 - \bar{x})T_4$$

Note that $\sum_{i=1}^{4} (x_i - \bar{x}) = 0$ so that L_R is a contrast. And

$$\text{SSR} = S_R^2 = L_R^2 / r \sum_{i=1}^{4} (x_i - \bar{x})^2$$

We now compute SSL = SST − SSR. Mean squares are obtained by dividing the sum of squares by their associated degrees of freedom. All of the F ratios have MSE as a denominator. These F ratios serve as test statistics for the following hypotheses:

1. F_T is a test statistic for

$$H_0 : \mu_1 = \mu_2 = \mu_3 = \mu_4$$

against

$$H_a : \mu_1 \neq \mu_2 \neq \mu_3 \neq \mu_4$$

The hypothesis is that all four treatment means are the same against the alternative that at least one mean differs from the others.

2. F_R is a test statistic for

$$H_0:\beta_1 = 0$$

against

$$H_a:\beta_1 \neq 0$$

The hypothesis is that there is no linear relationship between x and y against the alternative that a linear relation exists.

3. F_L is a test of

$$H_0:E(y) = \beta_0 + \beta_1 x$$

against

$$H_a:E(y) \neq \beta_0 + \beta_1 x$$

The hypothesis is that the simple linear regression model describes the data.

The last test might be illustrated as in Fig. 5.2. Under the regression model the true means of y should fall on the line

$$E(y) = \beta_0 + \beta_1 x$$

SSL is the sum of squares of deviations of the observed y means from the best fitting regression line. The higher are the deviations the larger will be SSL and the stronger is the evidence that a simple regression function does not describe the x, y relation. If F_L is significant there are two courses open:

1. Fit a more complex model.
2. Forget the regression approach.

If F_L is not significant but F_R is, then the conclusion is that linear relationship between y and x does exist. In this case we might want to find the equation of the relationship. Here the problem is to compute b_1, the sample estimate of β_1. We proceed as follows:

$$b_1 = \frac{(x_1 - \bar{x})T_1 + (x_2 - \bar{x})T_2 + (x_3 - \bar{x})T_3 + (x_4 - \bar{x})T_4}{r \sum_i (x_i - \bar{x})^2}$$

The sample regression equation is

$$\hat{y} = \bar{\bar{y}} + b_1(x_i - \bar{x})$$

where

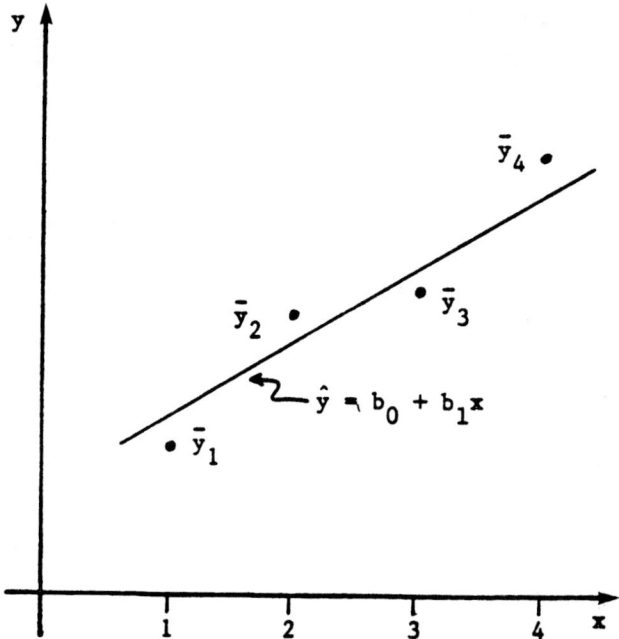

Fig. 5.2 Relationship between yield (y) and temperature (x).

$$\bar{\bar{y}} = (\sum_i T_i)/rp = \text{grand mean of } y$$

x_i = value of x (treatment level) at which y is to be estimated

5.10.2 Orthogonal Polynomials

Under some conditions we can simplify the fitting of curves within the ANOVA. If

1. There are equal numbers of observations at each treatment level
2. The levels of the treatment variable, x, are equally spaced (e.g. 20, 40, 60, 80 or 3, 6, 9, 12, 15)

we can replace ($x_i - \bar{x}$) by a simple set of multipliers called orthogonal polynomial coefficients. These coefficients are given in Table 5.25 for up to eight levels of treatment, x, and for polynomials of up to the fourth order. Other tables are available for x to more levels ($p \leq 100$) and for polynomials of higher order.

In general these orthogonal polynomial coefficients are used to define contrasts of totals. In particular these are contrasts associated with the linear (first-order), quadratic (second-order), cubic (third-order), etc. terms of the polynomial

TABLE 5.25 Orthogonal Polynomial Coefficients

No. of levels	Order	Ordered treatment number								D_i Divisor	λ
		1	2	3	4	5	6	7	8		
3	1	−1	0	+1						2	1
	2	+1	−2	+1						6	3
4	1	−3	−1	+1	+3					20	2
	2	+1	−1	−1	+1					4	1
	3	−1	+3	−3	+1					20	10/3
5	1	−2	−1	0	+1	+2				10	1
	2	+2	−1	−2	−1	+2				14	1
	3	−1	+2	0	−2	+1				10	5/6
	4	+1	−4	+6	−4	+1				70	35/12
6	1	−5	−3	−1	+1	+3	+5			70	2
	2	+5	−1	−4	−4	−1	+5			84	3/2
	3	−5	+7	+4	−4	−7	+5			180	5/3
	4	+1	−3	+2	+2	−3	+1			28	7/12
7	1	−3	−2	−1	0	+1	+2	+3		28	1
	2	+5	0	−3	−4	−3	0	+5		84	1
	3	−1	+1	+1	0	−1	−1	+1		6	1/16
	4	+3	−7	+1	+6	+1	−7	+3		154	7/12
8	1	−7	−5	−3	−1	+1	+3	+5	+7	168	2
	2	+7	+1	−3	−5	−5	−3	+1	+7	168	1
	3	−7	+5	+7	+3	−3	−7	−5	+7	264	2/3
	4	+7	−13	−3	+9	+9	−3	−13	+7	616	7/12

Note: 1. Order 1 = linear; order 2 = quadratic; order 3 = cubic; order 4 = quartic.

2. $D_i = \sum_{j=1}^{p} k_{ij}^2$, $SS_i = (\sum_{j=1}^{p} k_{ij}T_j)^2/rD_i = L_i^2/rD_i$.

equation relating yield, y, and treatment level, x. With these contrasts we can follow the usual procedures to compute a single-degree-of-freedom sum of squares associated with the contrast, and use an F ratio to test the hypothesis that the contrast is zero. This is equivalent to a test of the hypothesis that the corresponding polynomial coefficient is zero.

In the following development suppose that

p = number of equally spaced quantitative treatments

r = number of observations per treatment

$$x_j = j\text{th treatment level}$$

$$\bar{x} = \text{mean treatment level} = (1/p) \sum_j x_j$$

$$T_j = \text{sum of the } y\text{'s on the } j\text{th treatment}$$

$$k_{ij} = i\text{th-order polynomial coefficient for the } j\text{th treatment}$$

Then

$$L_i = \sum_{j=1}^{p} k_{ij} T_j$$

is the contrast of totals associated with the ith-order term of the polynomial, and

$$S_i^2 = (\sum_{j=1}^{p} k_{ij} T_j)^2 / r \sum_{j=1}^{p} k_{ij}^2 = L_i^2 / r \sum_j k_{ij}^2$$

is the single-degree-of-freedom sum of squares associated with the ith-order term. The ratio

$$F = S_i^2 / \text{MSE}$$

provides a test of the hypothesis that the regression coefficient for the ith order term in the polynomial is zero.

The usual procedure is to fit successive terms of the polynomial sequentially. Suppose that SST is the treatment sum of squares with $p - 1$ d.f. The procedure is:

1. Compute S^2 linear.
2. Compute $S_{\text{LOF}}^2 = \text{SST} - S_{\text{lin}}^2$ with $p - 2$ d.f.
3. Test "lack of fit" against MSE.
4. Compute S_{quad}^2.
5. Compute $S_{\text{LOF}}^2 = \text{SST} - S_{\text{lin}}^2 - S_{\text{quad}}^2$ with $p - 3$ d.f.
6. Test "lack of fit."

$\vdots$

We would continue in this way, adding one term at a time, until lack of fit is no longer significant. The final equation should contain all terms up to and including the term at which lack of fit first becomes nonsignificant.

To illustrate the computation of the individual sums of squares suppose we want to look at the following model:

$$y = \beta_0 + \beta_1 x + \beta_{11} x^2 + \beta_{111} x^3 + \beta_{1111} x^4$$

Assume we have conducted an experiment with five equally spaced levels of x ($p = 5$) each replicated four times ($r = 4$). The individual sums of squares would be computed as

1. For $H_0: \beta_1 = 0$

$$S_{\text{linear}}^2 = \frac{(-2T_1 - T_2 + 0T_3 + T_4 + 2T_5)^2}{(4)(10)}$$

2. For $H_0: \beta_{11} = 0$

$$S_{\text{quadratic}}^2 = \frac{(2T_1 - T_2 - 2T_3 - T_4 + 2T_5)^2}{(4)(14)}$$

3. For $H_0: \beta_{111} = 0$

$$S_{\text{cubic}}^2 = \frac{(-T_1 + 2T_2 + 0T_3 - 2T_4 + T_5)^2}{(4)(10)}$$

4. For $H_0: \beta_{1111} = 0$

$$S_{\text{quartic}}^2 = \frac{(T_1 - 4T_2 + 6T_3 - 4T_4 + T_5)^2}{(4)(70)}$$

5.10.3   Estimation

Having fit the equation and tested the terms for significance, we might now like to find the expected values of the y's at the several treatment levels. For this purpose we first compute

$$b_i = \left(\sum_{j=1}^{p} k_{ij} T_j \right) / r \sum_{j=1}^{p} k_{ij}^2$$

then

$$\hat{y}_j = \bar{y} + b_1 k_{1j} + b_2 k_{2j} + b_3 k_{3j} + b_4 k_{4j}$$

where

$\hat{y}_j$ = expected yield at the jth treatment level

$\bar{y}$ = observed mean yield

k_{ij} = orthogonal polynomial coefficient

To illustrate the procedure consider, again, four observations on each of five equally spaced treatments. The orthogonal polynomial coefficients are given in Table 5.26. First compute $\bar{y}$, b_1, b_2, b_3, b_4 as above. Then

$$\hat{y}_1 = \bar{y} + b_1(-2) + b_2(+2) + b_3(-1) + b_4(+1)$$
$$\hat{y}_2 = \bar{y} + b_1(-1) + b_2(-1) + b_3(+2) + b_4(-4)$$
$$\hat{y}_3 = \bar{y} + b_1(\ 0) + b_2(-2) + b_3(\ 0) + b_4(+6)$$

TABLE 5.26 Orthogonal Polynomial Coefficients for Five Treatments

	Treatment (j)				
Order (i)	1	2	3	4	5
1	−2	−1	0	+1	+2
2	+2	−1	−2	−1	+2
3	−1	+2	0	−2	+1
4	+1	−4	+6	−4	+1

$$\hat{y}_4 = \bar{y} + b_1(+1) + b_2(-1) + b_3(-2) + b_4(-4)$$
$$\hat{y}_5 = \bar{y} + b_1(+2) + b_2(+2) + b_3(+1) + b_4(+1)$$

If we want to obtain the equation in terms of the original x's the problem is more complicated. In addition to the symbols previously defined let

$\Delta = x_j - x_{j-1} = $ difference between treatment levels

$\lambda_i = $ ith-order multiplier from the table of orthogonal polynomial coefficients

First compute

$$a_i = \lambda_i b_i$$

Then compute

$$\zeta_{1j} = (x_j - \bar{x})/\Delta$$
$$\zeta_{2j} = [(x_j - \bar{x})/\Delta]^2 - (p^2 - 1)/12$$
$$\zeta_{3j} = [(x_j - \bar{x})/\Delta]^3 - (3p^2 - 7)[(x_j - \bar{x})/\Delta]/20$$
$$\zeta_{4j} = [(x_j - \bar{x})/\Delta]^4 - (3p^2 - 13)[(x_j - \bar{x})/\Delta]^2/14 + 3(p^2 - 1)(p^2 - 9)/560$$

Finally,

$$\hat{y}_j = \bar{y} + a_1\zeta_{1j} + a_2\zeta_{2j} + a_3\zeta_{3j} + a_4\zeta_{4j}$$

The combination of regression and ANOVA is illustrated in the numerical example presented in Section 5.10.4.

5.10.4 Numerical Example: Regression in the ANOVA

An experiment was conducted to determine the effect of storage temperature on the potency of an antibiotic. Fifteen samples of the antibiotic were obtained and three samples, selected at random, were stored at each of five temperatures: 10°,

30°, 50°, 70°, and 90°. After 30 days of storage the samples were tested for potency. The results are given in Table 5.27.

We first compute the preliminary ANOVA shown in Table 5.28.

Since there are highly significant differences among temperature effects we should partition SST into meaningful components. We see that temperature is a

TABLE 5.27 Data from an Antibiotic Potency Experiment

	Temperature					
	10°	30°	50°	70°	90°	
	62	26	16	10	13	
	55	36	15	11	11	
	57	31	23	18	9	Total
Sum	174	93	54	39	33	393
Mean	58	31	18	13	11	26.2

TABLE 5.28 ANOVA of Antibiotic Potency

Source	d.f.	SS	MS	F
Total	14	4680.40		
Temperature	4	4520.40	1130.10	70.63**
Error	10	160.00	16.00	

**Significant at the 1% level.

TABLE 5.29 Orthogonal Partition of Temperature Effects

		Treatment							
		10°	30°	50°	70°	90°			
Order	Sum	174	93	54	39	33	L_i	rD_i	S_i^2
Linear		-2	-1	0	$+1$	$+2$	-336	30	3763.20
Quadratic		$+2$	-1	-2	-1	$+2$	174	42	720.86
Cubic		-1	$+2$	0	-2	$+1$	-33	30	36.30
Quartic		$+1$	-4	$+6$	-4	$+1$	3	210	.04

Note: $L_i = \sum_{j=1}^{5} k_{ij} T_j$, $rD_i = r \sum_{j=1}^{5} k_{ij}^2$, $S_i^2 = L_i^2 / rD_i$.

TABLE 5.30 Sequential Test of Polynomial Temperature Effects

	Source	d.f.	SS	MS	F
(1)	Temperature	4	4520.40	1130.10	70.63**
(2)	Linear	1	3763.20	3763.20	235.20**
	Deviation from linear	3	757.20	252.40	15.78**
(3)	Quadratic	1	720.86	720.86	45.05**
	Deviation from linear and quadratic	2	36.34	18.17	1.14 N.S.

Note: MSE = 16.00, with 10 d.f., is the divisor for all F ratios.
**Significant at the 1% level.

quantitative variable, that the levels are equally spaced, a difference of 20° between levels, and that each level is equally replicated. We can use the orthogonal polynomial coefficients to partition SST into components associated with successive terms of a polynomial equation relating potency to storage temperature. This is displayed in Table 5.29.

We now add one term at a time to the polynomial and test for lack of fit after each addition. As soon as lack of fit becomes nonsignificant we stop. The terms added to this point define the relation we seek. In tabular form the sequence of tests is presented in Table 5.30.

After the quadratic term is added lack of fit is no longer significant. We conclude that potency is a quadratic function of temperature. We can now compute the expected potency under the assumed quadratic function. To do this we can use the polynomial coefficients. We compute

$$\bar{y} = \sum_j T_j/rp = \quad 393/15 = \quad 26.20$$

$$b_1 = L_1/rD_1 \quad = -336/30 = -11.20$$

$$b_2 = L_2/rD_2 \quad = \quad 174/42 = \quad 4.14$$

Then

$$\hat{y} = \bar{y} + b_1 k_{1j} + b_2 k_{2j} = 26.20 + (-11.20)k_{1j} + (4.14)k_{2j}$$

For example, at 10°

$$\hat{y}_{10} = 26.20 + (-11.20)(-2) + (4.14)(+2) = 56.88$$

In a similar way we can compute the expected potency at the other temperature levels:

Temperature	10°	30°	50°	70°	90°
Observed mean	58	31	18	13	11
Expected mean	56.9	33.3	17.9	10.9	12.1

We can go one step farther and derive the equation relating potency and storage temperature. For this we use the expression

$$\hat{y}_j = \bar{y} + a_1\zeta_{1j} + a_2\zeta_{2j}$$
$$= \bar{y} + \lambda_1 b_1(x_j - \bar{x})/\Delta + \lambda_2 b_2\{[(x_j - \bar{x})/\Delta]^2 - (p^2 - 1)/12\}$$

We see first that $\bar{x} = 50$, $\Delta = 20$, and that $\lambda_1 = \lambda_2 = 1$. Then

$$\hat{y} = 26.20 + (1)(-11.20)(x_j - 50)/20$$
$$+ (1)(4.14)\{[(x_j - 50)/20]^2 - (5^2 - 1)/12]$$
$$= 71.7950 - 1.5950x_j + .0104x_j^2$$

where:

$\hat{y}_j$ = expected potency

x_j = storage temperature

Report of statistical analysis. Over the 30-day storage period used in this study temperature had a highly significant effect on the potency of the antibiotic. Table 5.31 shows that potency is reduced as storage temperature is increased.

Further analysis of the data indicates that, over the temperature range 10° to 90°, potency may be described by the equation

$$\hat{y} = 71.80 - 1.60x + .01x^2$$

which

$\hat{y}$ = expected potency

x = 30-day storage temperature

TABLE 5.31 Mean Potency of Antibiotic Stored for 30 Days at Various Temperatures

Temperature	10°	30°	50°	70°	90°	Standard error
Mean potency	58	31	18	13	11	2.31

FURTHER READING*

Anderson, V. L., and R. A. McLean (1974), Ch. 1 and 2.
Cochran, W. G., and G. M. Cox (1957), Ch. 3.
Snedecor, G. W., and W. G. Cochran (1980), Ch. 12 and 19.
Steel, R. G. D., and J. H. Torrie (1980), Ch. 8 and 15.

REFERENCES

Carmer, S. G., and M. R. Swanson (1973). An Evaluation of Ten Pairwise Multiple Comparison Procedures by Monte Carlo Methods, *J. Amer. Stat. Assoc.* 68:66–74.

Carmer, S. G., and W. M. Walker (1982). Baby Bear's Dilemma: A Statistical Tale, *Agron. J.* 74:122–124.

Chew, V. (1976). Comparing Treatment Means: A Compendium, *HortScience* 11:348–357.

Cochran, W. G., and G. M. Cox (1957). *Experimental Designs,* 2nd ed., Wiley, New York, pp. 73–76.

Duncan, D. B. (1975). *t* Tests and Intervals for Comparisons Suggested by the Data, *Biometrics* 31:339–359.

Fisher, R. A. (1966). *The Design of Experiments,* 8th ed., Hafner, New York, Ch. 10.

Little, T. M. (1978). If Galileo Had Published in HortScience, *HortScience* 13:504–506.

Miller, R. G., Jr. (1966). *Simultaneous Statistical Inference,* McGraw-Hill, New York.

Petersen, R. G. (1977). Use and Misuse of Multiple Comparison Procedures, *Agron. J.* 69:205–208.

Steel, R. G. D., and J. H. Torrie (1980). *Principles and Procedures of Statistics,* 2nd ed., McGraw-Hill, New York, pp. 186–187.

Tukey, J. W. (1951). Quick and Dirty Methods in Statistics, Part II: Simple Analyses for Standard Designs, *Proc. 5th Annu. Conv. American Society for Quality Control,* 189–197.

Waller, R. A., and D. B. Duncan (1969, 1972). A Bayes Rule for the Symmetric Multiple Comparison Problem, *J. Amer. Stat. Assoc.* 64:1484–1503; Corregenda, *J. Amer. Stat. Assoc.* 67:253–255.

---

*See Bibliography for complete citation.

6

Factorial Experiments

6.1 INTRODUCTION

We have considered in some detail the assignment of treatments to the experimental units (the experimental design) and the analysis of the data at the conclusion of the trial. So far we have not given much attention to the selection of treatments to be included in the experiment. We might introduce this topic by considering an example. Suppose we want to introduce a new crop, triticale for example, into an area in which it has never before been grown, and we want to conduct an experiment to see how it will yield. There are a number of questions which must be answered in planning the experiment:

1. When should the crop be planted?
2. What should be the seeding rate?
3. Should the seed be drilled or broadcast?
4. Must we use fertilizer?
5. If so, how much of the major elements are needed?
6. Do we have to add minor elements?
7. Is irrigation necessary?
   ⋮

The traditional approach is to vary one factor at a time. For example, in one trial we could use different planting dates while holding other factors constant. Another experiment would be used to study seed rate, another for nitrogen fertilizer, another for phosphorus, etc. Although at first glance the idea is appealing, there are a number of problems with the one-factor-at-a-time approach. In order to study the effect of varying planting date, for instance, it is necessary to choose a seeding rate, fertilizer schedule, etc. Further, a planting date which produces a maximum yield for one variety might not be the date which would produce a maximum yield with a different variety.

6.1.1 Interaction

Sometimes the factors act independently of each other. By this we mean that changing the level of one factor produces the same effect at all levels of another factor. For example, the effect of different row spacings is the same at all planting dates if spacing and date are independent of each other. Often, however, the effects of two or more factors are not independent. A tall wheat variety might outyield a short variety in closely spaced rows while, because of lodging, the short variety might outyield the tall one in widely spaced rows. When two factors do not behave independently they are said to interact. *Interaction* is the failure of the differences in response to changes in levels of one factor to be the same at all levels of another factor. When two factors interact the response to changes in one factor is conditioned by the level of the other.

These concepts might be illustrated by considering some possible outcomes of an experiment in which three washday products are tested at four water temperatures. In one case a graph of the results might appear as in Fig. 6.1. In this case there is no product × temperature interaction. The differences between products are the same at all temperatures. Hence these factors are acting independently.

On the other hand, the outcome might be as shown in Fig. 6.2. In this instance there is a product × temperature interaction. As the level of temperature is increased the differences among products also increase. We cannot say anything about the difference between two products without at the same time specifying the temperature.

To illustrate these ideas further, consider the results of a trial designed to measure the effect of two gasoline additives, T and P, both alone and in combination, on gas mileage. Suppose that the mileages given in Table 6.1 were obtained.

1. We can compute the differences in mileage between added T and no T in the presence and absence of P. These are called the *simple effects* of T. We see that the simple effect of T in the absence of P is 10, while it is 15 in the presence of P. Similarly, we can compute the simple effects of P.
2. The average of the simple effects is called the *main effect*. For this trial the main effect of T is 12.5, while the main effect of P is 7.5.
3. The difference between simple effects is the *interaction*. In this example the $T \times P$ interaction is 5. If there were no interaction the simple effect of T would be the same at both levels of P.

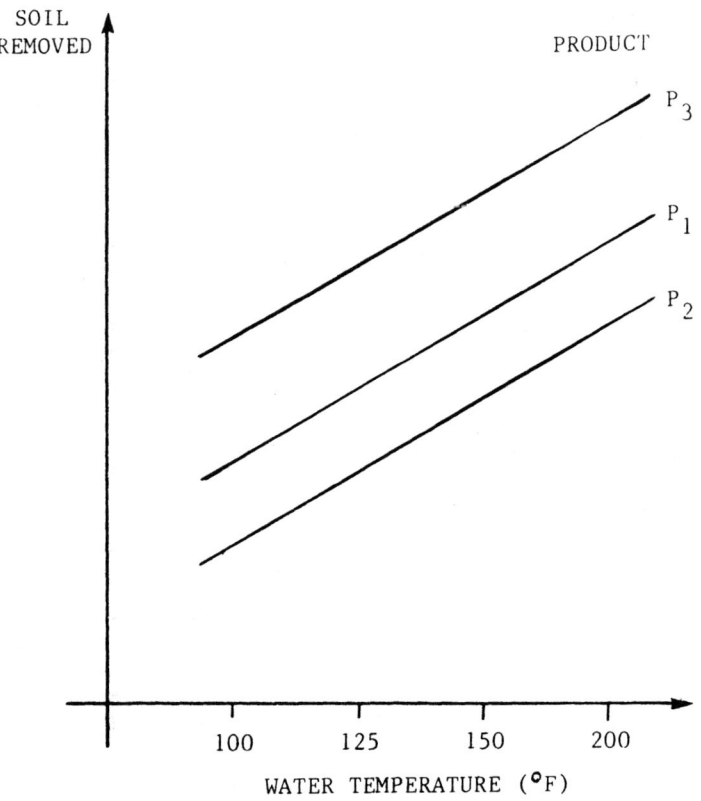

SOIL REMOVED

PRODUCT

P_3

P_1

P_2

WATER TEMPERATURE (°F)

100   125   150   200

Fig. 6.1   Soil removed by three washday products at different temperatures.

If interactions exist, and they are fairly common, we should plan our experiments in such a way that they can be estimated and tested. It is clear that we cannot do this if we vary only one factor at a time. For this purpose we must use multilevel, multifactor experiments.

6.2   FACTORIAL EXPERIMENTS

The most commonly used type of multifactor experiment is the factorial experiment. In a factorial experiment the treatments consist of combinations of two or more factors each at two or more levels. The combinations are such that each level of every factor occurs together with each level of every other factor. The number of treatments is the product of the number of levels of all factors.

We have already seen some examples of factorial sets of treatments. In the "paint life" example (Section 5.5) there are two factors, climate and wood type,

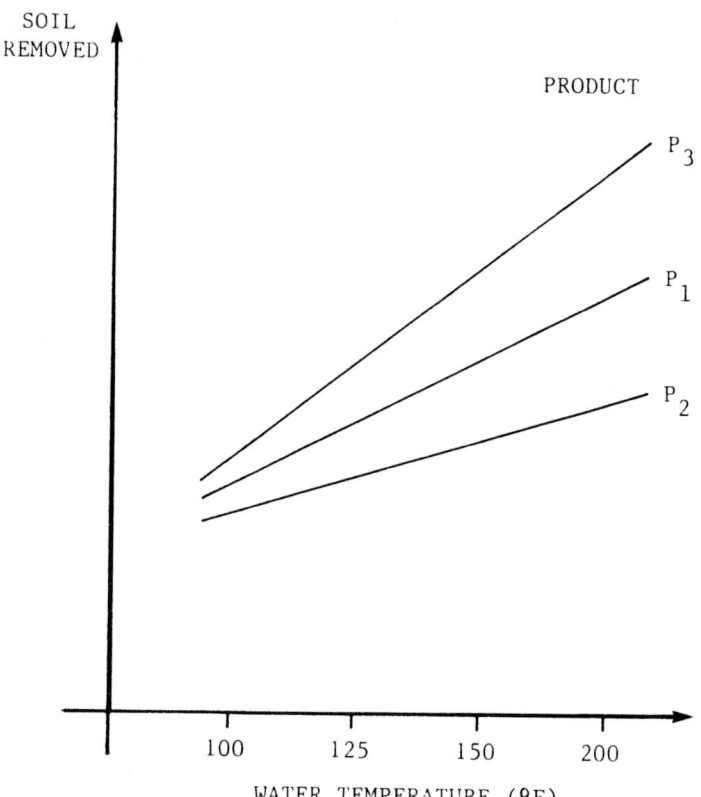

Fig. 6.2 Soil removed by three washday products at different temperatures.

TABLE 6.1 Simple Effects, Main Effects, and Interactions of Gasoline Additives on Mileage

| | *T* | | | Simple | Main | |
P	None	Some	Mean	effect	effect	Interaction
None	15	25	20.0	10	12.5	5
Some	20	35	27.5	15		
Mean	17.5	30				
Simple effect	5	10				
Main effect		7.5				
Interaction		5				

each at two levels. This is called a 2 × 2 factorial experiment. The gasoline additive example is another 2 × 2 factorial with *T* and *P*, each at two levels, as factors. The levels of each factor were absence and presence. There were 2 × 2 = 4 treatment combinations:

1. Both absent
2. *T* absent, *P* present
3. *T* present, *P* absent
4. Both present

We should point out that the term "factorial experiment" refers to the treatment combinations, not the type of experimental design. The factorial set of treatments can be used in any design: completely randomized, randomized block, Latin square, etc.

6.2.1 Advantages

There are a number of advantages to factorial experiments. When the factors are *independent* there are two principal advantages:

1. All of the simple effects of a factor are equal to the main effect. Hence, main effects are all that are needed to describe the action of a factor.
2. Hidden replication: Each main effect is estimated with the same precision as if the entire trial had been devoted to that factor alone. In the gasoline additive example, for illustration, half of the treatments contain *T* while half do not. The same holds true for *P*. Two single-factor experiments would require twice the number of units to attain the same precision as the factorial.

When there is *interaction* factorial experiments provide a systematic set of factor combinations for estimating all interactions, each with equal precision.

6.2.2 Disadvantages

Factorial experiments have two primary disadvantages.

1. As the number of factors increases the size of the experiment becomes very large. With eight factors each at two levels, for example, there are 256 combinations in the full factorial set. Not only are experiments with this many treatments costly to run, but it is also difficult to find sufficient uniform material to form blocks big enough to accommodate a full replication.
2. Large factorials may be difficult to interpret, particularly when interactions are present.

6.2.3 Uses

Factorial experiments are useful in a number of situations:

1. In exploratory experiments, where the aim is to examine a large number of factors to determine which are important and which are not.
2. To study relationships among several factors, in particular to determine the presence and magnitude of interactions.
3. In experiments designed to lead to recommendations over a wide variety of conditions. Some of the conditions can be included as factors in the trial even though they are not the principal factors of interest.

6.3 TWO-FACTOR EXPERIMENTS

The simplest kind of factorial experiment is one with only two factors. For example, a food processor might be interested in the effects of storage temperature and the length of storage on the quality of frozen strawberries. Suppose he chose two temperatures:

$$t_1 = -10°C, \qquad t_2 = -20°C$$

and four storage times:

$$s_1 = 1 \text{ mo}, \qquad s_2 = 2 \text{ mo}, \qquad s_3 = 3 \text{ mo}, \qquad s_4 = 4 \text{ mo}$$

The factorial set of treatments in this case would contain $2 \times 4 = 8$ treatments:

1. $t_1 s_1$ 5. $t_2 s_1$
2. $t_1 s_2$ 6. $t_2 s_2$
3. $t_1 s_3$ 7. $t_2 s_3$
4. $t_1 s_4$ 8. $t_2 s_4$

The experiment could be run with an experimental design chosen to fit the conditions. More often than not, a randomized block design, with blocks of eight experimental units, would be used. At the conclusion of the experiment the data would be analyzed in such a way that the main effects of temperature and time, and the time $\times$ temperature interaction, could be estimated and tested.

6.3.1 Data Analysis

To illustrate the process in general, suppose we have two factors: factor A at a levels and factor B at b levels. Suppose that the trial is run using a randomized block design with r blocks each containing ab units. We could describe an observation from this experiment with the following model:

$$y_{ijk} = \mu + \rho_i + \alpha_j + \beta_k + (\alpha\beta)_{jk} + \epsilon_{ijk}$$

where

y_{ijk} = yield of the jth level of factor A, kth level of factor B in the ith block

μ = overall mean yield

ρ_i = effect of the ith block, $\rho_i \sim N(0, \sigma_r^2)$

α_j = added effect of the jth level of factor A measured as a deviation from μ, $\sum_j \alpha_j = 0$

β_k = added effect of the kth level of factor B measured as a deviation from μ, $\sum_k \beta_k = 0$

$(\alpha\beta)_{jk}$ = added effect of the combination of the jth level of A with the kth level of B, the $A_j \times B_k$ interaction effect, $\sum_j (\alpha\beta)_{jk} = \sum_k (\alpha\beta)_{jk} = 0$

ϵ_{ijk} = random error, $\epsilon_{ijk} \sim N(0, \sigma^2)$

To begin the data analysis we first compute

$$T_{jk} = y_{.jk} = \sum_i y_{ijk}$$

Then we construct two tables of totals, Table 6.2 and 6.3.

TABLE 6.2   Factor A × Factor B Totals

Factor A	Factor B				
	1	2	$\cdots$	b	Sum
1	T_{11}	T_{12}	$\cdots$	T_{1b}	A_1
2	T_{21}	T_{22}	$\cdots$	T_{2b}	A_2
$\vdots$	$\vdots$	$\vdots$		$\vdots$	$\vdots$
a	T_{a1}	T_{a2}	$\cdots$	T_{ab}	A_a
Sum	B_1	B_2	$\cdots$	B_b	G

Note: $A_j = \sum_k T_{jk} = y_{.j.}$

$B_k = \sum_j T_{jk} = y_{..k}$

$G = \sum_j A_j = \sum_k B_k = y_{...}$

TABLE 6.3   Block Totals

Block	1	2	$\cdots$	r	Sum
Sum	R_1	R_2	$\cdots$	R_r	G

Note: $R_i = \sum_j \sum_k y_{ijk} = y_{i..}$

TABLE 6.4   Format for ANOVA of Two-Factor Factorial Experiment

Source	d.f.	SS	MS	F
Total	$rab - 1$	SSTot		
Block	$r - 1$	SSR	MSR	F_R
A	$a - 1$	SSA	MSA	F_A
B	$b - 1$	SSB	MSB	F_B
AB	$(a - 1)(b - 1)$	SSAB	MSAB	F_{AB}
Error	$(r - 1)(ab - 1)$	SSE	MSE	

From these tables we can compute the entries in an analysis of variance shown in Table 6.4. The sums of squares in Table 6.4 are computed using the following:

1. Correction term, $C = G^2/rab$

2. $\text{SSTot} = \sum_i \sum_j \sum_k y_{ijk}^2 - C$

3. $\text{SSR} = (1/ab) \sum_i R_i^2 - C$

4. $\text{SSA} = (1/rb) \sum_j A_j^2 - C$

5. $\text{SSB} = (1/ra) \sum_k B_k^2 - C$

6. $\text{SSAB} = (1/r) \sum_j \sum_k T_{jk}^2 - C - \text{SSA} - \text{SSB}$

7. $\text{SSE} = \text{SSTot} - \text{SSR} - \text{SSA} - \text{SSB} - \text{SSAB}$

Mean squares are obtained by dividing the sums of squares by their associated degrees of freedom. *F* ratios are computed by dividing the appropriate mean squares by MSE. Finally, we usually compute the coefficient of variation, CV, as

$$CV = (\sqrt{MSE}/\bar{\bar{y}})100$$

where

$$\bar{\bar{y}} = G/rab$$

6.3.2 Significance Tests

1. $F_A = MSA/MSE$, with $a - 1$, $(r - 1)(ab - 1)$ d.f., is a test statistic for the hypothesis that the main effect of factor A is zero. That is, it tests $H_0: \mu_1 = \mu_2 = \cdots = \mu_a = 0$. If we reject this hypothesis we conclude that there are significant differences among the means of the A factor levels.
2. $F_B = MSB/MSE$, with $b - 1$, $(r - 1)(ab - 1)$ d.f., is a test statistic for the hypothesis that the main effect of factor B is zero.
3. $F_{AB} = MSAB/MSE$, with $(a - 1)(b - 1)$, $(r - 1)(ab - 1)$ d.f., is a test statistic for the hypothesis that there is no $A \times B$ interaction. If we reject this hypothesis we conclude that factor A and factor B are not acting independently of each other.
4. $F_R = MSR/MSE$, with $(r - 1)$, $(r - 1)(ab - 1)$ d.f., is an approximate test statistic for the significance of differences among blocks. If F_R is larger than $F_{(table)}$ we conclude that blocking was effective in reducing experimental error.

6.3.3 Standard Errors

There are three types of means in a two-factor factorial experiment:

1. A factor means
2. B factor means
3. Treatment (AB) means

Associated with each type of mean and with differences between each type of mean is a standard error:

1. Factor A means, $\bar{y}_{.j.} = A_j/rb$.
   a. The standard error, $s_{\bar{y}A}$, of a factor A mean is $s_{\bar{y}A} = \sqrt{MSE/rb}$. Hence, the $(1 - \alpha)100\%$ confidence interval estimate of μ_j is

   $$L(\mu_j) = L(\mu + \alpha_j) = \bar{y}_{.j.} \pm t_\alpha \sqrt{MSE/rb}$$

   b. The standard error, $s_{\bar{d}A}$, of the difference between two factor A means is

   $$s_{\bar{d}A} = \sqrt{2MSE/rb}$$

   If we want to test the significance of the difference between two factor A means we can use

$$t = \frac{\bar{y}_{.j.} - \bar{y}_{.j'}}{\sqrt{2MSE/rb}}$$

with $(r - 1)(ab - 1)$ d.f.

2. Factor B means, $\bar{y}_{..k} = B_k/ra$.

a. The standard error, $s_{\bar{y}_B}$, of a factor B mean is

$$s_{\bar{y}_B} = \sqrt{MSE/ra}$$

b. The standard error, $s_{\bar{d}_B}$, of the difference between two factor B means is

$$s_{\bar{d}_B} = \sqrt{2MSE/ra}$$

These standard errors can be used in interval estimation and hypothesis tests in the same way as for the factor A means.

3. Treatment (AB) means, $\bar{y}_{.jk} = T_{jk}/r$.

a. The standard error, $s_{\bar{y}_{AB}}$, of the mean of any combination of factor A and factor B is

$$s_{\bar{y}_{AB}} = \sqrt{MSE/r}$$

b. The standard error, $s_{\bar{d}_{AB}}$, of the difference between two treatment means is

$$s_{\bar{d}_{AB}} = \sqrt{2MSE/r}$$

Note: The preceding interval estimates and significance tests hold under the normality assumption about ϵ_{ijk}. They are special cases of the more general formulation:

1. If $\hat{\theta}$ is a normally distributed estimator of the parameter θ, and if $s_{\hat{\theta}}$ is the sample standard error of $\hat{\theta}$, then

$$t = (\hat{\theta} - \theta)/s_{\hat{\theta}}$$

follows the t distribution with d.f. equal to the d.f. for $s_{\hat{\theta}}$.

2. The $(1 - \alpha)100\%$ confidence interval estimate of θ, $L(\theta)$, is given by

$$L(\theta) = \hat{\theta} \pm t_\alpha s_{\hat{\theta}}$$

3. A test statistic for $H_0{:}\theta = \theta_0$, where θ is the true value and θ_0 is the hypothetical value, is

$$t = (\theta - \theta_0)/s_{\hat{\theta}}$$

6.3.4 Interpretation

The interpretation of the results of a factorial experiment depends on the outcome of the significance tests. If the $A \times B$ interaction is significant then the main effects have no real meaning whether they are significant or not. This is because with interaction the response to changes in factor *A* depends on the level of factor *B*, and the mean for a level of *A* averaged over all levels of *B* has little meaning. In this case the results of the experiment are best summarized in a two-way table of means of the various *A* × *B* combinations, along with the standard error of these means.

If interaction is not significant then all of the information in the trial is contained in the significant main effects. The factors are independent and the difference between two *A* levels is essentially the same at all levels of *B*. In this case the results may be summarized in tables of means for factors with the significant main effects.

Details of the analysis and interpretation of a two-factor factorial experiment are presented in the accompanying numerical example.

6.3.5 Numerical Example: Two-Factor Factorial

An agronomist wanted to study the effect of different rates of phosphorus fertilizer on two types of broad bean (*Vicia faba*) plants. He thought that the plant types might respond differentially to fertilization so he decided to do a factorial experiment with two factors:

1. Plant type at two levels:

    T_1 = short, bushy

    T_2 = tall, erect

2. Phosphorus rate at three levels:

    P_1 = none

    P_2 = 25 kg/ha

    P_3 = 50 kg/ha

Using the full factorial set of combinations he had six treatments:

T_1P_1, T_1P_2, T_1P_3, T_2P_1, T_2P_2, T_2P_3

He conducted the experiment using a randomized block design with four blocks of six plots each. The field layout and yields (kilograms per hectare) were as shown in Fig. 6.3.

We begin the analysis by constructing the two tables of totals, Tables 6.5 and 6.6. The analysis of variance is given in Table 6.7.

Note that even though the main effect *F*'s exceed the 1% critical level there is a highly significant interaction. In this case the significance of the main effects has little meaning because the factors are not acting independently.

BLOCK

I	II	III	IV
T_2P_2 8.3	T_2P_1 11.2	T_1P_2 17.6	T_1P_3 18.9
T_2P_1 11.0	T_2P_2 10.5	T_1P_1 14.3	T_2P_2 12.8
T_1P_1 11.5	T_2P_3 16.7	T_2P_1 12.1	T_2P_3 17.5
T_2P_3 15.7	T_1P_2 17.6	T_1P_3 18.2	T_2P_1 12.6
T_1P_3 18.2	T_1P_1 13.6	T_2P_3 16.6	T_1P_2 18.1
T_1P_2 17.1	T_1P_3 17.6	T_2P_2 9.1	T_1P_1 14.5

Fig. 6.3 Field layout and yields for a 2 × 3 factorial experiment on broad beans.

TABLE 6.5 Block Totals

Block	I	II	III	IV	Sum
Total	81.8	87.2	87.9	94.4	351.3

TABLE 6.6 Type × Phosphorus Totals

Type	Phosphorus			
	P_1	P_2	P_3	Sum
T_1	53.9	70.4	72.9	197.2
T_2	46.9	40.7	66.5	154.1
Sum	100.8	111.1	139.4	351.3

TABLE 6.7   ANOVA of Broad Bean Yields

Source	d.f.	SS	MS	*F*
Total	23	243.3763		
Block	3	13.3213	4.4404	7.68**
Plant type	1	77.4004	77.4004	133.82**
Phosphorus	2	99.8725	49.9363	86.33*
$T \times P$	2	44.1059	22.0529	38.13**
Error	15	8.6762	.5784	

**Significant at the .01 level.

Note: CV $= (\sqrt{MSE/\bar{y}})100 = (\sqrt{.5784}/14.64)100 = 5.2\%$.

TABLE 6.8   Mean Yields (kg/ha) of Two
Types of Broad Bean (*Vicia faba*) Under
Different Levels of Phosphorus Fertilization

	Phosphorus (kg/ha)		
Plant type	0	25	50
Short, bushy	13.48	17.60	18.22
Tall, erect	11.72	10.18	16.62

Note: Std. error $= .38$.

Report of statistical analysis.   An experiment was conducted to measure the effect of different rates of phosphorus fertilizer on the yield of two types of broad bean. The results are summarized in Table 6.8.

It is apparent that while short, bushy plants outyield tall, erect plants at every phosphorus level, the response to increasing phosphorus differs from one type to the other. Short, bushy plants show their greatest yield increase with the first increment of added phosphorus, while tall, erect plants show no yield increase with 25 kg/ha of phosphorus.

The data indicate that blocking was effective and that the precision of the experiment was reasonably good (CV $= 5.2\%$).

6.4   THREE-FACTOR EXPERIMENTS

When a third factor is included in a factorial experiment the principles of design selection and randomization remain unchanged. The number of treatment combinations, however, increases fairly rapidly, and the analysis of the resulting data becomes somewhat more complicated. We now have to estimate and test three main effects, three two-factor (first-order) interactions and one three-factor (second-order) interaction.

6.4.1 Data Analysis

The basic pattern for analysis is the same as it is with the two-factor factorial. To illustrate, suppose we have three factors: factor A at a levels, factor B at b levels, and factor C at c levels. Assume that the experiment is run using a randomized block design with r blocks each containing abc experimental units. Let

y_{ijkl} = yield of the jth level of factor A, kth level of factor B, lth level of factor C in the ith block

We first construct a table of block totals, Table 6.9. We then form three two-way tables of factor totals. These are shown as Tables 6.10, 6.11, and 6.12.

TABLE 6.9 Block Totals

Block	1	2	$\cdots$	r	Sum
Sum	R_1	R_2	$\cdots$	R_r	G

Note: $R_i = \sum_j \sum_k \sum_l y_{ijkl} = y_{i...}$

$G = \sum_i R_i = y_{....}$

TABLE 6.10 $A \times B$ Totals

Factor A	Factor B				
	1	2	$\cdots$	b	Sum
1	$T_{.11.}$	$T_{.12.}$	$\cdots$	$T_{.1b.}$	A_1
2	$T_{.21.}$	$T_{.22.}$	$\cdots$	$T_{.2b.}$	A_2
$\vdots$	$\vdots$	$\vdots$		$\vdots$	$\vdots$
a	$T_{.a1.}$	$T_{.a2.}$	$\cdots$	$T_{.ab.}$	A_a
Sum	B_1	B_2	$\cdots$	B_b	G

Note: $T_{.jk.} = \sum_i \sum_l y_{ijkl} = y_{.jk.}$

$A_j = \sum_k T_{.jk.} = y_{.j..}$

$B_k = \sum_j T_{.jk.} = y_{..k.}$

$G = \sum_j A_j = \sum_k B_k = y_{....}$

TABLE 6.11 $A \times C$ Totals

	Factor C				
Factor A	1	2	· · ·	c	Sum
1	$T_{.1.1}$	$T_{.1.2}$	· · ·	$T_{.1.c}$	A_1
2	$T_{.2.1}$	$T_{.2.2}$	· · ·	$T_{.2.c}$	A_2
⋮	⋮	⋮		⋮	⋮
a	$T_{.a.1}$	$T_{.a.2}$	· · ·	$T_{.a.c}$	A_a
Sum	C_1	C_2	· · ·	C_c	G

Note: $T_{.j.l} = \sum_i \sum_k y_{ijkl} = y_{.j.l}$

$C_l = \sum_j T_{.j.l} = y_{...l}$

TABLE 6.12 $B \times C$ Totals

	Factor C				
Factor B	1	2	· · ·	c	Sum
1	$T_{..11}$	$T_{..12}$	· · ·	$T_{..1c}$	B_1
2	$T_{..21}$	$T_{..22}$	· · ·	$T_{..2c}$	B_2
⋮	⋮	⋮		⋮	⋮
b	$T_{..b1}$	$T_{..b2}$	· · ·	$T_{..bc}$	B_b
Sum	C_1	C_2	· · ·	C_c	G

Note: $T_{..kl} = \sum_i \sum_j y_{ijkl} = y_{..kl}$

We also require a three-way table of totals for the A, B, C factors. (Note, it is often easiest to start with this three-way table and then construct the two-way tables from the entries in the three-way table.) This is given in Table 6.13.

These tables form the basis for the computations in the analysis of variance, which takes the general form as shown in Table 6.14. The computations for the sums of squares in Table 6.14 are as follows:

TABLE 6.13 $A \times B \times C$ Totals

Factor A	Factor B	Factor C				
		1	2	$\cdots$	c	Sum
1	1	$T_{.111}$	$T_{.112}$	$\cdots$	$T_{.11c}$	$T_{.11.}$
	2	$T_{.121}$	$T_{.122}$	$\cdots$	$T_{.12c}$	$T_{.12.}$
	$\vdots$	$\vdots$	$\vdots$		$\vdots$	$\vdots$
	b	$T_{.1b1}$	$T_{.1b2}$	$\cdots$	$T_{.1bc}$	$T_{.1b.}$
	Sum	$T_{.1.1}$	$T_{.1.2}$	$\cdots$	$T_{.1.c}$	A_1
2	1	$T_{.211}$	$T_{.212}$	$\cdots$	$T_{.21c}$	$T_{.21.}$
	2	$T_{.221}$	$T_{.222}$	$\cdots$	$T_{.22c}$	$T_{.22.}$
	$\vdots$	$\vdots$	$\vdots$		$\vdots$	$\vdots$
	b	$T_{.2b1}$	$T_{.2b2}$	$\cdots$	$T_{.2bc}$	$T_{.2b.}$
	Sum	$T_{.2.1}$	$T_{.2.2}$	$\cdots$	$T_{.2.c}$	A_2
	$\vdots$	$\vdots$	$\vdots$	$\vdots$	$\vdots$	$\vdots$
a	1	$T_{.a11}$	$T_{.a12}$	$\cdots$	$T_{.a1c}$	$T_{.a1.}$
	2	$T_{.a21}$	$T_{.a22}$	$\cdots$	$T_{.a2c}$	$T_{.a2.}$
	$\vdots$	$\vdots$	$\vdots$		$\vdots$	$\vdots$
	b	$T_{.ab1}$	$T_{.ab2}$	$\cdots$	$T_{.abc}$	$T_{.ab.}$
	Sum	$T_{.a.1}$	$T_{.a.2}$	$\cdots$	$T_{.a.c}$	A_a
Sum		C_1	C_2	$\cdots$	C_c	G

Note: $T_{.jkl} = \sum_i y_{ijkl} = y_{.jkl}$

TABLE 6.14 Format for ANOVA of Three-Factor Factorial Experiment

Source	d.f.	SS	MS	F
Total	$rabc - 1$	SSTot		
Block	$r - 1$	SSR	MSR	F_R
A	$a - 1$	SSA	MSA	F_A
B	$b - 1$	SSB	MSB	F_B
AB	$(a - 1)(b - 1)$	SSAB	MSAB	F_{AB}
C	$c - 1$	SSC	MSC	F_C
AC	$(a - 1)(c - 1)$	SSAC	MSAC	F_{AC}
BC	$(b - 1)(c - 1)$	SSBC	MSBC	F_{BC}
ABC	$(a - 1)(b - 1)(c - 1)$	SSABC	MSABC	F_{ABC}
Error	$(r - 1)(abc - 1)$	SSE	MSE	

1. $C = G^2/rabc$

2. $\text{SSTot} = \sum_i \sum_j \sum_k \sum_l y_{ijkl}^2 - C$

3. $\text{SSR} = (1/abc) \sum_i R_i^2 - C$

4. $\text{SSA} = (1/rbc) \sum_j A_j^2 - C$

5. $\text{SSB} = (1/rac) \sum_k B_k^2 - C$

6. $\text{SSAB} = (1/rc) \sum_j \sum_k T_{.jk.}^2 - C - \text{SSA} - \text{SSB}$

7. $\text{SSC} = (1/rab) \sum_l C_l^2 - C$

8. $\text{SSAC} = (1/rb) \sum_j \sum_l T_{.j.l}^2 - C - \text{SSA} - \text{SSC}$

9. $\text{SSBC} = (1/ra) \sum_k \sum_l T_{..kl}^2 - C - \text{SSB} - \text{SSC}$

10. $\text{SSABC} = (1/r) \sum_j \sum_k \sum_l T_{.jkl}^2 - C - \text{SSA} - \text{SSB} - \text{SSAB}$

    $\quad - \text{SSC} - \text{SSAC} - \text{SSBC}$

11. $\text{SSE} = \text{SSTot} - \text{SSR} - \text{SSA} - \text{SSB} - \text{SSAB} - \text{SSC} - \text{SSAC} -$
    $\text{SSBC} - \text{SSABC}$

Mean squares, as usual, are obtained by dividing the sums of squares by their associated degrees of freedom. Similarly, MSE is an estimate of σ^2 and serves as the denominator of all F ratios.

6.4.2 Significance Tests

F_A, F_B, and F_C are used to test the significance of the main effects. F_{AB}, F_{AC}, and F_{BC} are test statistics for the first-order interactions. F_{ABC} tests the second-order ($A \times B \times C$) interaction. Again, F_R provides an approximate test of the effectiveness of blocking.

6.4.3 Standard Errors

The means, their standard errors, and the standard errors of differences between means in a three-factorial experiment may be summarized as shown in Table 6.15.

All of these standard errors may be used to construct interval estimates or to compute the sample t for test statistics. In either case all of the standard errors have $(r - 1)(abc - 1)$ degrees of freedom.

TABLE 6.15 Summary of Means and Standard Errors for a Three-Factor Factorial

Factor	Mean	Standard error Mean	Difference
A	$\bar{y}_{j..} = A_j/rbc$	$s_{\bar{y}A} = \sqrt{MSE/rbc}$	$s_{dA} = \sqrt{2MSE/rbc}$
B	$\bar{y}_{.k.} = B_k/rac$	$s_{\bar{y}B} = \sqrt{MSE/rac}$	$s_{dB} = \sqrt{2MSE\,rac}$
AB	$\bar{y}_{jk.} = T_{jk.}/rc$	$s_{\bar{y}AB} = \sqrt{MSE/rc}$	$s_{dAB} = \sqrt{2MSE/rc}$
C	$\bar{y}_{...l} = C_l/rab$	$s_{\bar{y}C} = \sqrt{MSE/rab}$	$s_{dC} = \sqrt{2MSE/rab}$
AC	$\bar{y}_{j.l} = T_{j.l}/rb$	$s_{\bar{y}AC} = \sqrt{MSE/rb}$	$s_{dAC} = \sqrt{2MSE/rb}$
BC	$\bar{y}_{.jl} = T_{.jl}/ra$	$s_{\bar{y}BC} = \sqrt{MSE/ra}$	$s_{dBC} = \sqrt{2MSE/ra}$
ABC	$\bar{y}_{jkl} = T_{jkl}/r$	$s_{\bar{y}ABC} = \sqrt{MSE/r}$	$s_{dABC} = \sqrt{2MSE/r}$

6.4.4 Interpretation

We begin the interpretation by first considering the significance of the three-factor interaction ($A \times B \times C$). If it is significant none of the factors is acting independently. In this case the results may be summarized in a three-way table of $A \times B \times C$ means along with their standard error.

If the *ABC* interaction is not significant we then look at the first-order (two-factor) interactions. If any two-factor interaction is significant then neither of the main effects has meaning, and we use two-way tables of means to summarize these results.

If two-factor interactions are not significant we look at the main effects. Factors with significant main effects are summarized in one-way tables of factor means.

6.4.5 Numerical Example: Three-Factor Factorial

An automobile manufacturer wanted to examine the effect of three welding rod variables: maker, diameter, and chromium content, on the strength of spot welds made by assembly line robots. He selected two makers: M_1, M_2; three rod diameters: $D = 30$ mm, $D_2 = 60$ mm, $D_3 = 90$ mm; and three chromium contents: $C_1 = 1.00\%$, $C_2 = 1.50\%$, $C_3 = 2.00\%$. He decided to use a $2 \times 3 \times 3$ factorial set of treatments ($M \times D \times C$). He randomly assigned the treatments to 18 robots on each of two assembly lines, which he treated as blocks in a randomized block design. The assignment of treatments and tensile strength of welds (psi $\times 10^{-3}$) were as shown in Fig. 6.4.

To begin the analysis we construct a number of summary tables: Tables 6.16, 6.17, 6.18, 6.19, and 6.20. The analysis of variance is presented in Table 6.21.

LINE I

$M_1D_1C_2$ 79.2	$M_1D_3C_2$ 78.3
$M_1D_1C_1$ 76.2	$M_2D_3C_1$ 72.2
$M_1D_2C_3$ 80.4	$M_2D_2C_3$ 79.8
$M_2D_3C_2$ 74.8	$M_2D_1C_3$ 78.8
$M_1D_3C_1$ 76.7	$M_2D_3C_3$ 77.8
$M_2D_1C_2$ 76.0	$M_1D_2C_1$ 75.0
$M_2D_2C_1$ 74.0	$M_1D_3C_3$ 80.6
$M_1D_1C_3$ 80.7	$M_2D_1C_1$ 75.1
$M_1D_2C_2$ 77.8	$M_2D_2C_2$ 76.2

LINE II

$M_2D_3C_1$ 73.0	$M_1D_3C_3$ 81.6
$M_1D_1C_1$ 77.6	$M_1D_2C_3$ 80.6
$M_1D_1C_2$ 78.8	$M_2D_2C_3$ 79.2
$M_1D_3C_1$ 76.5	$M_1D_2C_1$ 75.7
$M_1D_2C_2$ 78.2	$M_2D_2C_1$ 74.4
$M_2D_1C_2$ 76.4	$M_2D_1C_3$ 79.4
$M_2D_2C_2$ 77.1	$M_1D_3C_2$ 79.0
$M_2D_3C_2$ 74.8	$M_2D_1C_1$ 75.8
$M_1D_1C_3$ 81.8	$M_2D_3C_3$ 78.1

Fig. 6.4 Experimental plan and tensile strengths for a $2 \times 3 \times 3$ factorial experiment with welding rods.

TABLE 6.16 Block (Assembly Line) Totals

Block	I	II	Sum
Sum	1389.6	1398.0	2787.6

TABLE 6.17 Maker $\times$ Diameter $\times$ Chrome $(M \times D \times C)$ Totals

Maker	Diameter	Chrome			Sum
		C_1	C_2	C_3	
M_1	D_1	153.8	158.0	162.5	474.3
	D_2	150.7	156.0	161.0	467.7
	D_3	153.2	157.3	162.2	472.7
	Sum	457.7	471.3	485.7	1414.7
M_2	D_1	150.9	152.4	158.2	461.5
	D_2	148.4	153.3	159.0	460.7
	D_3	145.2	149.6	155.9	450.7
	Sum	444.5	455.3	473.1	1372.9
Sum		902.2	926.6	958.8	2787.6

TABLE 6.18 Maker $\times$ Diameter $(M \times D)$ Totals

Maker	Diameter			Sum
	D_1	D_2	D_3	
M_1	474.3	467.7	472.7	1414.7
M_2	461.5	460.7	450.7	1372.9
Sum	935.8	928.4	923.4	2787.6

TABLE 6.19 Maker × Chrome ($M \times C$) Totals

Maker	Chrome			Sum
	C_1	C_2	C_3	
M_1	457.7	471.3	485.7	1414.7
M_2	444.5	455.3	473.1	1372.9
Sum	902.2	926.6	958.8	2787.6

TABLE 6.20 Diameter × Chrome ($D \times C$) Totals

Diameter	Chrome			Sum
	C_1	C_2	C_3	
D_1	304.7	310.4	320.7	935.8
D_2	299.1	309.3	320.0	928.4
D_3	298.4	306.9	318.1	923.4
Sum	902.2	926.6	958.8	2787.6

TABLE 6.21 ANOVA of Tensile Strength of Welding Rods

Source	d.f.	SS	MS	F
Total	35	206.5600		
Line (block)	1	1.9600	1.9600	14.18**
Maker	1	48.5344	48.5344	351.10**
Diameter	2	6.4867	3.2434	23.46**
$M \times D$	2	9.5356	4.7678	34.49**
Chrome	2	134.3267	67.1634	485.86**
$M \times C$	2	.5489	.2745	1.98
$D \times C$	4	1.9816	.4954	3.58*
$M \times D \times C$	4	.8361	.2090	1.51
Error	17	2.3500	.1382	

*Significant at the 5% level.
**Significant at the 1% level.

Note: CV = $\sqrt{.1382/77.43})100$ = 0.5%.

Note that neither the three-factor interaction nor the maker × chrome interaction is significant. However, the diameter × chrome interaction is significant at the 5% level while the maker × diameter interaction is significant at the 1% level. We can summarize the results in two 2-way tables of means: $M \times D$ and $D \times C$. Since the F value for chrome is larger than the F value for $D \times C$ by a factor of over 100, we should also say something about the main effect of chrome.

Report of statistical analysis. The results of an experiment designed to test the effect of chromium content, diameter, and maker of welding rods on the strength of spot welds are summarized in Tables 6.22 and 6.23.

In Table 6.22 it is seen that there is a tendency for welds to be stronger for rods from maker M_1 than for those from maker M_2. Differences between makers, however, are not consistent from one diameter to another. The data in Table 6.23 indicate that spot weld strength increases as the chromium content increases.

TABLE 6.22 Mean Tensile Strength (p.s.i. $\times 10^{-3}$) of Spot Welds Made by Rods of Three Diameters Manufactured by Two Makers

	Diameter		
Maker	30 mm	60 mm	90 mm
M_1	79.05	77.95	78.78
M_2	76.92	76.78	75.12

Note: Standard error = 0.15.

TABLE 6.23. Mean Tensile Strength (p.s.i. $\times 10^{-3}$) of Spot Welds Made by Rods of Three Diameters with Three Chromium Contents

Diameter	Chromium (%)		
(mm)	1.00	1.50	2.00
30	76.18	77.60	80.18
60	74.78	77.32	80.00
90	74.60	76.72	79.52
Mean	75.18	77.22	79.90

Note: Standard error (diameter × chrome) = 0.18.
       Standard error (chrome) = 0.11.

There is a tendency for this increase to be inconsistent from one diameter to another.

As indicated by a coefficient of variation of 0.5%, the precision of this experiment was quite high. Blocking on assembly line contributed effectively to this high precision.

6.5 EXTENSION

More than three factors may be included in a factorial experiment. The computations involved become more cumbersome in this case, and the interpretation may become more difficult. However, no new principles are involved so we will not consider more than three factors at this point.

6.6 SPLIT-PLOT DESIGNS

The most common procedure in many areas of research is to use a factorial set of treatments and a randomized block experimental design. In this design the experimental units are grouped into blocks of units which are more or less homogeneous with respect to some criteria. Treatments are then assigned at random to the units within the blocks. With some types of treatments, however, the requirement that all treatments be assigned to the units at random within the blocks may cause some difficulty. Consider some examples:

1. A ceramics engineer wants to study the effect of firing temperature and clay composition on the strength of bricks. In a randomized block design each brick would be randomly assigned to a temperature, an expensive, if not impossible, procedure.
2. A food scientist wants to study the effect of storage time and type of container on keeping quality. Here it would be convenient to handle time as a unit.
3. An agronomist is interested in tillage treatments and fertilizers. Tillage machinery requires large plots while fertilizers can be applied to small plots.

These are only a few examples of experiments involving two or more factors requiring different sizes of experimental unit for their efficient application. For this reason, among others, it would be convenient to have an experimental design in which some of the factors could be applied to large units which are made up of groups of the small units. This is possible with split-plot designs or variants of them.

In the split-plot design the levels of one factor are assigned at random to large experimental units within blocks of such units. The large units are then divided into smaller units, and the levels of the second factor are assigned at random to the small units within the larger units. In the terminology of agricultural research, where these designs were developed, the large units are called whole plots or main plots, while the small units are called split-plots or subplots.

As an example, the whole-plot factor might be three methods of seedbed preparation using farm-scale machinery, S_1, S_2, S_3, while the subplot factor might be four rates of nitrogen fertilizer applied by hand, N_0, N_1, N_2, N_3. If the experiment were run in two replicates with the whole plots in a randomized block design the field plan might be as shown in Fig. 6.5.

In the split-plot design the whole-plot factor effects are estimated from the large units, while the subplot effects and the interaction of the whole-plot and subplot factors are estimated from the small units. Because there are two sizes of unit there are two experimental errors, one for each type of unit. Generally the error associated with the subplots is smaller than that for the whole plots. This is because

1. Small units within the large units tend to be positively correlated. This has the effect of reducing experimental error.
2. Error degrees of freedom for the whole plots are usually less than those for the subplots. This has the effect of increasing the whole-plot error relative to that of the subplots.

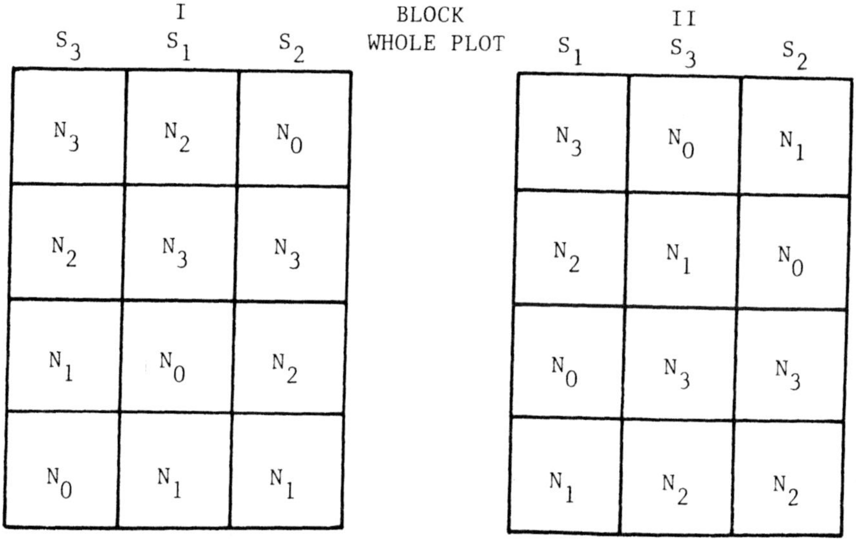

Fig. 6.5 Experimental plan for a split-plot design with S on the whole plots and N on the subplots.

The net effect is that the whole-plot factor is less precisely estimated than is the subplot factor and its interaction with the whole-plot factor. This should be taken into consideration when deciding to use a split-plot design.

6.6.1   Advantages

There are a number of advantages to be gained with the split-plot design:

1. It permits the efficient use of some factors which require large experimental units in combination with other factors which require small experimental units.
2. It provides increased precision in the comparison of some factors.
3. It permits the introduction of new treatments into an experiment which is already in progress.

6.6.2   Disadvantages

The split-plot design has a few disadvantages:

1. Statistical analysis is complicated because different comparisons have different error variances.
2. Low precision on the whole plots can result in large differences being nonsignificant, while small differences on the subplots may be statistically significant even though they are of no practical significance.

6.6.3   Uses

Split-plot designs are most effectively used for:

1. Experiments in which one factor requires larger experimental units than another factor.
2. Introduction of a new factor into an experiment which is already in progress.

6.6.4   Comment

Care should be taken in the analysis of data from an experiment which is said to have been conducted as an ordinary factorial experiment. Often the way in which the factors were assigned to the experimental units, or the way in which the experiment was conducted, creates an unintentional split-plot experiment. This is particularly true where time is one of the treatment factors.

6.6.5 Data Analysis

Note, first, that in their simplest form split-plot designs contain two factors with the factor combinations assigned to the subplots. We will want to estimate and test the two main effects and the interaction. To illustrate the analysis in general, suppose we have factor A at a levels assigned at random to the whole plots, and factor B at b levels assigned at random to the subplots within the whole plots. Assume that the whole plots are arranged in a randomized block design with r blocks. Let

y_{ijk} = the yield of the jth level of factor A, kth level of factor B in the ith block

To begin the analysis we construct two tables of totals, Table 6.24 and Table 6.25. The analysis of variance takes the form shown in Table 6.26.

The sources of variation and degrees of freedom are the same as those for a two-factor factorial experiment in a randomized block design except that the error is partitioned into two parts:

1. Error (A), with $(r - 1)(a - 1)$ d.f., the experimental error associated with the whole plots
2. Error (AB), with $a(r - 1)(b - 1)$ d.f., the subplot error.

TABLE 6.24 Factor A by Block Totals

Block	Factor A				
	1	2	$\cdots$	a	Sum
1	$y_{11.}$	$y_{12.}$	$\cdots$	$y_{1a.}$	R_1
2	$y_{21.}$	$y_{22.}$	$\cdots$	$y_{2a.}$	R_2
$\vdots$	$\vdots$	$\vdots$		$\vdots$	$\vdots$
r	$y_{r1.}$	$y_{r2.}$	$\cdots$	$y_{ra.}$	R_r
Sum	A_1	A_2	$\cdots$	A_a	G

Note: $y_{ij.} = \sum_k y_{ijk}$

$R_i = \sum_j y_{ij.} = y_{i..}$

$A_j = \sum_i y_{ij.} = y_{.j.}$

$G = \sum_i R_i = \sum_j A_j = y_{...}$

TABLE 6.25 *A by B* Totals

Factor A	Factor B				
	1	2	$\cdots$	b	Sum
1	$T_{.11}$	$T_{.12}$	$\cdots$	$T_{.1b}$	A_1
2	$T_{.21}$	$T_{.22}$	$\cdots$	$T_{.2b}$	A_2
$\vdots$	$\vdots$	$\vdots$		$\vdots$	$\vdots$
a	$T_{.a1}$	$T_{.a2}$	$\cdots$	$T_{.ab}$	A_a
Sum	B_1	B_2	$\cdots$	B_b	G

Note: $T_{.jk} = \sum_i y_{ijk} = y_{.jk}$

$B_k = \sum_j T_{.jk} = y_{..k}$

and

$A_j = \sum_k T_{.jk} = y_{.j.}$

$G = \sum_j A_j = \sum_k B_k = y_{...}$

TABLE 6.26 Format for the ANOVA of a Split-Plot Design

Source	d.f.	SS	MS	F
Total	$rab - 1$	SSTot		
Block	$r - 1$	SSR	MSR	F_R
A	$a - 1$	SSA	MSA	F_A
Error (A)	$(r - 1)(a - 1)$	SSEA	MSEA	
B	$b - 1$	SSB	MSB	F_B
AB	$(a - 1)(b - 1)$	SSAB	MSAB	F_{AB}
Error (AB)	$a(r - 1)(b - 1)$	SSEAB	MSEAB	

The sums of squares in Table 6.26 are computed as

1. Correction term, $C = G^2/rab$

2. $\text{SSTot} = \sum_i \sum_j \sum_k y_{ijk}^2 - C$

3. $\text{SSR} = (1/ab) \sum_i R_i^2 - C$

4. $\text{SSA} = (1/rb) \sum_j A_j^2 - C$

5. $\text{SSEA} = (1/b) \sum_i \sum_j y_{ij.}^2 - C - \text{SSR} - \text{SSA}$

6. $\text{SSB} = (1/ra) \sum_k B_k^2 - C$

7. $\mathrm{SSAB} = (1/r) \sum_j \sum_k T_{.jk}^2 - C - \mathrm{SSA} - \mathrm{SSB}$

8. $\mathrm{SSEAB} = \mathrm{SSTot} - \mathrm{SSR} - \mathrm{SSA} - \mathrm{SSEA} - \mathrm{SSB} - \mathrm{SSAB}$

Mean squares are computed, in the usual way, by dividing the sums of squares by their associated degrees of freedom. The F ratios are computed somewhat differently, however, since there are two errors in a split-plot design. The proper ratios are indicated by arrows in the ANOVA table. The point of the arrow indicates the mean square in the numerator while the tail indicates the mean square in the denominator of the F ratio. We have

$$F_R = \mathrm{MSR/MSEA}$$

$$F_A = \mathrm{MSA/MSEA}$$

and

$$F_B = \mathrm{MSB/MSEAB}$$

$$F_{AB} = \mathrm{MSAB/MSEAB}$$

6.6.6 Significance Tests

1. F_A, with $(a - 1)$, $(r - 1)(a - 1)$ d.f., is used to test the significance of differences among A factor means (main effect of factor A).
2. F_B, with $(b - 1)$, $a(r - 1)(b - 1)$ d.f., is a test statistic for the significance of the main effect of factor B.
3. F_{AB}, with $(a - 1)(b - 1)$, $a(r - 1)(b - 1)$ d.f., provides a test of the significance of the $A \times B$ interaction.
4. F_R, with $(r - 1)$, $(r - 1)(a - 1)$ d.f., provides an approximate test of the effectiveness of blocking in reducing the whole-plot error.

6.6.7 Standard Errors

Standard errors for the split-plot design are somewhat complicated, particularly when comparing one mean with another. For the individual means there are three standard errors:

1. Factor A means, $\bar{y}_{.j.} = A_j/rb$

$$s_{\bar{y}A} = \sqrt{\mathrm{MSEA}/rb}$$

2. Factor B means, $\bar{y}_{..k} = B_k/ra$

$$s_{\bar{y}B} = \sqrt{\mathrm{MSEAB}/ra}$$

3. Treatment ($A \times B$) means, $\bar{y}_{.jk} = T_{.jk}/r$

    $$s_{\bar{y}AB} = \sqrt{MSEAB/r}$$

For differences between means there are four types of comparison, each with its own standard error:

1. Difference between two A means, $\bar{y}_{A2} - \bar{y}_{A1}$

    $$s_{\bar{d}A} = \sqrt{2MSEA/rb}$$

2. Difference between two B means, $\bar{y}_{B2} - \bar{y}_{B1}$

    $$s_{\bar{d}B} = \sqrt{2MSEAB/ra}$$

3. Difference between two B means at same level of A, $\bar{y}_{A1B2} - \bar{y}_{A1B1}$

    $$s_{\bar{d}AB} = \sqrt{2MSEAB/r}$$

4. Difference between two A means at same or different level of B, $\bar{y}_{A2B1} - \bar{y}_{A1B1}$ or $\bar{y}_{A2B2} - \bar{y}_{A1B1}$

    $$s_{\bar{d}} = (2[(b - 1)MSEAB + MSEA]/rb)^{1/2}$$

These standard errors may be used to compute interval estimates just as in ordinary factorial experiments. Standard errors involving only MSEA have $(r - 1)(a - 1)$ d.f., while those involving only MSEAB have $a(r - 1)(b - 1)$ d.f. The standard error which contains both MSEA and MSEAB has no exact value for the d.f. associated with it. To obtain an approximate value we can use a procedure due to Satterthwaite (1946). This approximation is

$$\text{d.f.} = \frac{a(r - 1)(a - 1)[(b - 1)MSEAB + MSEA]^2}{[(a - 1)(b - 1)(MSEAB)^2] + [a(MSEA)^2]}.$$

Note: Because there are two sizes of experimental unit and two experimental errors a coefficient of variation is not ordinarily computed for split-plot experiments.

6.6.8 Interpretation

The interpretation of the analysis of data from a split-plot experiment proceeds in much the same way as it does for an ordinary factorial experiment. First, we test the $A \times B$ interaction. If it is significant the main effects have no meaning whether they are significant or not. In this case the results may be summarized in a two-way table of means.

If the $A \times B$ interaction is not significant then the main effects which are significant have some meaning. Here the results may be summarized in one-way tables of means for the significant factors.

The analysis and interpretation of data from a split-plot experiment are illustrated in the accompanying numerical example.

6.6.9 Numerical Example: Split-Plot Design

A metallurgist wanted to determine the effect of annealing temperature on the breaking strength of three experimental metal alloys. In his research laboratory he had four laboratory ovens each capable of annealing three metal samples. He decided to use a split-plot design with temperatures assigned to ovens as whole plots and metal samples within ovens as subplots. The temperature levels assigned to the whole plots were: $T_1 = 675°F$, $T_2 = 700°F$, $T_3 = 725°F$, $T_4 = 750°F$. The alloys assigned to the subplots were designated: A_1, A_2, A_3.

He could make only one run per day in his experimental ovens. He decided to use a randomized block design for the ovens (whole plots) with three blocks, using days as a blocking factor.

The experimental plan and the gauge reading for the alloy samples are shown in Fig. 6.6. In this trial the higher the gauge reading the higher is the breaking strength of the alloy.

The summary tables for the totals needed to complete the analysis of variance are given as Table 6.27 and Table 6.28. The analysis of variance of the data from this trial is presented in Table 6.29.

Because there are two error terms, a number of different standard errors for the means and for differences between means are required:

1. Temperature means, $s_{\bar{y}A}$

$$s_{\bar{y}A} = \sqrt{MSEA/rb} = \sqrt{51.42/9} = 2.39,\ 6\ \text{d.f.}$$

2. Alloy means, $s_{\bar{y}B}$

$$s_{\bar{y}B} = \sqrt{MSEAB/ra} = \sqrt{9.88/12} = .91,\ 16\ \text{d.f.}$$

3. Temperature × alloy means, $s_{\bar{y}AB}$

$$s_{\bar{y}AB} = \sqrt{MSEAB/r} = \sqrt{9.88/3} = 1.81,\ 16\ \text{d.f.}$$

4. Difference between two temperature means, $s_{\bar{d}A}$

$$s_{\bar{d}A} = \sqrt{2MSEA/rb} = \sqrt{(2)(51.42)/9} = 3.38,\ 6\ \text{d.f.}$$

5. Difference between two alloy means, $s_{\bar{d}B}$

$$s_{\bar{d}B} = \sqrt{2MSEAB/ra} = \sqrt{(2)(9.88)/12} = 1.28,\ 16\ \text{d.f.}$$

6. Difference between two alloy means at same temperature, $s_{\bar{d}AB}$

$$s_{\bar{d}AB} = \sqrt{2MSEAB/r} = \sqrt{(2)(9.88)/3} = 2.57,\ 16\ \text{d.f.}$$

BLOCK

T_4		T_1		T_2		T_3	
A_3	58	A_3	34	A_2	69	A_1	73
A_1	38	A_2	30	A_3	79	A_3	80
A_2	45	A_1	18	A_1	57	A_2	82

(I)

T_1		T_3		T_4		T_2	
A_3	38	A_2	79	A_2	53	A_1	62
A_1	24	A_1	65	A_1	45	A_3	73
A_2	36	A_3	82	A_3	60	A_2	56

(II)

T_2		T_3		T_4		T_1	
A_1	51	A_3	88	A_3	50	A_2	33
A_2	59	A_2	88	A_1	39	A_3	32
A_3	70	A_1	72	A_2	45	A_1	17

(III)

Fig. 6.6 Experimental plan and breaking strength (gauge reading) for a split-plot experiment on alloy heat treatment.

TABLE 6.27 Block × Temperature Totals

	Temperature				
Block	T_1	T_2	T_3	T_4	Sum
I	82	205	235	141	663
II	98	191	226	158	673
III	82	180	248	134	644
Sum	262	576	709	433	1980

TABLE 6.28 Temperature × Alloy Totals

	Temperature				
Alloy	T_1	T_2	T_3	T_4	Sum
A_1	59	170	210	122	561
A_2	99	184	249	143	675
A_3	104	222	250	168	744
Sum	262	576	709	433	1980

TABLE 6.29 ANOVA of Breaking Strength of Alloys Under Different Annealing Temperatures

Source	d.f.	SS	MS	F
Total	35	14,368.00		
Day (block)	2	36.17	18.08	
Temperature	3	12,276.67	4,092.22	79.58**
Error(A)	6	308.49	51.42	
Alloy	2	1,423.50	711.75	72.04**
Temperature × alloy	6	165.16	27.52	2.78*
Error(AB)	16	158.01	9.88	

*Significant at the 5% level.
**Significant at the 1% level.

7. Difference between two temperature means for same or different alloy, $s_{\bar{d}}$

$$s_{\bar{d}} = (2[(b - 1)\text{MSEAB} + \text{MSEA}]/rb)^{1/2}$$

$$= (2[(2)(9.88) + 51.42]/9)^{1/2} = 3.98$$

For this standard error the approximate d.f., d.f., are (Satterthwaite, 1946)

$$\text{d.f.} = \frac{a(r - 1)(a - 1)[(b - 1)\text{MSEAB} + \text{MSEA}]^2}{[(a - 1)(b - 1)(\text{MSEAB})^2] + [a(\text{MSEA})^2]}$$

$$= \frac{(4)(2)(3)[(2)(9.88) + 51.42]^2}{[(3)(2)(9.88)^2] + [(4)(51.42)^2]}$$

$$= 10.89 = 11 \text{ d.f.}$$

On considering the ANOVA (Table 6.29) it is seen that the temperature $\times$ alloy interaction is barely significant at the 5% level while both main effects, temperature and alloy, are significant at the 1% level. In this case, even though the two factors are not totally independent, the main effects are so much larger than the interaction that something probably should be said about them.

Report of statistical analysis. The results of an experiment to examine the effect of annealing temperature on the breaking strength of three experimental metal alloys are presented in Table 6.30.

Although there is a slight indication that the differences between alloys depend on temperature, alloy differences and temperature differences are highly significant. Breaking strength was highest at all temperatures for alloy A_3 and lowest at all temperatures for alloy A_1. For all alloys the breaking strength

TABLE 6.30 Mean Breaking Strength (Gauge Reading) of Three Alloys Annealed at Four Different Temperatures

Alloy	Temperature				Mean
	675°	700°	725°	750°	
A_1	19.67	56.67	70.00	40.67	46.75
A_2	33.00	61.33	83.00	47.67	56.25
A_3	34.67	74.00	83.33	56.00	62.00
Mean	29.11	64.00	78.78	48.11	55.00

Note: Standard errors: temperature, 2.39; alloy, .91; temperature $\times$ alloy, 1.81.

increases as temperature is increased until it reaches a maximum at 725°. Above 725° there is a decrease in breaking strength.

FURTHER READING*

Anderson, V. L., and R. A. McLean (1974), Ch. 2 and 7.
Cochran, W. G., and G. M. Cox (1957), Ch. 5 and 7.
Federer, W. T. (1955), Ch. 8 and 10.
Kempthorne, O. (1973), Ch. 13.
Snedecor, G. W., and W. G. Cochran (1980), Ch. 16.
Steel, R. G. D., and J. H. Torrie (1980), Ch. 15 and 16.
Yates, F. (1937).

REFERENCES

1. Satterthwaite, F. E. (1946). An Approximate Distribution of Estimates of Variance Components, *Biometrics Bull.* 2:110–114.

---

*See Bibliography for complete citation.

7

Data Interpretation: Some Examples

7.1 INTRODUCTION

We have now had a brief look at the principles of modern experimental design. We have also considered the basic kinds of designs along with their advantages, disadvantages, and uses. In addition, we have touched on the analysis and interpretation of the data resulting from an experiment. Before going into more sophisticated and specialized designs, however, it might be well to pause at this point to consider a few examples of data interpretation.

Many of the scientists who use designed experiments in their research have had little or no exposure to more than elementary statistical methods. The exposure they have had has been concerned largely with the process of applying statistical techniques in the analysis of data. Little attention has been paid to the consideration of what the data can tell the scientist about whether the experiment has met its objectives or about what the results really mean within the context of the problem. As a consequence, too often the data are subject to procedures with which the scientist is familiar, whether they are appropriate or not. This is particularly true of pairwise multiple comparison procedures, which are frequently used in situations where they are entirely inappropriate (Petersen, 1977).

The scientist should look at statistical analysis as a tool to guide his interpretation of the experimental results. In this regard statistical analyses should be considered on the same basis as chemical analyses. They support the conduct and interpretation of the experiment. They are not usually an end in themselves.

There are a number of questions that can be answered by a statistical analysis. These include, but are not limited to:

1. To what extent do the results throw light on the original objectives of the experiment? Are the questions which were asked in formulating the objectives answered by the experimental results?

2. Do the results indicate anything unexpected or unusual in the response to the treatments? If so, is this an artifact of the conduct of the experiment or is it something that should be investigated further?
3. What are the estimates of the treatment effects and the differences between treatment effects? What is the precision of these estimates?
4. Did the experimental design provide satisfactory control of extraneous variation consistent with the desired precision and the cost of conducting the experiment?
5. How can the conduct of similar experiments be improved in the future?

We have looked at some examples of data interpretation in the numerical examples used to illustrate the experimental designs presented to this point. It might be well, however, to consider some more complicated examples before going on.

7.2 EXAMPLE I: WASHDAY PRODUCTS

An industrial chemist wanted to study the effectiveness of a number of washday products in removing soil from cloth during a typical washing process. Specifically, he wanted to compare a synthetic detergent with soap, to see whether the form of the detergent had an effect and whether or not the addition of phosphate made a difference. Accordingly, he selected the following treatments for study:

1. Detergent, liquid
2. Detergent, granules
3. Detergent, flakes

4. Detergent + phosphate, liquid
5. Detergent + phosphate, granules
6. Detergent + phosphate, flakes
7. Soap

He obtained 21 bed sheets, soiled each in a standard way, then washed three randomly selected sheets with each product. Because he had no basis for blocking he used a completely randomized design. At the conclusion of each washing he measured the amount of dirt removed. The resulting data are summarized in Table 7.1.

A preliminary analysis of variance of the data in Table 7.1 is given in Table 7.2.

Since the preliminary analysis of variance indicates significant differences among the treatment means, we now look for a set of contrasts, preferably mutually orthogonal, which correspond to the original objectives of the chemist. Note that these contrasts can be constructed whether or not the F test for treatments is significant so long as they are not based on an examination of the means. Often, however, a nonsignificant F test provides a reason to discontinue the analysis. One set of orthogonal contrasts for these data is shown in Table 7.3.

TABLE 7.1 Amount of Dirt (mg) Removed by Seven Washday Products

Product	Dirt removed (mg)			Sum	Mean
1	32.7	32.3	31.5	96.5	32.17
2	32.5	31.1	29.7	93.3	31.10
3	32.1	29.7	29.1	90.9	30.30
4	38.2	37.8	31.9	107.9	35.97
5	35.7	35.9	33.1	104.7	34.90
6	36.0	34.2	31.2	101.4	33.80
7	31.8	28.0	29.2	89.0	29.67

TABLE 7.2 ANOVA (Preliminary) of Dirt Removed

Source	d.f.	SS	MS	F
Total	20	161.9314		
Treatment	6	103.1514	17.1919	4.09*
Error	14	58.7800	4.1986	

*Significant at the 5% level.

Note: $CV = (\sqrt{4.1986}/32.56)100 = 6.3\%$.

TABLE 7.3 Orthogonal Contrasts Among Treatment Totals for Washday Products

		Treatment							
		1	2	3	4	5	6	7	
Contrast	Total	96.5	93.3	90.9	107.9	104.7	101.4	89.0	SS
Soap vs. detergent		+	+	+	+	+	+	−6	29.2420*
Phosphate vs. no phosphate		−	−	−	+	+	+	0	61.6050**
Solid vs. liquid		−2	+	+	−2	+	+	0	9.5069 NS
Granules vs. flake		0	−	+	0	−	+	0	2.7075 NS
$P \times$ (solid vs. liquid)		2	−	−	−2	+	+	0	.0225 NS
$P \times$ (granules vs. flake)		0	+	−	0	−	+	0	.0675 NS

*Significant at the 5% level.
**Significant at the 1% level.
Note: NS = Not significant.

TABLE 7.4 Analysis of Variance of Dirt Removed by Various Washday Products

Source	d.f.	SS	MS	F
Total	20	161.9314		
Soap vs. detergent	1	29.2420	29.2420	6.96*
Phosphate vs. none	1	61.6050	61.6050	14.67**
Form of detergent	2	12.2144	6.1072	1.45
Phosphate × form	2	.0900	.0450	.01
Error	14	58.7800	4.1986	

*Significant at the 5% level.
**Significant at the 1% level.

These results may be summarized in the final analysis of variance displayed in Table 7.4.

Notes on analysis:

1. The sums of squares, with 2 d.f., for "form of detergent" and for "phosphate × form" interaction are obtained by pooling appropriate sums of squares from the table of contrasts.
2. The error mean square, with 14 d.f., serves as denominator for all *F* tests.

Report of statistical analysis. An experiment was conducted to evaluate the ability of a number of washday products to remove soil from cloth. Under study was a comparison of soap with synthetic detergent. Also under consideration was the form of the detergent and the effect of the addition of phosphate to the detergent.

Statistical analysis of the resulting data indicated a real difference between soap and detergent as well as a significant effect of the addition of phosphate to the detergent. On the average, detergent removed about 3.4 mg more dirt than did soap (standard error = 1.28). Similarly, the addition of phosphate to the detergent increased the average amount of dirt removed by about 3.7 mg (standard error = .96) compared to detergent without phosphate. There was no real effect of form of the detergent, nor did the phosphate effect depend on the form.

The coefficient of variation, 6.3%, was satisfactorily low. There is little indication from the data that an experimental design more complex than a completely randomized design would be useful in future work of this type.

7.3 EXAMPLE II: FROZEN ORANGE JUICE

Early attempts to produce acceptable frozen orange juice failed because the juice separated into a liquid and a solid fraction shortly after water was added to the frozen concentrate. A number of additives were tried in an attempt to retard separation. The goal of the experiments with the additives was to determine whether the addition of the particular compound retarded separation and, if it did, to estimate the relationship between amount of compound added and time to separation.

In one experiment the treatments consisted of four levels: 1, 2, 3, 4 ppm, of a certain enzyme plus a control with no added enzyme. Each treatment was replicated four times in a completely randomized design. Water was added to the 20 samples, and the time to separation (minutes) was measured on each sample. The basic results are tabulated in Table 7.5.

The preliminary analysis of variance is given in Table 7.6.

The F test indicates that there are highly significant differences among the treatment means. Accordingly, we should begin a meaningful decomposition of the treatment sum of squares. As a start we consider two questions:

1. Does the presence of enzyme retard separation as compared to the absence of enzyme?
2. Is there a differential effect of the level of added enzyme?

We can answer the first question by a t test of the difference between two means: (1) the mean for the control without enzyme and (2) the mean of all treatments with added enzyme. For the first mean we have

TABLE 7.5 Results of a Trial Designed to Measure the Effect of Added Enzyme on Separation Time of Frozen Orange Juice

Enzyme level (ppm)	Time to separation (min)				Sum	Mean
0	3.96	6.24	5.42	11.11	26.73	6.68
1	27.31	26.96	32.21	30.13	116.61	29.15
2	35.30	34.71	36.75	38.38	145.14	36.28
3	41.09	43.99	48.70	41.78	175.56	43.89
4	48.90	47.61	50.38	49.61	196.50	49.12

TABLE 7.6 ANOVA of Juice Separation Time at Various Enzyme Concentrations

Source	d.f.	SS	MS	F
Total	19	4482.20		
Treatment	4	4387.24	1096.81	173.27**
Error	15	94.96	6.33	

**Significant at the 1% level.

Note: CV = $\sqrt{6.33}/33.03)100 = 7.6\%$.

$$\bar{y}_0 = 26.73/4 = 6.68$$

and for the second

$$\bar{y}_E = (116.61 + 145.14 + 175.56 + 196.50)/16 = 39.61$$

then

$$t = (\bar{y}_E - \bar{y}_0)/\sqrt{[(1/4) + (1/16)]MSE}$$
$$= (39.61 - 6.68)/\sqrt{(20)(6.33)/64} = 32.93/1.4065 = 23.41$$

since $t_{.01(15)} = 2.947 < 23.41$ the difference is highly significant. We conclude that the addition of enzyme does affect the separation time.

To answer the second question we can use an F test on the subset of means with added enzyme. We compute, first, the sum of squares, S^2, for the four enzyme treatments:

$$S^2 = (1/4)(116.61^2 + 145.14^2 + 175.56^2 + 196.50^2)$$
$$- (116.61 + 145.14 + 175.56 + 196.50)^2/16$$
$$= 26,024.27 - 25,107.19 = 917.08$$

Then the mean squares, MS, is

$$MS = 9.17.08/3 = 305.69$$

and

$$F = 305.69/6.33 = 48.29$$

Since $F_{.01(3,15)} = 5.42 < 48.29$ we conclude that there are highly significant differences among the means of the enzyme treatments.

We would now like to see if we can find a relationship between enzyme concentration (with added enzyme) and time to separation. Because the enzyme levels are equally spaced, and because they are equally replicated, we can use orthogonal polynomial coefficients to fit and test successive terms of a polynomial relating time to level. The polynomial contrasts are shown in Table 7.7.

To perform this sequential test we add one term at a time, then compute and test lack of fit after each addition. The procedure is illustrated in Table 7.8.

We see that "lack of fit" is not significant after fitting the linear term. Further, the quadratic term is not significant. This would indicate that the relationship between separation time (y) and concentration of added enzyme (x) may be described by an equation of the form

$$\hat{y} = b_0 + b_1 x = \bar{y}_E + a_1 \zeta_1$$

TABLE 7.7 Orthogonal Polynomial Contrasts Among Enzyme Levels

| | \multicolumn{4}{c}{Concentration (ppm)} | | |
| | 1 | 2 | 3 | 4 | | |
Total	116.61	145.14	175.56	196.50	L_i	S_i^2
Linear	−3	−1	1	3	270.09	911.86
Quadratic	1	−1	−1	1	−7.59	3.60
Cubic	−1	3	−3	1	−11.37	1.62

Note: $L_i = \sum_{j=1}^{4} k_{ij}T_j, \ S_i^2 = L_i^2/r \sum_{j=1}^{4} k_{ij}^2.$

TABLE 7.8 Sequential Test of Successive Polynomial Coefficients

Source	d.f.	SS	MS
Enzyme	3	917.08	305.69**
Linear	1	911.86	911.86**
Deviation from linear	2	5.22	2.61 NS
Quadratic	1	3.60	3.60 NS
Deviation from linear + quadratic	1	1.62	1.62 NS

**Significant at the 1% level.
NS, not significant.
Note: Experimental error mean square (6.33) is used as the denominator in all of the significance (F) tests.

where

$$a_1 = b_1'\lambda_1$$

$$\zeta_1 = (x_j - \bar{x})/\Delta$$

$$\bar{y}_E = \text{mean separation time for enzyme treatments}$$

$$\bar{x} = \text{mean enzyme level}$$

(For details of this computation see Section 5.10.)
    For the present data we have

$$\bar{y}_E = (116.61 + 145.14 + 175.56 + 196.50)/16 = 39.61$$

$$\bar{x} = (1 + 2 + 3 + 4)/4 = 2.50$$

$$b_1' = L_1/(r \sum_{j=1}^{4} k_{1j}^2) = 270.09/(4)(20) = 3.38$$

$$\lambda = 2$$

$$a = b_1'\lambda = (3.38)(2) = 6.76$$

$$\zeta_i = (x_j - 2.50)/1 = (x_j - 2.50)$$

Then

$$\hat{y}_i = 39.61 + 6.76(x_j - 2.50)$$

$$= 6.76x_j + 22.71$$

Report of statistical analysis. An experiment was conducted to determine whether or not the addition of a certain enzyme would retard the separation of reconstituted frozen orange juice. Statistical analyses of the resulting data indicate that, indeed, separation time is increased by the addition of small amounts of the enzyme. A graph of the results is given in Fig. 7.1.

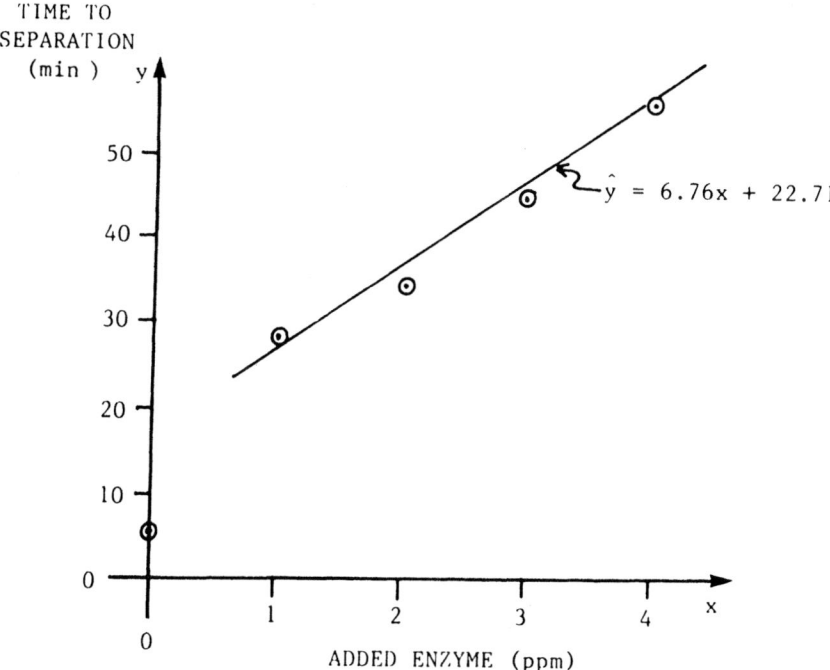

Fig. 7.1 Separation time of reconstituted frozen orange juice as affected by the amount of added enzyme.

Consideration of Fig. 7.1 indicates that the first added increment of enzyme increases the separation time substantially (about 22 min). Increasing increments of enzyme produce an additional delay in time to separation. Further analysis indicates that, once enzyme is added, the relationship between time to separation and amount of added enzyme may be described by the equation

Separation time $= 6.76$ (enzyme concentration) $+ 22.71$

That is, time to separation is increased by about 6.75 min for each added part per million of enzyme over the range of concentrations used in this study. The graph of this equation is plotted in Fig. 7.1.

7.4   EXAMPLE III: BUSH BEAN SPACING

A horticultural scientist wanted to determine the effect of row spacing on the yield of bush beans. Because of differing growth habits among varieties he suspected that the effect of spacing might differ from one variety to another. Accordingly, he decided to use a 4×3, variety $\times$ spacing factorial set of treatments. For the four varieties he chose two, "New Era" and "Big Green," which form low, bushy plants, and two varieties, "Little Gem" and "Red Lake," which form erect plants with few branches. The three row spacings he selected were 20, 40, and 60 cm between rows. After emergence the plants were thinned to a 20-cm spacing within rows.

The experiment was to be conducted in a field which had a gentle slope. To overcome any variation due to slope he decided to use a randomized block design with four blocks of 12 plots each. Blocks were laid out as a line of 12 plots oriented perpendicular to the gradient. At harvest time the yield of dried beans on each plot was determined.

The yields from the trial are presented in Table 7.9. To begin the analysis two summary tables of totals, Table 7.10 and 7.11, are computed. The standard factorial analysis of variance of these data is given in Table 7.12.

On considering the ANOVA we see that there is a highly significant interaction between variety and spacing. For this reason the significance of the differences among variety means has no meaning. It would now be of interest to see if further analysis can shed some light on the source of the interaction. Two courses are open. Since spacing is a quantitative variable we might get some information by graphing yield against spacing for each variety separately. A second alternative is to further decompose the sums of squares by looking at a reasonable set of mutually orthogonal contrasts among totals.

A graph of the means, shown in Fig. 7.2, shows that the yield of two of the varieties, "New Era" and "Big Green," increased as spacing increased. On the other hand, the yield of the other two varieties, "Little Gem" and "Red Lake," decreased as the spacing increased.

TABLE 7.9 Yield of Dried Beans (kg/plot)

Variety	Spacing (cm)	Block I	II	III	IV
New Era	20	32	21	19	22
	40	36	26	21	24
	60	42	33	26	26
Big Green	20	37	38	27	30
	40	39	45	44	37
	60	50	54	54	42
Little Gem	20	35	32	29	30
	40	34	33	28	28
	60	33	29	25	26
Red Lake	20	40	36	35	38
	40	35	33	31	35
	60	28	28	23	30

TABLE 7.10 Block Totals

Block	I	II	III	IV	Σ
Total	441	408	362	368	1579

TABLE 7.11 Variety × Spacing Totals

Variety	Spacing 20 cm	40 cm	60 cm	Σ
New Era	94	107	127	328
Big Green	132	165	200	497
Little Gem	126	123	113	362
Red Lake	149	134	109	392
Σ	501	529	549	1579

TABLE 7.12 ANOVA of Yield of Dried Beans

Source	d.f.	SS	MS	F
Total	47	3046.48		
Block	3	341.90	113.96	8.78**
Variety	3	1332.56	444.19	34.22**
Spacing	2	72.67	36.33	2.80 NS
Variety × spacing	6	871.00	145.17	11.18**
Error	33	428.35	12.98	

**Significant at the 1% level.
NS, not significant.
Note: CV = $(\sqrt{12.98}/32.89)100$ = 10.9%.

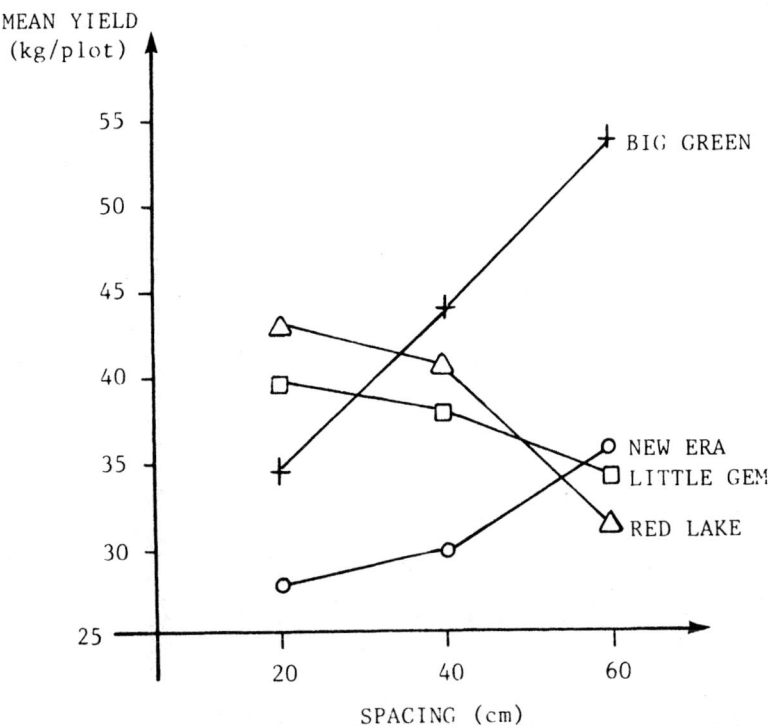

Fig. 7.2 Relationship between row spacing and yield of four bush bean varieties.

TABLE 7.13 Orthogonal Contrasts Among Treatment Totals of Bush Bean Yields

	New Era			Big Green			Little Gem			Red Lake				
Variety	20	40	60	20	40	60	20	40	60	20	40	60		
Space	94	107	127	132	165	200	126	123	113	149	134	109		
Contrast													L_i	S_i^2
Variety														
1) Tall vs. bushy	−1	−1	−1	−1	−1	−1	+1	+1	+1	+1	+1	+1	−71	105.02**
2) Within bushy	−1	−1	−1	+1	+1	+1	0	0	0	0	0	0	169	1190.04**
3) Within tall	0	0	0	0	0	0	−1	−1	−1	+1	+1	+1	30	37.50
Spacing														
4) Linear	−1	0	+1	−1	0	+1	−1	0	+1	−1	0	+1	48	72.00*
5) Quadratic	+1	−2	+1	+1	−2	+1	+1	−2	+1	+1	−2	+1	−8	.67
Variety × spacing														
6) (Tall vs. bushy) × linear	+1	0	−1	+1	0	−1	−1	0	+1	−1	0	+1	−154	741.13**
7) (Tall vs. bushy) × quadratic	−1	+2	−1	−1	+2	−1	+1	−2	+1	+1	−2	+1	−26	7.04
8) (Within bushy) × linear	+1	0	−1	−1	0	+1	0	0	0	0	0	0	35	76.56
9) (Within bushy) × quadratic	−1	+2	−1	+1	−2	+1	0	0	0	0	0	0	−5	.52
10) (Within tall) × linear	0	0	0	0	0	0	+1	0	−1	−1	0	+1	−27	45.56
11) (Within tall) × quadratic	0	0	0	0	0	0	−1	+2	−1	+1	−2	+1	−3	.19

*Significant at the 5% level.
**Significant at the 1% level.

The orthogonal contrasts among totals provide detail to the information presented in the graph. A useful set of these orthogonal contrasts is given in Table 7.13. There is a different linear relationship between yield and spacing for the tall varieties as compared to the bushy ones. Further, this linear relationship appears to differ between the bushy varieties but not between the tall varieties.

Questions answered by the contrasts in Table 7.13 are:

1. Is there a difference in mean yield between tall and bushy varieties?
2. Is there a varietal difference in mean yield within the bushy varieties?
3. Is there a varietal difference in mean yield within the tall varieties?
4. When averaged over varieties is there a linear relationship between yield and spacing?
5. When averaged over varieties is there curvature in the relationship between yield and spacing?
6. Does the linear relationship between yield and spacing differ between tall and bushy varieties?
7. Is the curvature the same for tall as for bushy varieties?
8. Within bushy varieties is the linear relation the same for both varieties?
9. Within bushy varieties is the curvature the same for both varities?

Contrasts 10, 11 are the same for tall varieties as are contrasts 8, 9 for bushy varieties.

Report of statistical analysis. An experiment was conducted to determine the effect of row spacing on the yield of selected varieties of two types of bush beans. A 3 × 4 factorial combination of spacing × variety was used for the 12 treatments in the experiment. The trial was run using a randomized block design with four blocks. Mean yields (kilograms per plot) of dried beans are given in Table 7.14. Note that with the bushy varieties the yield increases as spacing is increased. Yield of the erect varieties, on the other hand, decreases with increasing spacing.

TABLE 7.14 Mean Yield (kg/plot) of Four Varieties of Bush Beans at Three Row Spacings

		Row spacing		
Variety	Type	20 cm	40 cm	60 cm
New Era	Bushy	23.50	26.75	31.75
Big Green	Bushy	33.00	41.25	50.00
Little Gem	Erect	31.50	30.75	28.25
Red Lake	Erect	37.25	33.50	27.25

Note: Standard error = 1.80.

Further analyses indicate that the relationship between yield and spacing may be described by the following equations:

1. New Era

   $\hat{y} = 19.08 + .21x$

2. Big Green

   $\hat{y} = 24.42 + .42x$

3. Average of the tall varieties

   $\hat{y} = 38.04 - .16x$

In all of the equations

$\hat{y}$ = expected yield (kg/plot)

x = distance between rows (cm)

Blocking was effective in increasing the precision of this experiment. The relative efficiency was 150% with blocks as compared to a completely randomized design for the same experiment.

7.5   EXAMPLE IV: WHEAT YIELD TRIAL

In the cereal breeding programs of the International Agricultural Research Centers one method of evaluating promising selections is through the "yield nursery" program. Under this program seeds from a number of selections are sent to cooperators at national or university research centers throughout the world. Here they are grown in a yield trial using a standard experimental design, normally a randomized block design with four blocks. At harvest time yield data are collected, tabulated, and returned to the breeder at the International Center. The results are then analyzed to select the best varieties for each location and to measure the adaptability of the selections to a variety of growing conditions.

The results of one such trial grown in Shawbak, Jordan, in 1978 are given in Table 7.15.

We begin the analysis by first computing the block totals shown in Table 7.16. Then the analysis of variance of the data is as presented in Table 7.17.

Since the breeder wants to separate the highest yielding selections from the others, and since there is no underlying relationship or structure connecting the selections, this is an example of one of the few situations in which a multiple comparison procedure makes sense. For a number of reasons, primarily the work done by Carmer and Swanson (1973), we prefer to use Fisher's protected LSD for this purpose. Alternatively we would use the Waller-Duncan BLSD (1975). The basic procedure is:

TABLE 7.15 Yield of Wheat (kg/ha) in Shawbak, Jordan, in 1978

Selection	Block				Sum
	1	2	3	4	
1	2,546	2,139	2,050	2,083	8,818
2	2,843	2,369	2,117	1,808	9,137
3	3,546	2,865	2,908	1,750	11,069
4	2,611	2,610	2,575	1,542	9,338
5	2,792	2,209	2,000	1,892	8,893
6	1,913	1,491	2,033	1,925	7,362
7	1,754	1,796	1,850	1,517	6,917
8	2,315	2,537	2,158	2,375	9,385
9	3,366	2,759	2,992	2,450	11,567
10	2,463	2,162	1,842	2,308	8,775
11	2,352	2,148	1,950	1,917	8,367
12	2,412	2,310	2,150	2,250	9,122

TABLE 7.16 Block Totals

Block	1	2	3	4	Sum
Total	30,913	27,395	26,625	23,817	108,750

TABLE 7.17 ANOVA of Wheat Yield Data

Source	d.f.	SS	MS	F
Total	47	9,510,745		
Block	3	2,133,257	711,086	8.63**
Selection	11	4,659,329	423,575	5.14**
Error	33	2,718,159	82,368	

**Significant at the 1% level.

Note: CV = $(\sqrt{82,368}/2,265.62)100 = 12.7\%$.

Relative efficiency = $[(3)(711,086) + (33)(82,368)]/(47)(82,368) = 1.25$.

1. Note that in the ANOVA the F test for "selections" is significant so we can proceed with the means-separation procedure.

2. Compute the LSD at the chosen significance level. For this example we will use the 5% level:

$$\text{LSD} = t_\alpha\sqrt{2\text{MSE}/r} = 2.042\sqrt{(2)(82,368)/4}$$

$$= 414.40$$

3. Arrange the means in order of descending yield; then either look for differences between adjacent ranked means which are larger than the LSD, or use the LSD to group the means into groups within which the means are not significantly different. For the present example the grouping is given in Table 7.18.

Report of statistical analysis. As part of a wheat breeding program 12 new selections were grown in a yield trial in Shawbak, Jordan, to select possible varieties for use in that area. Mean yields of the selections, arranged in order of decreasing yield, are given in Table 7.19.

Statistical analysis indicates that although the difference in yield between the top two selections, 9 and 3, is not significant, there is a significant gap in yield between these two and the other selections. On the basis of this trial these two would warrant further evaluation for use in this area.

As far as the trial itself is concerned the coefficient of variation (12.7%) is somewhat higher than usual. This may have been the result of greater than

TABLE 7.18 Grouping of Selections Using the FPLSD Procedure

Selection	Rank (i)	Mean yield	Gap $\bar{y}_i - \bar{y}_{i+1}$	Grouping
9	1	2891.75		
3	2	2767.25	124.70	
8	3	2346.25	421.00*	
4	4	2334.50	11.75	
2	5	2284.25	50.25	
12	6	2280.50	3.75	
5	7	2223.25	57.25	
1	8	2204.50	18.75	
10	9	2193.75	10.75	
11	10	2091.75	102.00	
6	11	1840.50	251.25	
7	12	1729.25	111.25	

*Gap significant at the 5% level.

Note: Means enclosed by the same line are not significantly different.

TABLE 7.19 Mean Yield (kg/ha) of
Wheat Selections Grown in Shawbak,
Jordan, in 1978

Selection	Mean yield
9	2891.75 *a*
3	2767.25 *a*
8	2346.25 *b*
4	2334.50 *b*
2	2284.25 *b*
12	2280.50 *b*
5	2223.25 *b c*
1	2204.50 *b c*
10	2193.75 *b c*
11	2091.75 *b c d*
6	1840.50 *c d*
7	1729.25 *d*
Standard error	143.50

Note: Means designated by the same letter
are not significantly different at the
5% level.

normal variation among plots at the test site. Some of this variation was controlled
by blocking. Analysis of the data indicates that it would have required five
replications of a completely randomized design to reach the same precision
attained by four replications of the randomized block design.

7.6   STATISTICAL COMPUTING PACKAGES

With the advent of high-speed, electronic computers and their increasingly wide-
spread availability some mention should be made about the use of computers in
the analysis of experimental data. A number of packaged programs are available
for using computers to do statistical analysis. Among these packages the more
well known and widely used include SPSS, SAS, BMDP, MINITAB, GLIM,
GENSTAT, and many others. All of them include programs for the description
of data sets, for correlation analysis, simple and multiple regression, analysis
of variance, multivariate analysis, etc. For data from designed experiments these
packages include programs to handle most of the standard experimental designs.

For the most part the experimental design programs are based on the decomposition of data from factorial experiments into sums of squares associated with
the main effects and interactions among the various factors. With the blocked

designs the blocking criterion is introduced as a factor, and the interaction of blocks with treatment factors is used as experimental error. The general idea might be illustrated with some examples:

1. Randomized block design with p treatments in r blocks. The analysis of variance for this design is outlined in Table 7.20.
2. Two-factor factorial with factor A at a levels and factor B at b levels run in a randomized block design with r blocks. The analysis of variance for this design is shown in Table 7.21.

    If "block" is taken to be a factor, then Table 7.22 gives the interactions which may be computed.

TABLE 7.20 ANOVA of RBD with p Treatments in r Blocks

Source	d.f.
Total	$rp - 1$
Block	$r - 1$
Treatment	$p - 1$
Error	$(r - 1)(p - 1)$

TABLE 7.21 ANOVA of $A \times B$ Factorial in RBD with r Blocks

Source	d.f.
Total	$rab - 1$
Block	$r - 1$
A	$a - 1$
B	$b - 1$
$A \times B$	$(a - 1)(b - 1)$
Error	$(r - 1)(ab - 1)$

TABLE 7.22 Block $\times$ Treatment Interactions

Source	d.f.
Block $\times A$	$(r - 1)(a - 1)$
Block $\times B$	$(r - 1)(b - 1)$
Block $\times A \times B$	$(r - 1)(a - 1)(b - 1)$

These interactions (involving blocks) are pooled by adding the sums of squares to obtain an "experimental error" sum of squares. The degrees of freedom for this sum of squares is obtained by adding the individual d.f. We have

$$(r - 1)(a - 1) + (r - 1)(b - 1) + (r - 1)(a - 1)(b - 1)$$

$$= ra - r - a + 1 + rb - r - b + 1 + rab - ra - rb$$
$$- ab + r + a + b - 1$$

$$= rab - ab - r + 1$$

$$= (r - 1)(ab - 1)$$

3. Split-plot design with factor A at a levels on the whole plots in a randomized block design with r blocks, and factor B at b levels on the subplots. The analysis of variance for this design is presented in Table 7.23.

   With "block" as a factor and "A" estimated from the whole plots, the "block $\times$ A" interaction, with $(r - 1)(a - 1)$ d.f., serves as "error(a)". The subplot error, "error(ab)", is obtained by pooling the "block $\times$ B" and the "block $\times$ A $\times$ B" interactions. That is, the interactions of blocks with effects estimated on the subplots are pooled to obtain the subplot error. The pooled d.f. is

$$(r - 1)(b - 1) + (r - 1)(a - 1)(b - 1)$$

$$= rb - r - b + 1 + rab - ra - rb - ab + r + a + b$$
$$- 1$$

$$= rab - ra - ab + a$$

$$= a(r - 1)(b - 1)$$

TABLE 7.23 ANOVA of a Split-Plot Design for Two Factors

Source	d.f.
Total	$rab - 1$
Block	$r - 1$
A	$a - 1$
Error(a)	$(r - 1)(a - 1)$
B	$b - 1$
A $\times$ B	$(a - 1)(b - 1)$
Error(ab)	$a(r - 1)(b - 1)$

The general procedure illustrated here is often useful in developing the error terms to be used with unusual designs. Restricting criteria are entered as factors. The interactions of these factors with treatment factors are then used as error terms.

Many of the computing packages contain programs for pairwise multiple comparison procedures. These should be used only where they are appropriate. Too often a multiple comparison test is used simply because it is included in the computing package, whether or not its use makes sense within the context of the problem or of the treatments used in the experiment (Petersen, 1977). Some computing packages also contain procedures for constructing and testing single-degree-of-freedom contrasts. In many cases, however, these contrasts can be computed and tested by using a hand calculator in less time than it takes to determine how to instruct the computer to do what is wanted.

Computers are useful in the analysis of data from designed experiments in a number of situations:

1. To compute the preliminary analysis of variance for standard experimental designs
2. To analyze data from very large experiments
3. To analyze experiments where a number of variables are measured on each experimental unit

For many situations, however, most of the analysis of data from designed experiments can be quickly and conveniently handled with a hand-held calculator. This is particularly true for the detailed examination of data after the initial analysis of variance has been completed.

Because computer systems and their associated software are changing rapidly we will not go into greater detail here. The computations required to analyze the data from the designs presented here are given in detail with each design. They can be readily adapted to any computing system or statistical analysis package.

FURTHER READING

Cox, D. R., and E. J. Snell (1981). Applied Statistics, Principles and Examples, *Methuen*, New York.

REFERENCES

Carmer, S. G., and M. R. Swanson (1973). An Evaluation of Ten Pairwise Multiple Comparison Procedures by Monte Carlo Methods, *J. Amer. Stat. Assoc.* 68:66–74.

Duncan, D. B. (1975). *t* Tests and Intervals for Comparisons Suggested by the Data, *Biometrics* 31:339–359.

Peterson, R. G. (1977). Use and Misuse of Multiple Comparison Procedures, *Agron. J.* 69:205–208.

8

Multifactor Experiments

8.1. INTRODUCTION

Many research problems go through a number of stages before a satisfactory solution is reached. The first problem is to identify those factors which affect response and to separate them from factors whose level is unimportant (the initial or exploratory stage). Next comes the problem of determining how the response changes with changes in the level of the more important factors, and whether the response to changes in one factor is affected by the level of other factors (the intermediate stages). Finally comes the problem of finding limitations, maxima and minima, both physical and economic, and, in general, finding the precise relationship between response and levels of the production factors (the final stages). Different types of designs are useful at each stage. For the most part, however, some type of multifactor experiment should be employed in most of the stages.

8.2 THE 2^k FACTORIAL SERIES

One of the most useful types of multifactor experiment is the 2^k factorial series. In this series there are k factors each at two levels. Hence there is a total of 2^k treatments in the full factorial set. This series is particularly useful in the exploratory stages of an investigation because it permits the examination of a fairly large number of factors and their interactions in a trial of reasonable size.

To illustrate the 2^k series and the standard symbolism that goes along with it suppose we are interested in three factors: A, B, and C. A full factorial set would contain $2^3 = 8$ treatments. These are commonly designated:

1.	(1)	5.	c
2.	a	6.	ac
3.	b	7.	bc
4.	ab	8.	abc

These are listed in so-called standard order. Presence of a letter in a combination indicates that the factor is present at the high level. Absence of a letter indicates that the factor is present at the low level. The symbol (1) is used to designate the combination in which all factors are present at the low level. Often, but not necessarily, the low level is the absence of the factor while the high level is its presence. When the factor is a qualitative variable one category is arbitrarily designated the high level while the other is called the low.

Two other systems of symbols, one of which we will use later, are also in use:

1.	$a_1b_1c_1$	$a_1b_1c_2$	2.	000	001
	$a_2b_1c_1$	$a_2b_1c_2$		100	101
	$a_1b_2c_1$	$a_1b_2c_2$		010	011
	$a_2b_2c_1$	$a_2b_2c_2$		110	111

8.2.1 Analysis

In a factorial experiment we want to estimate and test the main effects and interactions. In the 2^k series there are k main effects, C_2^k two-factor (first-order) interactions, C_3^k three-factor (second-order) interactions, . . . , and one k-factor interaction. Each of these factorial effects carries with it one degree of freedom.

In the analysis of variance the treatment sum of squares is invariably partitioned into the main effect and interaction components. For a 2^3 factorial with factors A, B, and C conducted using a randomized block design with r blocks the key-out for the analysis of variance would be as shown in Table 8.1.

Note: Main effects and interactions are usually designated by capital letters while small letters are used to identify treatment combinations.

There are several ways to compute the sums of squares for the main effects and interactions. They can be computed from two-way, three-way, etc. tables of totals. However, an easier way is to construct contrasts of totals and to compute the sums of squares from these contrasts. For the 2^3 system, for example, the standard table of coefficients for the factorial contrasts is illustrated in Table 8.2. Then, if

T_j = the yield total for the jth treatment

r = number of replications

TABLE 8.1 Format for the
ANOVA of a 2^3 Factorial
Experiment

Source	d.f.
Total	$8r - 1$
Blocks	$r - 1$
A	1
B	1
AB	1
C	1
AC	1
BC	1
ABC	1
Error	$7(r - 1)$

TABLE 8.2 Table of Coefficients for Factorial Contrasts in a 2^3 Factorial

	Treatment							
Contrast	(1)	a	b	ab	c	ac	bc	abc
A	−	+	−	+	−	+	−	+
B	−	−	+	+	−	−	+	+
AB	+	−	−	+	+	−	−	+
C	−	−	−	−	+	+	+	+
AC	+	−	+	−	−	+	−	+
BC	+	+	−	−	−	−	+	+
ABC	−	+	+	−	+	−	−	+

k_{ij} = coefficient for the jth treatment for the ith contrast (entry in the body of Table 8.2)

$L_i = \sum_j k_{ij}T_j$ is the ith contrast

and

$S_i^2 = L_i^2/r2^k = (\sum_j k_{ij}T_j)^2/r2^k$ is the sum of squares for the ith contrast

8.2.2 Contrasts in the 2^k Series

There are a number of procedures for writing down the coefficients of the contrasts in the 2^k factorial system. These are given by the following rules, any of which may be used:

I. The "product rule"
 1. For each main effect use a + for those treatment combinations in which the letter for the effect occurs. All other combinations receive a − .
 2. The signs for the interaction contrasts are the product of the signs of the main effects whose letters appear in the interaction.
II. The "odds vs. evens" rule
 1. In every factorial contrast half the combinations receive a + sign and half a − sign.
 2. Combinations which receive one sign are those which contain an even number (zero is considered to be even) of the letters in the factorial effect. Combinations which contain an odd number of the letters receive the other sign.
 3. By convention the + sign is assigned to combinations which contain all of the letters of the effect.
III. The "algebraic system"
 1. Write the following expression:

$$(a \pm 1)(b \pm 1)(c \pm 1) \ldots$$

 using a − sign for those factors having a letter in the effect and a + sign for the other factors.
 2. Expand the expression algebraically to obtain the contrast.

8.2.3 Yates Algorithm for 2^k Factorials

An algorithm for computing the effects, sums of squares, and treatment means in the 2^k system was developed by Yates (1937). The procedure may be illustrated, in Table 8.3., using data from a 2^3 factorial trial with four replications.

TABLE 8.3 Yates Algorithm for Analysis of a 2^3 Factorial

Treatment	Total yield	I	II	III	Effect	SS
(1)	121	302	663	1362	Total	—
a	181	361	699	408	A	5202.00
b	104	296	213	166	B	861.12
ab	257	403	195	188	AB	1104.50
c	123	60	59	36	C	40.50
ac	173	153	107	−18	AC	10.12
bc	129	50	93	48	BC	72.00
abc	274	145	95	2	ABC	.12

1. Treatments are listed in standard order beginning with (1) and adding one letter at a time followed by all combinations with letters which have previously been added.
2. In the yield column enter the total yields for the treatment combinations listed in the first column.
3. Fill in as many columns (headed by Roman numerals) as there are factors in the experiment in the following way:
   a. Add the first two yields and enter the sum as the first entry in column I. Add the third and fourth yields to obtain the second entry in column I. Continue in this way, adding successive pairs of yields, until the top half of column I has been filled (all pairs have been added).
   b. Subtract the first yield from the second to obtain the next entry in column I. Subtract the third yield from the fourth to obtain the next entry in column I. Continue in this way, subtracting the first from the second member of each successive pair, until the bottom half of column I has been filled. (Differences of all pairs have been obtained.)
4. To obtain the entries in column II repeat steps 3a and 3b using the entries in column I.
5. To obtain the entries in column III repeat steps 3a and 3b using the entries in column II.
6. Continue in this way filling successive columns until as many columns have been filled as there are factors in the experiment.

The first entry in the last column headed by a Roman numeral (III) is the grand total of all yields in the trial. The mean yield of the trial is obtained by dividing this entry by $r2^k$, where r is the number of replications. The remaining entries in this column (III) are effect totals for the main effects and interactions in standard order (identified by the symbols in the "effect" column). The effect means are computed by dividing the remaining entries in this column by $r2^{k-1}$.

The sums of squares for the main effects and interactions are obtained as follows:

1. Square the effect total (entry in the last column headed by a Roman numeral).
2. Divide this square by the number of observations in the experiment $(r2^k)$.

Often it is desirable to compute summary tables of means for combinations of factors. These can be obtained by direct averaging of appropriate plot yields. However, a faster method for large trials involves further application of the Yates algorithm. The procedure for obtaining mean yields for all combinations of any set of factors is as follows:

1. Take the grand total and all main effect and interaction effect totals for the factors to be studied from the last column headed by a Roman numeral in the Yates analysis. Write these in a "total" column in reverse of standard order.
2. Perform additions and subtractions according to the Yates scheme. Fill as many columns as there are factors in the table of means.
3. Divide each entry in the last column filled under the Yates scheme by the number of observations in the experiment ($r2^k$).
4. Read the results as means listed in the order in which the effect totals were entered.

For example, a two-way table of means for factors A and C may be obtained from the preceding Yates analysis (Table 8.3) as shown in Table 8.4.

8.2.4 Numerical Example: Factorial Experiment

An agronomist wanted to determine the effect of three fertilizer elements: nitrogen (N), phosphorus (P), and potassium (K), on the yield of corn. He believed that the fertilizers used in combination might produce different effects than when they are used alone. He decided to use a $2 \times 2 \times 2$ factorial combination of the three elements. His treatments were:

1.	No fertilizer	(1)	5.	20# K		k
2.	50# N	n	6.	50# N + 20# K		nk
3.	20# P	p	7.	20# P + 20# K		pk
4.	50# N + 20# P	np	8.	50# N + 20# P + 20# K		npk

The area in which the trial was to be conducted was furrow irrigated. He suspected that water volume might be greater near the furrow than farther away. He decided to use a randomized block design with four blocks. These were

TABLE 8.4 Computation of Means Using the Reverse Yates Algorithm

Effect	Total	I	II	÷ 32	Mean
AC	− 18	18	1788	55.88	ac
C	36	1770	1008	31.50	c
A	408	54	1752	54.75	a
(1)	1362	954	900	28.12	(1)

arranged to permit elimination of differences due to water volume if such differences existed. The plot plan and yields (grams per plot) of corn were as shown in Fig. 8.1. A preliminary analysis of variance of the yield data is presented in Table 8.5.

The treatment differences are highly significant. We now look at the main effects and interactions by using the contrasts of totals given in Table 8.6.

Alternatively, the main effects and interactions may be computed by using the Yates method shown in Table 8.7. The final ANOVA of the yield data is summarized in Table 8.8.

BLOCK

I		II		III		IV	
nk	(1)	pk	n	p	nk	np	k
291	101	407	89	323	334	361	302
pk	k	p	npk	(1)	k	nk	(1)
398	265	324	449	87	279	272	131
p	n	k	np	np	n	n	npk
312	106	272	338	324	128	103	437
np	npk	nk	(1)	pk	npk	p	pk
373	450	306	106	423	471	324	445

———————▶   WATER FLOW   ———————▶

Fig. 8.1 Field plan and yields for an NPK fertilizer experiment on corn.

TABLE 8.5 Preliminary ANOVA of Corn Yield Data

Source	d.f.	SS	MS	F
Total	31	466,779.7		
Blocks	3	774.1	258.03	.74
Treatments	7	458,718.0	65,531.14	188.83*
Error	21	7,287.6	347.03	

*Significant at the 1% level.

Note: CV $= (100\sqrt{347.03}/291.59) = 6.4\%$.

Standard error of treatment mean $= \sqrt{347.03/4} = 9.31$.

TABLE 8.6 Factorial Effect Contrasts and Sums of Squares

Contrast	(1) 425	n 426	p 1283	np 1396	k 1118	nk 1203	pk 1673	npk 1807	L_i	S_i^2
N	−	+	−	+	−	+	−	+	333	3,465.28
P	−	−	+	+	−	−	+	+	2,987	278,817.78
NP	+	−	−	+	+	−	−	+	161	810.03
K	−	−	−	−	+	+	+	+	2,271	161,170.03
NK	+	−	+	−	−	+	−	+	105	344.53
PK	+	+	−	−	−	−	+	+	− 669	13,986.28
NPK	−	+	+	−	+	−	−	+	− 63	124.03

Note: $L_i = \Sigma k_{ij} T_j.$

$S_i^2 = (\Sigma k_{ij} T_j)^2 / r \Sigma k_{ij}^2 = L_i^2/32.$

TABLE 8.7 Yates Algorithm for Factorial Effects and Sums of Squares

Treatment	Total yield	I	II	III	Effect	SS
(1)	425	851	3,530	9,331	Total	—
n	426	2,679	5,801	333	N	3,465.28
p	1,283	2,321	114	2,987	P	278,817.78
np	1,396	3,480	219	161	NP	810.03
k	1,118	1	1,828	2,271	K	161,170.03
nk	1,203	113	1,159	105	NK	344.53
pk	1,673	85	112	− 669	PK	13,986.28
npk	1,807	134	49	− 63	NPK	124.03

Note: SS $= (III)^2/r2^k = (III)^2/(4)(8)$

We see that neither the NP, the NK, nor the NPK interactions are significant. So we can say something about the significant main effect on *N*. On the other hand, the PK interaction is significant. This means that neither the *P* nor the *K* main effects have a meaningful interpretation.

Report of statistical analysis. The results of an NPK fertilizer trial on corn are summarized in Tables 8.9 and 8.10.

From Table 8.9 it is seen that the addition of 50 lb/A (A = acre) of nitrogen results in a mean yield increase of about 20 g of corn per plot. This increase occurs regardless of the level of phosphorus or potassium.

Consideration of Table 8.10 shows that the yield response to potassium depends on whether or not phosphorus is present. In the absence of phosphorus

TABLE 8.8 Final ANOVA of Corn Yields
(g/plot)

Source	d.f.	MS	F
Total	31		
Blocks	3	258.03	.74
N	1	3,465.28	9.98**
P	1	278,817.78	803.44**
NP	1	810.03	2.33
K	1	161,170.03	464.43**
NK	1	344.53	.99
PK	1	13,986.28	40.30**
NPK	1	124.03	.36
Error	21	347.03	

**Significant at the 1% level.

TABLE 8.9 Mean Yield of Corn (g/plot) at
Different Levels of Nitrogen

N (lb/A)	0	50	Standard error
Mean yield	281.19	302.00	4.66

TABLE 8.10 Mean Yield of Corn (g/plot)
at Different Combinations of Phosphorus and
Potassium

	K (lb/A)		Standard error
P (lb/A)	0	20	
0	106.38	290.13	6.59
20	334.88	435.00	

the addition of 20 lb/A of potassium results in a mean increase of about 185 g of corn per plot. In the presence of 20 lb/A of phosphorus, however, the addition of 20 lb/A of potassium results in an average increase of only about 100 g of corn per plot.

As might be expected, the minimum yield, about 105 g/plot, occurred when no fertilizer was added. The maximum yield, about 450 g/plot, was obtained when all three fertilizer elements were used together.

Blocking had little effect on the experimental error in this trial. Apparently there was little difference in water volume from one end of the furrow to the other. Further, the relative variation among plots in this area is not excessive (CV = 6.4%). For future trials in this area it is suggested that a completely randomized design be used in place of a randomized block design.

Note: Using the reverse Yates procedure, the pertinent means may be computed as shown in Tables 8.11 and 8.12.

8.3 SINGLE REPLICATION

One of the problems with factorial experiments is that the number of treatments increases rapidly as the number of factors is increased, even if the factors are included at only two levels. This leads to difficulty because the time and expense involved often put a limit on the total number of units which can be included in an experiment. For this reason consideration must be given to finding ways of examining a large number of factors without using an excessive number of units.

Suppose there are 64 units available and it is desired to study five factors. It would be possible to run two replications of a 2^5 factorial, with the analysis of variance format given in Table 8.13.

This experiment provides 32 d.f. for testing all main effects and interactions. Consider, however, the interpretation of the results. Main effects and low-order interactions can be interpreted in a meaningful way. High-order interactions, on the other hand, are difficult to interpret. Further:

TABLE 8.11 Nitrogen Main Effect Means

Effect	Total	I	÷ 32	Mean
N	333	9664	302.00	n
Total	9331	8998	281.19	(1)

TABLE 8.12 P × K Means

Effect	Total	I	II	÷ 32	Mean
PK	− 669	1,602	13,920	435.00	pk
K	2,271	12,318	9,284	290.12	k
P	2,987	2,940	10,716	334.88	p
Total	9,331	6,344	3,404	106.38	(1)

TABLE 8.13 ANOVA of a 2^5
Factorial Experiment

Source	d.f.
Total	63
Main effects	5
Two-factor interactions	10
Three-factor interactions	10
Four-factor interactions	5
Five-factor interactions	1
Error	32

TABLE 8.14 ANOVA for a Single Replication of a 2^6 Factorial

Source	d.f.	
Total	63	
Main effects	6	
Two-factor interactions	15	
Three-factor interactions	20	
Four-factor interactions	15	⎫
Five-factor interactions	6	⎬ pooled error, 22 d.f.
Six-factor interactions	1	⎭

1. Experience has shown that high-order interactions are usually negligible.
2. In the analysis of variance a negligible interaction will have a mean square about equal to the error mean square.
3. An experiment can be designed with only one unit per treatment with a mean square obtained by pooling the high-order interactions being used as an estimate of error.
4. Such a procedure will provide conservative tests of main effects and low-order interactions because any effects declared significant under this test would certainly be significant with the proper error.

Let us reconsider the preceding example in which 64 units are used for two replications of a 2^5 factorial. Suppose, instead, we choose to add an additional factor at two levels and run a single replication of a 2^6 factorial. Our ANOVA would be as shown in Table 8.14. Based on the previous argument we could pool the four, five, and six-factor interactions to obtain an estimate of error with 22 d.f.

By this procedure information can be obtained on an additional factor without losing any of the relevant information, nor reducing the precision, about the original five factors. From this standpoint, then, the best design is often one which "saturates" the experimental units with factors. A few points should be noted:

1. If some high-order interactions are of particular interest they need not be pooled in error but may be isolated and tested along with the low-order effects.
2. Interactions which are to be pooled for error must be selected before the experiment is run. It is not correct to do the ANOVA and then pool only those interactions with small mean squares. This would lead, falsely, to too many declarations of significance.
3. As a general rule it is usually safe to pool four-factor and higher interactions; two-factor interactions should not be pooled; and three-factor interactions should be pooled only if necessary.

8.3.1   Numerical Example: Single Replication

A research chemist wanted to study the effect of a number of factors on the yield of a new high-impact plastic. The plastic is produced by mixing a resin with an extender in a solvent. The process takes place in a heated reaction vat. The materials are allowed to react for a period of time, and the plastic settles to the bottom of the vat in the form of soft granules. These granules are removed from the vat, filtered on a screen, and dried.

The chemist worked in a laboratory which contained 32 experimental reaction vats and a filter and dryer with each vat. He knew that he could run one trial per day in each vat. He decided to study five factors using a single replication of a 2^5 factorial set of treatments. He wanted to look at main effects and first-order interactions. He assumed that second- and higher-order interactions were negligible and could be pooled to obtain an estimate of error.

He selected the following factors, each at two levels:

A = reaction temperature: 300°C, 150°C
B = reaction time: 4 hr, 2 hr
C = filter pressure: 2 atm, 1 atm
D = drying temperature: 200°C, 100°C
E = resin/extender ratio: 2/1, 1/1

The resulting yields (grams) and their analysis by the Yates method are given in Table 8.15.

Report of statistical analysis.   An experiment was conducted to determine the effect of the following five factors on the yield of a new high-impact plastic:

TABLE 8.15 Analysis of Yields from a Single Replication of a 2^5 Factorial Experiment on Plastic Production Using the Yates Method

Treatment	Yield	I	II	III	IV	V	Effect	SS$_i$
(1)	246	549	1,161	2,341	4,544	10,106	Total	—
a	303	612	1,180	2,203	5,562	358	A	4,005.125
b	276	602	1,240	2,792	330	212	B	1,404.500
ab	336	578	963	2,770	28	− 64	AB	128.000
c	258	559	1,398	251	202	− 202	C	1,275.125
ac	344	681	1,394	79	10	30	AC	28.125
bc	265	461	1,355	2	− 152	− 248	BC	1,922.000
abc	313	502	1,415	26	88	− 40	ABC	50.000 +
d	249	705	117	39	− 258	− 160	D	800.000
ad	310	693	134	163	56	− 148	AD	684.500
bd	318	748	106	− 114	− 116	362	BD	4,095.125
abd	363	646	− 27	124	146	− 330	ABD	3,403.125 +
cd	212	649	− 48	− 35	− 168	− 232	CD	1,682.000
acd	249	706	50	− 117	− 80	− 200	ACD	1,250.000 +
bcd	283	674	− 11	168	− 126	106	BCD	351.125 +
abcd	219	741	37	− 80	86	− 62	ABCD	120.125 +
e	379	57	63	19	− 138	1,018	E	32,385.125**
ae	326	60	− 24	− 277	− 22	− 302	AE	2,850.125
be	344	86	122	− 4	− 172	− 192	BE	1,152.000
abe	349	48	41	60	24	240	ABE	1,800.000 +
ce	389	61	− 12	17	124	314	CE	3,081.125
ace	359	45	− 102	− 133	238	262	ACE	2,145.125 +
bce	283	37	57	98	− 82	88	BCE	242.000 +
abce	363	− 64	67	48	− 248	212	ABCE	1,404.500 +
de	313	− 53	3	− 87	− 296	116	DE	420.500
ade	336	5	− 38	− 81	64	196	ADE	1,200.500 +
bde	370	− 30	− 16	− 90	− 150	114	BDE	406.125 +
abde	336	80	− 101	10	− 50	− 166	ABDE	861.125 +
cde	322	23	58	− 41	6	360	CDE	4,050.000 +
acde	352	− 34	110	− 85	100	100	ACDE	312.500 +
bcde	367	30	− 57	52	− 44	94	BCDE	276.125 +
abcde	374	7	− 23	34	− 18	26	ABCDE	21.125 +
Sum	10,106							73,806.875

+ Sums of squares pooled to estimate experimental error.
**Effect significant at the 1% level.

Note: SS$_i$ = $(V)^2/32$.

$$\text{MSE} = (50.000 + 3403.125 + 1250.000 + 351.125 + 120.125 + 1800.000$$
$$+ 2145.125 + 242.000 + 1404.500 + 1200.500 + 406.125 + 861.125$$
$$+ 4050.000 + 312.500 + 276.125 + 21.125)/16$$
$$= 1118.344.$$

1. Reaction temperature
2. Reaction time
3. Filter pressure
4. Drying temperature
5. Resin/extender ratio

The only factor whose change had a significant effect on plastic yield was the resin/extender ratio. The results are given in Table 8.16.

For an exploratory experiment the relative variation in this trial was not excessive (CV = 10.6%).

8.4 FRACTIONAL REPLICATION

The arguments which lead to the choice of a single replication can be extended to permit the use of an experiment in which the number of units is less than the number of combinations in the full factorial set. The trick lies in the choice of the combinations to be used. For example, if $q < 2^k$ combinations are used they should be chosen so that the factorial contrasts of interest can be estimated and tested, and so that some negligible contrasts remain to be used for error. A suitable set of combinations may, with luck, be chosen by trial and error. However, a systematic procedure is available.

To illustrate the procedure (illustration only) consider the following four combinations from the eight required for a full factorial set with three factors each at two levels:

a b c abc

Now, in a 2^3 factorial we can estimate the main effects and interactions as contrasts. These contrasts can be estimated with the above four combinations using the coefficients shown in Table 8.17.

8.4.1 Defining Contrasts and Aliases

We see that all of the above combinations carry a + sign for the ABC contrast, so that this contrast cannot be estimated at all from the selected treatments. This contrast (ABC) is called the *defining contrast*. Note also that the following contrasts are equivalent to each other:

TABLE 8.16 Mean Yield (g) of Plastic at Two Resin/Extender Ratios

Ratio	2/1	1/1	Standard error
Mean yield	284.00	347.62	8.36

TABLE 8.17 Coefficients for
Factorial Effects Using Four
Treatments from a 2^3 Factorial Set

	Treatment			
Contrast	a	b	c	abc
A	$+$	$-$	$-$	$+$
B	$-$	$+$	$-$	$+$
AB	$-$	$-$	$+$	$+$
C	$-$	$-$	$+$	$+$
AC	$-$	$+$	$-$	$+$
BC	$+$	$-$	$-$	$+$
ABC	$+$	$+$	$+$	$+$

$$A = BC$$
$$B = AC$$
$$C = AB$$

That is, the contrast which estimates A is the same as the contrast which estimates BC. Such contrasts are said to be *aliases* of each other. If the contrast shows an apparent effect of A there is no way of knowing whether it is really an effect of A or an effect of BC or a mixture of both.

Whenever a fractional replication is used every contrast has one or more other contrasts as an alias. To find the aliases we make use of the following rule:

In the 2^k system the alias of any factorial contrast is its generalized interaction with the defining contrast.

To find the generalized interaction of two contrasts in the 2^k system combine all of the letters that appear in the two contrasts, then delete those which occur twice. The remaining letters symbolize the generalized interaction.

For example, with ABC as a defining contrast the generalized interaction of B with the defining contrast would be

$$A\cancel{B}\cancel{B}C = AC$$

Hence, AC is the alias of B.

8.4.2 Design Construction

We can use these concepts in the construction of a design containing only 1/2 replication of 2^k factorial. To do this we select a contrast, usually a high-order interaction, to serve as a defining contrast. Then we use either those combinations which carry a + sign or those combinations which carry a − sign under the defining contrast. For example, with four factors we could choose *ABCD* as a defining contrast. Then we could use either the set:

(1)   *ab*   *ac*   *bc*   *ad*   *bd*   *cd*   *abcd*   (+)

or the set:

a   *b*   *c*   *abc*   *d*   *abd*   *acd*   *bcd*   (−)

to give us 1/2 replication of the full set.

Proper use of fractional replication depends on choosing the defining contrast in such a way that:

1. No main effect or low-order interaction has another main effect or low-order interaction as an alias.
2. Some contrasts and their aliases are unimportant (negligible) so that they can be used as error.

Normally, we want to estimate the test main effects and first-order interactions. If we assume that second-order (three-factor) and higher interactions are negligible the implication is that six factors is the minimum number for which a 1/2 replication of a 2^k factorial is a useful design.

Suppose we want to run a 1/2 replication of a 2^6 factorial. We could choose *ABCDEF* as the defining contrast. This would give us the following alias pattern:

I	= *ABCDEF*	*D*	= *ABCEF*	*E*	= *ABCDF*	*DE*	= *ABCF*
A	= *BCDEF*	*AD*	= *BCEF*	*AE*	= *BCDF*	*ADE*	= *BCF*
B	= *ACDEF*	*BD*	= *ACEF*	*BE*	= *ACDF*	*BDE*	= *ACF*
AB	= *CDEF*	*ABD*	= *CEF*	*ABE*	= *CDF*	*ABDE*	= *CF*
C	= *ABDEF*	*CD*	= *ABEF*	*CE*	= *ABDF*	*CDE*	= *ABF*
AC	= *BDEF*	*ACD*	= *BEF*	*ACE*	= *BDF*	*ACDE*	= *BF*
BC	= *ADEF*	*BCD*	= *AEF*	*BCE*	= *ADF*	*BCDE*	= *AF*
ABC	= *DEF*	*ABCD*	= *EF*	*ABCE*	= *DF*	*ABCDE*	= *F*

All of the main effects have five-factor interactions as aliases; two-factor interactions have four-factor interactions as aliases; and three-factor interactions have other three-factor interactions as aliases. Under the assumption that second, and higher, order interactions are negligible we have the analysis of variance presented in Table 8.18.

8.4.3 Analysis of the 2^{k-p} Fractional Factorials

The quickest method of analyzing fractional replications of the 2^k factorials is through an adaptation of the Yates algorithm.

1. Since the design contains 2^{k-p} combinations it consists of a complete replication of $k - p$ factors. Arrange $k - p$ factors in standard order,

TABLE 8.18 ANOVA of 1/2
Replication of a 2^6 Factorial

Source	d.f.
Total	31
Main effects	6
Two-factor interaction	15
Error	10

in the usual Yates format, ignoring p of the factors. Symbols for the factors which are ignored are placed in brackets along with proper combinations of the $k - p$ factors to identify treatment combinations used in the design.

2. Apply the Yates method as if the experiment were a complete replication of a 2^{k-p} factorial set.
3. Reintroduce the ignored factors in the "effect column." When this is done each effect will have one or more aliases involving the ignored factors.
4. Select the main effects and low-order interactions from the alias set to be identified with the contrast.

For example, suppose we use *ABCD* as a defining contrast to obtain a 1/2 replication of 2^4 factorial and choose, as treatments,

(1)   *ab*   *ac*   *bc*   *ad*   *bd*   *cd*   *abcd*

Then, the Yates setup for the analysis of this 1/2 replication would be given in Table 8.19.

8.4.4 Numerical Example: Fractional Factorial

An investigation was conducted under manufacturing conditions to examine the effect of five factors on the quality of a basic dyestuff. The process involved three phases, and the factors were:

Phase I—Oxidation
    A—Temperature: low $(-)$, high $(+)$
    B—Quality of material: low $(-)$, high $(+)$
Phase II—Reduction
    C—Pressure: atmospheric $(-)$, increased $(+)$

TABLE 8.19 Format for a Yates Method Analysis of 1/2 Replication of a 2^4 Factorial

Treatment	Yield	I	II	III	Effect	
(1)					Total	
a [d]					$\underline{A}$	= BCD
b [d]					$\underline{B}$	= ACD
ab					$\overline{AB}$	= CD
c [d]					$\underline{C}$	= ABD
ac					$\overline{AC}$	= BD
bc					BC	= AD
abc [d]					ABC	= $\underline{D}$

TABLE 8.20 Treatment Combinations and Quality Readings for Dyestuff Experiment

Treatment	(1)	ab	ac	bc	ad	bd	cd	abcd
Yield	202	176	178	174	240	208	274	274
Treatment	ae	be	ce	abce	de	abde	acde	bcde
Yield	178	184	188	196	256	244	258	256

Phase III—Oven drying under pressure
  D—Pressure: low (−), high (+)
  E—Vacuum leak: low (−), high (+)

Technical considerations suggested that A and B might interact and so might C, D, and E. However, A and B would not interact with C, D, and E so that these interaction effects could be used for error.

The experiment was run as 1/2 replication of a 2^5 factorial with ABCDE as the defining contrast. It was assumed that interactions of second order and higher as well as first-order interactions of A and B with C, D, and E would be negligible. Quality was measured with a photoelectric spectrometer; the lower the reading the higher the quality.

Treatment combinations and their quality readings were as shown in Table 8.20.

We begin the analysis, using the Yates method, by arranging the treatments in standard order with respect to *A, B, C,* and *D*. We then supply *E* where appropriate and proceed as in Table 8.21.

We see that both the *AB* and the *CD* interactions are significant. We can use the Yates method to compute the two-way tables of means for these factors as shown in Tables 8.22 and 8.23.

TABLE 8.21 Yates Analysis of 1/2 Replication of 2^5 Experiment on Dyestuff Quality

Treatment	Yield	I	II	III	IV	Contrast	Alias	SS
(1)	202	380	740	1476	3486	Total	*ABCDE*	—
a(e)	178	360	736	2010	2	*A*	*BCDE*	.25
b(e)	184	366	948	− 20	− 62	*B*	*ACDE*	240.25
ab	176	370	1062	22	134	*AB*	*CDE*	1,122.25*
c(e)	188	496	− 32	− 16	110	*C*	*ABDE*	756.25*
ac	178	452	12	− 46	26	*AC* +	*BDE*	42.25
bc	174	532	20	48	66	*BC* +	*ADE*	272.25
abc(e)	196	530	2	86	− 2	*DE*	*ABC*	.25
d(e)	256	− 24	− 20	− 4	534	*D*	*ABCE*	17,822.25**
ad	240	− 8	4	114	42	*AD* +	*BCE*	110.25
bd	208	− 10	− 44	44	− 30	*BD* +	*ACE*	56.25
abd(e)	244	22	− 2	− 18	38	*CE*	*ABD*	90.25
cd	274	− 16	16	24	118	*CD*	*ABE*	870.25*
acd(e)	258	36	32	42	− 62	*BE* +	*ACD*	240.25
bcd(e)	256	− 16	52	16	18	*AE* +	*BCD*	20.25
abcd	274	18	34	− 18	− 34	*E*	*ABCE*	72.25

+ Contrasts used to estimate error.
*Significant at the 5% level.
**Significant at the 1% level.

Note: SS = (IV)^2/16.
     MSE = (42.25 + 272.25 + 110.25 + 56.25 + 240.25 + 20.25)/6 = 123.58.

TABLE 8.22 Computation of $A \times B$ Means

Effect	Total	I	II	÷ 16	Mean
AB	134	72	3560	222.50	*ab*
B	− 62	3488	3288	205.50	*b*
A	2	− 196	3416	213.50	*a*
(1)	3486	3484	3680	230.00	(1)

Report of statistical analysis. The results of an experiment to examine the factors affecting the quality of a basic dyestuff may be summarized in two tables of means, Table 8.24. and Table 8.25.

Consideration of Table 8.24 leads to the conclusion that high oxidation temperatures should be used with low-quality material, while low temperatures should be used on high-quality material. From Table 8.25 it is seen that high drying pressure results in reduced quality particularly if reduction pressure is increased. At low drying pressure the difference in reduction pressure has little effect on quality. Further consideration of the data indicates that differences in the rate of vacuum leak have little effect on the quality of the dyestuff.

TABLE 8.23 Computation of $C \times D$ Means

Effect	Total	I	II	÷ 16	Mean
CD	118	652	4248	265.50	cd
D	534	3596	3792	237.00	d
C	110	416	2944	184.00	c
(1)	3486	3376	2960	185.00	(1)

TABLE 8.24 Mean Quality at Different Oxidation Temperatures and Material Qualities

Material quality	Oxidation temperature		Standard error
	Low	High	
Low	230.00	213.50	5.56
High	205.50	222.50	

Note: Standard error = $\sqrt{MSE/4} = \sqrt{123.58/4} = 5.56$.

TABLE 8.25 Mean Quality at Different Reduction and Drying Pressures

Drying pressure	Reduction pressure		Standard error
	Atmospheric	Increased	
Low	185.00	184.00	5.56
High	237.00	265.50	

Note: Standard error = $\sqrt{MSE/4} = \sqrt{123.58/4} = 5.56$.

8.5 FURTHER FRACTIONATION

The principles developed thus far can be used to further subdivide the treatment combinations in a 2^k series design. To obtain a 1/2 replication we choose a defining contrast and then use those combinations which carry either a + or a − sign under the defining contrast. To select half of the combinations obtained in this way we choose a second defining contrast and use the combinations with a + or a − under this contrast. This procedure divides the original 2^k combinations into four groups, each with 2^{k-2} combinations. If X represents the first defining contrast and Y the second, the signs for the four groups are

X	Y		X	Y
+	+		−	+
+	−		−	−

As an example, suppose we use AC as a defining contrast to obtain a 1/2 replication of a 2^5 factorial. The combinations with a + sign under this contrast are:

(1)	b	ac	abc	d	bd	acd	abcd
e	be	ace	abce	de	bde	acde	abcde

Then, suppose we choose BD as a second defining contrast. The combinations in the first 1/2 replication which carry a + sign for the second defining contrast are:

(1)	ac	bd	abcd	e	ace	bde	abcde

These combinations would constitute a 1/4 replication of a 2^5 factorial set.

When constructing a 1/4 replication the following general rules apply:

1. In the 2^k system any two factorial effects may be used as defining contrasts. Their generalized interaction also acts as a defining contrast and cannot be estimated.
2. Any factorial effect which is not a defining contrast has three aliases: its generalized interaction with the three defining contrasts.

In the preceding example we chose AC and BD as defining contrast. Under rule 1 their generalized interaction, $ABCD$, is also a defining contrast. Under rule 2 we would have the following alias pattern:

I	$= AC$	$= BD$	$= ABCD$
A	$= C$	$= ABD$	$= BCD$
B	$= ABC$	$= D$	$= ACD$
AB	$= BC$	$= AD$	$= CD$
E	$= ACE$	$= BDE$	$= ABCDE$
AE	$= CE$	$= ABDE$	$= BCDE$

$$BE = ABCE = DE = ACDE$$
$$ABE = BCE = ADE = CDE$$

8.5.1 Restrictions

If we adhere to the restriction that all main effects and first-order interactions are to be estimated, and that all second- and higher-order interactions are negligible, then eight is the minimum number of factors for which a 1/4 replication design is available. With seven factors it is not possible to find two defining contrasts for which some two-factor interactions do not have other two-factor interactions as aliases.

With eight factors the problem is to find a set of defining contrasts which meet the restrictions. For this purpose each of the three defining contrasts must have at least five letters. With any defining contrast that has only four letters six two-factor interactions will form alias pairs. If any defining contrast has only three letters, three main effects will have two-factor interactions as aliases.

One permissible set for a 2^{8-2} fractional replication would be *ABDEF* and *ACGEH* along with their generalized interaction *BCDFGH*. The combinations, 64 out of the original 256, which form a 1/4 replication could be one out of four possible sets, those which have the following combinations of signs for two of the defining contrasts:

Set	ABDEF	ACGEH
1	+	+
2	+	−
3	−	+
4	−	−

8.6 SEQUENCES OF FRACTIONAL REPLICATIONS

When the results of an experiment can be obtained quickly, and when the effect of a large number of factors must be studied, it may be advantageous to begin with a fractional replication. The expectation is to make a decision or to add another fraction and continue after seeing the results from the first fraction. For example, we might run a 1/4 replication, examine the data, run another 1/4 replication, examine the data, etc.

The basic problem is: How do we choose the combinations to be included in later fractions in such a way that they will not duplicate what has already been done? The fundamental rule for this purpose is:

> To obtain another fractional replication multiply all of the combinations in the first fraction by any combination of letters which does not occur in the first fraction, deleting letters which occur twice in the product.

For example, suppose we have the following 1/4 replication of a 2^5 factorial:

(1)   *ab*   *cd*   *abcd*   *ace*   *bce*   *ade*   *bde*

To obtain a second 1/4 replication we could use, for example, *a* or *b* or *ac* or *abce* as a multiplier. It is easy to verify, however, that using *ab* as a multiplier and deleting letters according to the rule will reproduce the original 1/4 replication. Suppose we use *ac* as a multiplier. A second 1/4 replication would then be

ac   *bc*   *ad*   *bd*   *e*   *abe*   *cde*   *abcde*

To obtain a third 1/4 replication multiply the first set by any combination of letters which does not occur in either of the first two sets. For the preceding two 1/4 replications *a* or *b* or *abc* would be suitable multipliers, while *ab* or *e* would not. Using *a* as a multiplier we obtain

a   *b*   *acd*   *bcd*   *ce*   *abce*   *de*   *abde*

as a third 1/4 replication.

A more critical problem is to choose a second 1/4 replication which, together with the first, will give us a 1/2 replication having a desired defining contrast. Three contrasts are used to define the first 1/4 replication. One of these will be the defining contrast for the 1/2 replication obtained by adding a second 1/4 replication. The defining contrast should be one whose alias pattern will permit us to sort out the effects of interest.

In the preceding example the defining contrasts for the first 1/4 replication of the 2^5 factorial set were *ABE, CDE*, and *ABCD*. Hence, main effects have two-factor interactions as aliases ($A = BE, C = DE$, etc.). If we want to generate a 1/2 replication to sort out main effects we cannot have *ABE* or *CDE* as a defining contrast because three main effects will still have first-order interactions as aliases. The required defining contrast is *ABCD*.

To choose the combinations for the second 1/4 replication so that the 1/2 replication will have the desired defining contrast:

Multiply the combinations in the first 1/4 replication by any combination of letters which:

1.   Does not appear in the first fraction, and
2.   Contains an even number (including zero) of the letters in the desired defining contrast.

In our example, to generate a 1/2 replication with *ABCD* as the defining contrast we see that *e* or *ac* or *bd* would be suitable multipliers while *a* or *abc* would not. Using *e* as a multiplier we have *e   abe   cde   abcde   ac   bc   ad   bd*

as a second 1/4 replication which, together with the first 1/4 replication is a 1/2 replication with *ABCD* as a defining contrast.

It is clear that we could further fractionate a 2^k factorial using the procedures we have developed to this point. However, we will not go farther here. Catalogs of such designs are available (Cochran and Cox, 1957; Connor et al., 1956).

8.7 FORMAL CONSIDERATION OF EFFECTS

8.7.1 2^k Series

In order to lay a background for the construction of fractional replications with factors at more than two levels it is convenient to take a more formal look at the 2^k series. Consider the case with two factors *A* and *B* each at two levels, 0 and 1. The treatments which constitute a 2^2 factorial set may be symbolized as

$$00 = (1) \quad 01 = b$$
$$10 = a \quad 11 = ab$$

These may be represented by points in a plane which describes the two-dimensional factor space for the experiment. This representation is illustrated in Fig. 8.2. In

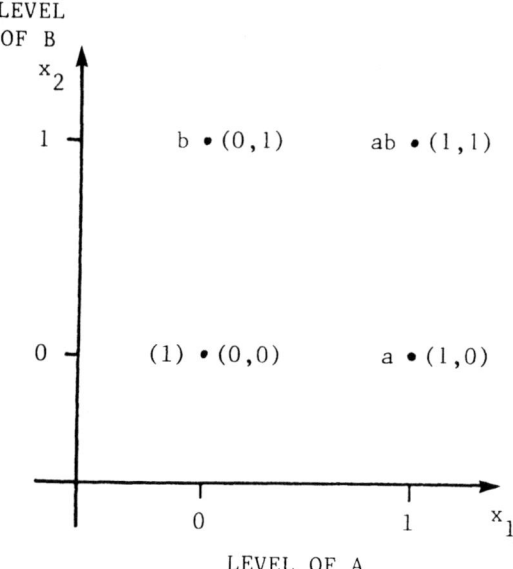

Fig. 8.2 Graphic representation of factor combinations in a 2 × 2 factorial set.

Fig. 8.2 the x_1 coordinate represents the level of A and the x_2 coordinate represents the level of B.

Now, the contrast which gives the main effect of A is

$$- (1) + a - b + ab = a + ab - (1) - b$$
$$= [(1,0) + (1,1)] - [(0,0) + (0,1)]$$

That is, it is the comparison of treatment combinations for which x_1 is zero with those for which x_1 is one. Similarly, the main effect of B compares those points for which $x_2 = 0$ with those for which $x_2 = 1$.

The AB interaction is the following comparison among treatments:

$$(1) - a - b + ab = [(0,0) + (1,1)] - [(1,0) + (0,1)]$$

For point $(0,0)$ the equation, in the coordinates of the factor space, $x_1 + x_2$ takes the value zero.

For point $(1,1)$, $x_1 + x_2 = 2$

For point $(0,1)$, $x_1 + x_2 = 1$

and

For point $(1,0)$, $x_1 + x_2 = 1$

Suppose we consider what happens if we work with numbers reduced modulo 2 (*mod 2*). That is, we replace any number by the remainder when it is divided by 2. To illustrate, we have

Number	Number reduced mod 2
1	1
2	0
3	1
4	0
5	1
⋮	⋮

Under this reduction the main effect of A compares those points for which the equation $x_1 = 0$ (mod 2) with those for which $x_1 = 1$ (mod 2). The main effect of B compares the points at which $x_2 = 0$ (mod 2) with those at which $x_2 = 1$ (mod 2). Finally, the AB interaction compares points for which

$$x_1 + x_2 = 0 \ (\text{mod } 2)$$

with those for which

$$x_1 + x_2 = 1 \pmod 2$$

8.7.2 Extension to More Than Two Factors

This concept can be extended to describe factorial contrasts for any number of factors each at two levels. Consider the case with three factors, A, B, C. The eight combinations may be represented by points in a three-dimensional factor space with coordinate axes x_1, x_2, and x_3.

$$
\begin{aligned}
(1) &= (0,0,0) & c &= (0,0,1) \\
a &= (1,0,0) & ac &= (1,0,1) \\
b &= (0,1,0) & bc &= (0,1,1) \\
ab &= (1,1,0) & abc &= (1,1,1)
\end{aligned}
$$

The main effects and interactions are the comparisons between groups of points given by the following equations:

Effect	Equation
A	x_1
B	x_2
AB	$x_1 + x_2$
C	x_3
AC	$x_1 + x_3$
BC	$x_2 + x_3$
ABC	$x_1 + x_2 + x_3$

The effect (contrast) is the comparison of those points for which the value of the equation is 0 (mod 2) with those for which the value of the equation is 1 (mod 2). For example, the ABC interaction is given by

$$ABC = -(1) + a + b - ab + c - ac - bc + abc$$
$$= [a + b + c + abc] - [(1) + ab + ac + bc]$$

Using the coordinates of the points in 3-space, we have

Combination	Point	$x_1 + x_2 + x_3 =$
a	$(1,0,0)$	$1 = 1 \pmod 2$
b	$(0,1,0)$	$1 = 1 \pmod 2$
c	$(0,0,1)$	$1 = 1 \pmod 2$
abc	$(1,1,1)$	$3 = 1 \pmod 2$
(1)	$(0,0,0)$	$0 = 0 \pmod 2$
ab	$(1,1,0)$	$2 = 0 \pmod 2$
ac	$(1,0,1)$	$2 = 0 \pmod 2$
bc	$(0,1,1)$	$2 = 0 \pmod 2$

The extension of this concept to two-level factorials with more than three factors is straightforward.

8.8 FACTORS AT THREE LEVELS

The concepts developed for the 2^k series can be extended to cover factorials in which the factors are at three levels. Consider the case with two factors, A and B, each at three levels. For the full factorial set there would be $3^2 = 9$ treatment combinations:

$$a_1b_1 \qquad a_1b_2 \qquad a_1b_3$$

$$a_2b_1 \qquad a_2b_2 \qquad a_2b_3$$

$$a_3b_1 \qquad a_3b_2 \qquad a_3b_3$$

The treatment sum of squares, with 8 d.f., may be partitioned as follows:

Source	d.f.
A main effect	2
B main effect	2
AB interaction	4

The approach used with the 2^k system suggests an approach which might be used here. The nine treatment combinations can be represented as points in a two-dimensional factor space as shown in Fig. 8.3.

The main effect of A represents a comparison of three sets of points: (1) points at which $x_1 = 0$, (2) points at which $x_1 = 1$, and (3) points at which $x_1 = 2$. Similarly, the main effect of B compares three sets of points: those at which $x_2 = 0$, 1, or 2. These main effects each have two degrees of freedom.

The AB interaction has four degrees of freedom. We can consider this interaction as being composed of two components, each with two degrees of freedom. We can construct this decomposition by superimposing a 3×3 Graeco-Latin square on the factor space as presented in Fig. 8.4.

In this case, comparisons among the columns give the main effect of A and comparisons among the rows give the main effect of B. Comparisons among the Latin letters and comparison among the Greek letters, each with two degrees of freedom, represent the AB interaction. [Note, the Latin letters correspond to the J components and the Greek letters correspond to the I components of Yates (1937).]

8.8.1 Modulo 3 Representation

Suppose we now work with numbers reduced modulo 3 (mod 3), replacing any number by the remainder when it is divided by 3. We have, analogous to the case with (mod 2):

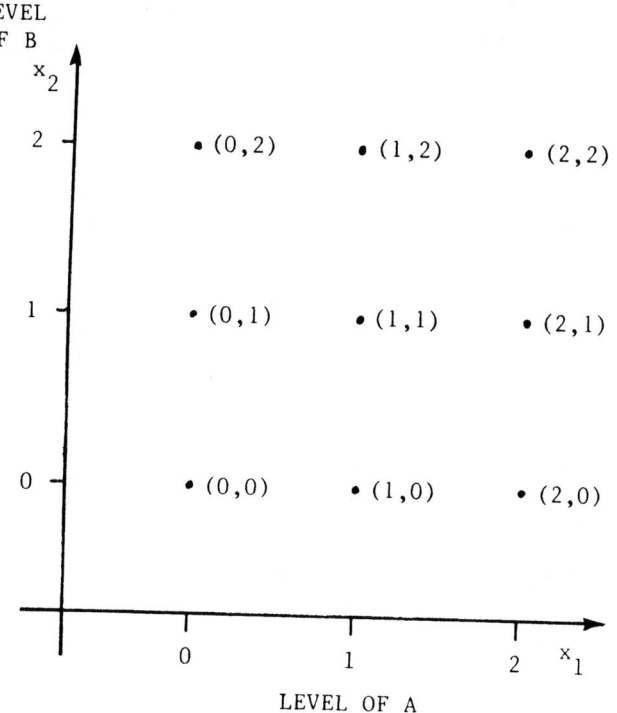

Fig. 8.3 Graphic representation of factor combinations in a 3 × 3 factorial set.

Number	Number reduced mod 3
0	0
1	1
2	2
3	0
4	1
5	2
6	0
⋮	⋮

The main effect of A is a comparison of the groupings of points

(0,0), (0,1), (0,2) for which $x_1 = 0$ (mod 3)

vs. (1,0), (1,1), (1,2) for which $x_1 = 1$ (mod 3)

vs. (2,0), (2,1), (2,2) for which $x_1 = 2$ (mod 3)

Similarly, the main effect of B is a comparison of groupings of points for which $x_2 = 0, 1, 2$ (mod 3)

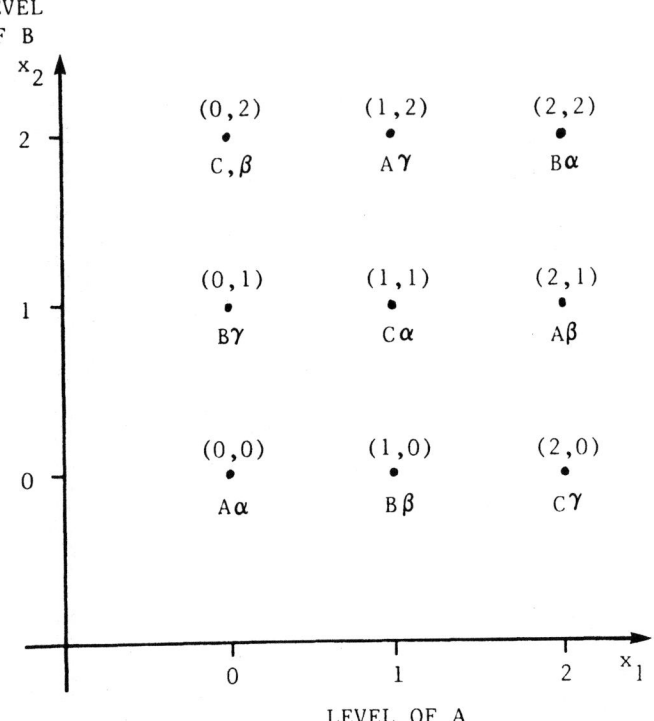

Fig. 8.4 Graeco-Latin square superimposed on a 3×3 factorial set.

Now consider the groupings given by the Latin letters:

A: (0,0), (1,2), (2,1)

B: (0,1), (1,0), (2,2)

C: (0,2), (1,1), (2,0)

For those points in the A groups,

$$x_1 + x_2 = 0, 3, 3$$
$$= 0 \pmod{3} \qquad \text{for each point}$$

For the points in the B group $x_1 + x_2 = 1, 1, 4 = 1 \pmod 3$, and for the C group $x_1 + x_2 = 2, 2, 2 = 2 \pmod 3$. Thus, two of the degrees of freedom for the AB interaction represent the comparison among the three groups of points for which the equation $x_1 + x_2 = 0, 1, 2 \pmod 3$.

The groupings given by the Greek letters are

α: (0,0), (1,1), (2,2)

β: (0,2), (1,0), (2,1)

γ: (0,1), (1,2), (2,0)

For points in the α group the equation $x_1 + 2x_2 = 0, 3, 6 = 0$ (mod 3). For points in the β group $x_1 + 2x_2 = 4, 1, 4 = 1$ (mod 3), and for points in the γ group $x + 2x_2 = 2, 5, 2 = 2$ (mod 3).

In summary, the main effects and interactions for a 3^2 design are formed by comparisons among three groups of treatments defined by the following equations:

$$\left. \begin{array}{ll} A: & x_1 \qquad\;\; = 0, 1, 2 \\[4pt] B: & x_2 \qquad\;\; = 0, 1, 2 \\[4pt] & x_1 + x_2 \;= 0, 1, 2 \\[4pt] & x_1 + 2x_2 = 0, 1, 2 \end{array} \right\} \text{(mod 3)}$$

Interaction

Each of these equations has associated with it two degrees of freedom. For convenience we designate the two d.f. corresponding to the equation $x_1 + x_2$ by the symbol AB, and the two d.f. associated with $x_1 + 2x_2$ by the symbol AB^2.

8.8.2 Analysis

The partition of the treatment sum of squares into main effects and interactions may be accomplished by using the totals of the groupings defined by the preceding equations. For each 2-d.f. effect there will be three totals: one for which the equation is 0 (mod 3), one for which it is 1 (mod 3), and one for which it is 2 (mod 3).

This procedure is illustrated by the accompanying numerical example.

8.8.3 Numerical Example: Partition of Treatment SS in a 3 × 3 Design

The procedure for partitioning the treatment sum of squares in a 3 × 3 design may be illustrated using data from an experiment designed to measure the effect of three levels of N and three levels of P on the emergence of lettuce plants. The treatments may be symbolized as shown in Fig. 8.5. The data, given in Table 8.26. are the total number of plants which emerged, and are totals over 12 plots.

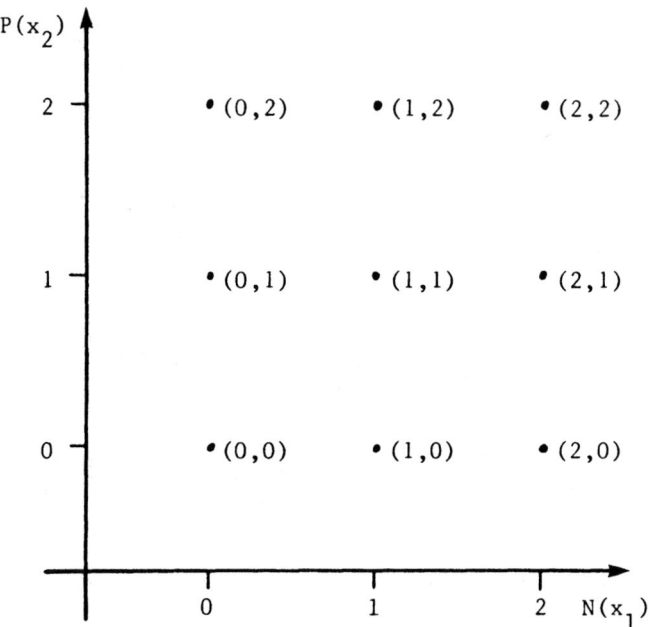

Fig. 8.5 Graphic representation of treatment combinations for an $N \times P$ experiment on lettuce emergence.

TABLE 8.26 Number of Plants Emerging (Total of 12 Replications) in Lettuce Experiment

	N (x_1)			
P (x_2)	0	1	2	Sum
0	449	413	326	1188
1	409	358	291	1058
2	341	278	312	931
Sum	1199	1049	929	3177

The treatment sum of squares, SST, is:

$$SST = (1/12)\Sigma T_{ij}^2 - (1/108)(\Sigma T_{ij})^2$$
$$= (1/12)(449^2 + 409^2 + \cdots + 312^2) - (1/108)(3177)^2$$
$$= 2333.34 \quad \text{with} \quad 8 \text{ d.f.}$$

SST may be partitioned into four components, each with two degrees of freedom: main effect of N, main effect of P, and two $N \times P$ interaction components. Each component is computed from totals of three sets of treatment combinations grouped according to whether the following equations take the value 0, 1, or 2 modulo 3:

$$\left. \begin{array}{llll} N: & x_1 = 0, 1, 2 \\ P: & x_2 = 0, 1, 2 \\ & x_1 + x_2 = 0, 1, 2 \\ N \times P: & x_1 + 2x_2 = 0, 1, 2 \end{array} \right\} (\text{mod } 3)$$

where x_1 and x_2 represent the coordinates of N and P in the design space. For the N main effect we have: $x_1 = 0, 1, 2$ (mod 3)

$$x_1(0) = (0,0) + (0,1) + (0,2) = 449 + 409 + 341 = 1199$$
$$x_1(1) = (1,0) + (1,1) + (1,2) = 413 + 358 + 278 = 1049$$
$$x_1(2) = (2,0) + (2,1) + (2,2) = 326 + 291 + 312 = 929$$

Then

$$\text{SSN} = (1/36)(1199^2 + 1049^2 + 929^2) - (3177)^2/108 = 1016.67$$

For the P main effect we have: $x_2 = 0, 1, 2$ (mod 3)

$$x_2(0) = (0,0) + (1,0) + (2,0) = 449 + 413 + 326 = 1188$$
$$x_2(1) = (0,1) + (1,1) + (2,1) = 409 + 358 + 291 = 1058$$
$$x_2(2) = (0,2) + (1,2) + (2,2) = 341 + 278 + 312 = 931$$

Then

$$\text{SSP} = (1/36)(1188^2 + 1058^2 + 931^2) - (3177)^2/108 = 917.39$$

For the component of $N \times P$ defined by $x_1 + x_2 = 0, 1, 2$ (mod 3) we have:

$$(x_1 + x_2)(0) = (0,0) + (1,2) + (2,1) = 449 + 278 + 291 = 1018$$
$$(x_1 + x_2)(1) = (0,1) + (1,0) + (2,2) = 409 + 413 + 312 = 1134$$
$$(x_1 + x_2)(2) = (0,2) + (1,1) + (2,0) = 341 + 358 + 326 = 1025$$

And

$$\text{SS} = 1/36(1018^2 + 1134^2 + 1025^2) - (3711)^2/108 = 235.06$$

Finally, for the $N \times P$ component defined by $x_1 + 2x_2 = 0, 1, 2 \pmod 3$

$$(x_1 + 2x_2)(0) = (0,0) + (1,1) + (2,2) = 449 + 358 + 312 = 1119$$

$$(x_1 + 2x_2)(1) = (0,2) + (1,0) + (2,1) = 341 + 413 + 291 = 1045$$

$$(x_1 + 2x_2)(2) = (0,1) + (1,2) + (2,0) = 409 + 278 + 326 = 1013$$

Then

$$SS = (1/36)(1119^2 + 1045^2 + 1013^2) - (3177)^2/108 = 164.22$$

Each of these components has 2 d.f. Note that

$$1016.67 + 917.39 + 235.06 + 164.22 = 2333.34 = SST$$

8.9 MORE THAN TWO FACTORS AT THREE LEVELS

We can extend the preceding exposition to situations involving more than two factors at three levels. Before doing this, however, we need to look at some rules about the symbols describing an effect, and about the formation of interactions.

1. In the symbol describing an effect the letters are placed in alphabetic order and the exponent on the first letter must be unity. If the exponent is not unity we take the square of the symbols and reduce the exponents modulo 3. For example, suppose that in the formation of an interaction we obtain the symbol A^2BC^2; we replace this with

$$(A^2BC^2)^2 = A^4B^2C^4 \stackrel{\text{mod } 3}{=} AB^2C$$

In the 3^k system the $3^k - 1$ degrees of freedom for treatment may be separated into $(3^k - 1)/2$ sets of pairs of degrees of freedom. To each pair is attached a symbol formed according to the preceding rule. Further, the interaction of two effects consists of two pairs of d.f. The interaction of A and B, for instance, is AB and AB^2. In general,

2. If X and Y represent two pairs of d.f. then their generalized interaction consists of two pairs of d.f. corresponding to XY and XY^2, with the describing symbol formed according to rule 1.

For example, the generalized interaction of AB, (X), and AC^2 (Y), is

$$XY = (AB) \times (AC^2) = A^2BC^2 \stackrel{\text{sq}}{=} A^4B^2C^4 \stackrel{\text{mod } 3}{=} AB^2C$$

and also

$$XY^2 = (AB) \times (AC^2)^2 = (AB) \times (A^2C^4) = A^3BC^4 \overset{\text{mod } 3}{=} BC$$

With this background we can begin to look at more than two factors at three levels. With three factors, A, B, and C, the 26 d.f. for treatment can be partitioned into 13 pairs of d.f. If (x_1, x_2, x_3) designated the coordinates of a point in the three-dimensional factor space of A, B, and C, each pair of d.f. represents a comparison among three groups of points. The equations for obtaining the groups are

Effect	Equation
A	x_1
B	x_2
AB	$x_1 + x_2$
AB^2	$x_1 + 2x_2$
C	x_3
AC	$x_1 + x_3$
AC^2	$x_1 + 2x_3$
BC	$x_2 + x_3$
BC^2	$x_2 + 2x_3$
ABC	$x_1 + x_2 + x_3$
ABC^2	$x_1 + x_2 + 2x_3$
AB^2C	$x_1 + 2x_2 + x_3$
AB^2C^2	$x_1 + 2x_2 + 2x_3$

The three groups to be compared for any effect are the three sets of points for which the equation for the effect takes the value 0, 1, 2 (mod 3).

The extension of the procedure to cover more than three factors is straightforward. The 16 d.f. for the four-factor interaction in a 3^4 factorial set break down into the following eight effects, each with two degrees of freedom:

Effect	Equation
$ABCD$	$x_1 + x_2 + x_3 + x_4$
$ABCD^2$	$x_1 + x_2 + x_3 + 2x_4$
ABC^2D	$x_1 + x_2 + 2x_3 + x_4$
ABC^2D^2	$x_1 + x_2 + 2x_3 + 2x_4$
AB^2CD	$x_1 + 2x_2 + x_3 + x_4$
AB^2CD^2	$x_1 + 2x_2 + x_3 + 2x_4$

Effect	Equation
AB^2C^2D	$x_1 + 2x_2 + 2x_3 + x_4$
$AB^2C^2D^2$	$x_1 + 2x_2 + 2x_3 + 2x_4$

The extension to cover additional factors at three levels is obvious, and we will pursue it no farther here.

8.10 SINGLE AND FRACTIONAL REPLICATION: FACTORS AT THREE LEVELS

As the number of factors increases the number of treatment combinations increases much more rapidly when the factors are at three levels than when they are at two levels. A 2^5 factorial, for example, has only 32 combinations, while a 3^5 has 243. We would like to find ways of reducing the size of three-level experiments on several factors in much the same way as we did with two-level experiments. For this reason we are tempted to explore the concept of single and fractional replication of three-level factorial experiments.

We can use the same criteria here as we used before:

1. We want to estimate and test main effects and two-factor interactions.
2. We assume that high-order interactions are negligible and use them to estimate error.

Under these criteria the minimum number of factors to be considered is three. With three factors, each at three levels, a single replication would contain 27 treatments. Of the 26 d.f. there would only be 8 d.f. for the three-factor interaction used to estimate error. Normally we would like to have a minimum of 10 degrees of freedom for error, so we would hesitate to use a single replication of a 3^3 factorial.

On the other hand, with four factors at three levels, for a total of 81 treatments in a single replication, there would be 48 d.f. for error if we pooled both three-factor and four-factor interactions. This is certainly ample. Further, we could use the four-factor interaction, with 16 d.f., as an estimate of error if we wanted to look at the three-factor interactions.

Of course, experiments with more than four factors have sufficient degrees of freedom in the high-order interactions to permit them to be run in a single replication. The basic problem here, however, is the large size of the experiment even with only a single replication: 243 treatments with five factors, 729 with six. With many factors we need to consider ways of reducing the size of the experiment to less than what is required for a complete replication.

8.10.1 Fractional Replication

The principles of fractional replication for the 3^k series are much the same as they are for the 2^k series. We choose a defining effect which, in this case, partitions the treatment combinations into three groups. These are the groups for which the equation for the defining effect takes the value 0 (mod 3), 1 (mod 3), or 2 (mod 3). Each of these groups is a 1/3 replication of the full factorial set. For example, suppose we take ABC as a defining effect for a 3^3 factorial. The equation for the effect is $x_1 + x_2 + x_3$, and the partition is:

$x_1 + x_2 + x_3 = 0$ (mod 3)

(0,0,0)	(1,0,2)	(2,0,1)
(0,1,2)	(1,1,1)	(2,1,0)
(0,2,1)	(1,2,0)	(2,2,2)

$x_1 + x_2 + x_3 = 1$ (mod 3)

(0,0,1)	(1,0,0)	(2,0,2)
(0,1,0)	(1,1,2)	(2,1,1)
(0,2,2)	(1,2,1)	(2,2,0)

$x_1 + x_2 + x_3 = 2$ (mod 3)

(0,0,2)	(1,0,1)	(2,0,0)
(0,1,1)	(1,1,0)	(2,1,2)
(0,2,0)	(1,2,2)	(2,2,1)

Any one of these three groups of nine combinations is a 1/3 replication defined by the ABC effect.

In a 1/3 replication of a 3^k each effect which is not a defining effect has as an alias two other effects. These are the generalized interaction of the effect with the defining effect and the square of the defining effect ($X \times Y = XY$ and XY^2). To illustrate, if ABC is the defining effect for a 1/3 replication of a 3^3 factorial we would have the following alias pattern:

$$
\begin{aligned}
I &= ABC &&= A^2B^2C^2 \\
A &= AB^2C^2 &&= BC \\
B &= AB^2C &&= AC \\
AB &= ABC^2 &&= C \\
AB^2 &= AC^2 &&= BC^2
\end{aligned}
$$

8.10.2 Restrictions

Note that each main effect has a two-factor interaction as an alias. Hence, this design would not meet the criteria we have set up. Further, there is no other defining effect for which main effects and two-factor interactions are not aliases. Thus, no useful fractional replication of a 3^3 factorial can be found.

With four factors at three levels we might choose one of the eight four-factor interaction effects as a defining effect. It is tedious, but not too difficult, to show that any of these will have main effects with three- and four-factor interactions as aliases. However, half of the two-factor interactions will have other two-factor interactions as aliases. Such a design might be useful for estimating main effects provided that all of the interactions are negligible.

With five factors any of the 32 five-factor effects may be used to give a 1/3 replication with 81 treatments. Further, all main effects and two-factor interactions will have only high-order interactions as aliases. If a 1/3 replication with five factors meets the criteria, then a 1/3 replication with more than five factors must meet the limitations we have set down. However, as the number of factors increases even a 1/3 replication contains a large number of treatments. To overcome this difficulty we can select a second defining effect to further subdivide the set into 1/9 of a replication. However, we will not go into the details here.

FURTHER READING*

Anderson, V. L., and R. A. McLean (1974), Ch. 10 and 11.
Box, G. E. B., and J. S. Hunter (1961).
Connor, W. S., and M. Zelan (1959).
Daniel, C. (1976), Ch. 11, 12, and 14.
Finney, D. J. (1960), Ch. 4.
Kempthorne, O. (1973), Ch. 16 and 20.

REFERENCES

Cochran, W. G., and G. M. Cox (1957). *Experimental Designs,* 2nd ed., Wiley, New York, Ch. 6A.
Connor, W. S., M. Zelan, and L. Deming (1956). Fractional Factorial Experiment Designs for Factors at Two Levels, *Natl. Bur. Stand. Appl. Math Ser., No. 48.*
Yates, F. (1937). The Design and Analysis of Factorial Experiments, *Common. Bur. Soil Sci. Tech. Commun. 35.*

---

*See Bibliography for complete citation.

9

Confounding

9.1 INTRODUCTION

One of the principles of modern experimental design is to group the experimental material into blocks of homogeneous units. Treatments are then assigned to units within blocks. The aim is to remove block-to-block variation from experimental error and to permit the comparisons among treatments under as nearly uniform conditions as possible. To the extent that we are able to maximize the variation among blocks and minimize the variation within blocks, we can often effect a substantial increase in the precision of an experiment.

Experience has shown, however, that experimental error variance is related to block size, increasing as block size is increased. Frequently, particularly with factorial experiments, it is impossible to find sufficient uniform material to construct blocks large enough to accommodate all of the treatment combinations. Further, the negative correlation commonly encountered between block size and homogeneity of the plots in the blocks may reduce or eliminate the advantage of blocking.

The evidence, then, points to the necessity of keeping the blocks small to gain the increase in precision due to blocking. No general rule can be stated as to what constitutes a "small" block. This will depend on the nature of the experimental material. However, we need to find ways of gaining the advantage of small blocks without, at the same time, limiting the use of factorial sets of treatments.

One device for this purpose is called *confounding*. To illustrate the basic idea of confounding, suppose we want to run a 2^3 factorial experiment with factors A, B, and C. But, suppose that the maximum feasible block size is four experimental units. We might form two blocks of four units each and assign treatments to the blocks as follows:

TABLE 9.1 Contrasts for a 2^3 Factorial Assigned to Blocks of Four Units

Contrast	Block Treatment	I				II			
		(1)	ab	ac	bc	a	b	c	abc
A		−	+	+	−	+	−	−	+
B		−	+	−	+	−	+	−	+
AB		+	+	−	−	−	−	+	+
C		−	−	+	+	−	−	+	+
AC		+	−	+	−	−	+	−	+
BC		+	−	−	+	+	−	−	+
ABC		−	−	−	−	+	+	+	+
Block		−	−	−	−	+	+	+	+

Block I	(1)	ab	ac	bc
Block II	a	b	c	abc

Now, consider Table 9.1, a table of contrasts for this experiment.

9.2 PRINCIPLES OF CONFOUNDING

Note that, except for *ABC*, all of the factorial contrasts are orthogonal to the block contrast. Hence, these contrasts can be estimated and tested exactly as they could be in blocks of eight units. On the other hand, the *ABC* contrast is exactly the same as the contrast used to estimate the difference between blocks. *ABC* is said to be *confounded* with blocks because both contrasts are the same. We have no way of separating the true interaction effect from the block effect. By confounding *ABC* with blocks we have sacrificed information on the *ABC* interaction to gain the increased precision of smaller blocks for our estimation of the other effects.

Note that confounding is used to reduce block size, not to reduce the size of the experiment. We can have any number of replications and still keep the blocks small by confounding a contrast in which we have no interest. Basically, the idea of confounding is to make a block contrast equivalent to a factorial contrast. We give up information on the confounded contrast to gain the precision of small blocks. Unlike fractional replication, which it superficially resembles, we have no alias pattern in a confounded design. All of the treatment combinations are present and, except for the confounded effect, all of the main effects and interactions are estimable.

9.3 ANALYSIS

Suppose we have enough material to run three replications of a 2^3 factorial experiment, but suppose that the maximum feasible block size is four units. We might construct six blocks of four units each and completely confound the *ABC* interaction with blocks. The experimental plan might appear as in Fig. 9.1.

The outline of the analysis of variance would take the form shown in Table 9.2. Note that *ABC* is not included in the ANOVA because it is the same as one of the block contrasts.

I		II		III		REP
1	2	3	4	5	6	BLOCK
bc	abc	ab	c	bc	abc	
ac	a	(1)	a	ac	c	
(1)	b	bc	abc	ab	a	
ab	c	ac	b	(1)	b	

Fig. 9.1 Experimental plan for a 2^3 factorial confounded in blocks of four.

TABLE 9.2 Format for the ANOVA of 3 Replications of a 2^3 Factorial in Blocks of Four with *ABC* Confounded

Source	d.f.
Total	23
Block	5
A	1
B	1
AB	1
C	1
AC	1
BC	1
Error	12

9.3.1 Interblock Estimates

It is possible to get some information about a completely confounded contrast by making proper comparisons among the block totals. This information is not very good, and is seldom computed unless the number of blocks is quite large. To illustrate the procedure, however, consider the preceding example for a 2^3 factorial in six blocks of four with *ABC* completely confounded. Combinations with the minus sign for *ABC* were placed in blocks 2, 4, and 6. One partition of the five degrees of freedom for blocks is shown in Table 9.3.

Table 9.4 gives one set of contrasts which can be used to obtain the sums of squares for this partition.

To make the test of interaction, first compute the mean square for error MSE(*ABC*) as

$$\text{MSE}(ABC) = (S_4^2 + S_5^2)/2$$

Then

$$F = S_3^2/\text{MSE}(ABC)$$

is a test of significance for *ABC* with 1, 2 d.f.

TABLE 9.3 ANOVA for Block Totals

Source	d.f.
Replication	2
ABC	1
Interblock error	2

TABLE 9.4 Contrasts for Partitioning the Block Sum of Squares

	Block total						
Contrast	B_1	B_2	B_3	B_4	B_5	B_6	Sum of squares
Replication (1)	-1	-1	$+1$	$+1$	0	0	S_1^2 } *Rep*
Replication (2)	-1	-1	-1	-1	$+2$	$+2$	S_2^2
ABC	-1	$+1$	-1	$+1$	-1	$+1$	S_3^2 *ABC*
Error (1)	$+1$	-1	-1	$+1$	0	0	S_4^2 } Error
Error (2)	$+1$	-1	$+1$	-1	-2	$+2$	S_5^2

9.4 FURTHER CONFOUNDING

We have used one confounding contrast to divide the 2^k possible combinations into two parts. This permits us to use blocks of size 2^{k-1}. We can reduce block size again by choosing a second confounding contrast and again dividing the treatments. With two confounding contrasts, say X and Y, we have four combinations of signs:

X	Y
−	−
−	+
+	−
+	+

We could then use four blocks of size 2^{k-2} for a full replication of the experiment. We would place combinations with $(-,-)$ in one block, $(-,+)$ in another block, $(+,-)$ in a third, and $(+,+)$ in a fourth block.

To illustrate, suppose we want to run a 2^3 factorial, but suppose the maximum usable block size is two (twins?). We could choose any confounding contrast, say AB, to divide the eight treatments into two sets of four, and another confounding contrast, say BC, to further divide them into four groups of two. We have:

Contrast	(1)	a	b	ab	c	ac	bc	abc
				Treatment				
AB	+	−	−	+	+	−	−	+
BC	+	+	−	−	−	−	+	+

The blocks in one replication would be

Block			
1	2	3	4
(1)	ab	a	b
abc	c	bc	ac

It is easily verified that the AB interaction is the comparison of blocks 1 and 2 with blocks 3 and 4, and that BC is the comparison of blocks 1 and 3 with blocks 2 and 4. These two comparisons among blocks are orthogonal to each other. A third comparison which is orthogonal to the first two is the comparison of blocks 1 and 4 with blocks 2 and 3. It is not difficult to verify that this third comparison is identical to the AC interaction contrast.

This illustrates the general rule that if any two effects are completely confounded their generalized interaction is also completely confounded. Again it
should be pointed out that all of the unconfounded contrasts are orthogonal to
the block contrasts and, hence, are estimated and tested as if no confounding
had been done. In a confounded experiment with full replication there is no alias
pattern to be concerned with.

9.5   PARTIAL CONFOUNDING

We have considered in some detail the concept of complete confounding in the
2^k series. We note that when confounding is complete we have no information
on the confounding contrast. If there are several replications in an experiment,
we need not sacrifice all of the information on one contrast to gain the advantage
of small blocks. As an alternative to complete confounding we can confound
one contrast in one replication, another contrast in another replication, etc. We
then compute these contrasts from the replications in which they are not confounded. This procedure is known as *partial confounding*.

9.5.1   Illustration

To illustrate the process, suppose we want to run four replications of a 2^3 factorial
in factors A, B, and C. Suppose, further, that we have ample experimental
material but we require no more than four units per block. (For example, perhaps
we can make a maximum of four runs per day.) Assume that we are interested
in looking at all of the main effects and all of the interactions. We could form
the material into eight blocks of four units each. We could then confound AB
in the first pair of blocks, AC in the second pair, BC in the third pair, and ABC
in the last pair. The assignment of treatments to blocks before randomization
would appear as shown in Fig. 9.2.

The analysis of variance would appear as in Table 9.5.

The degrees of freedom for the interactions are marked with a prime to
indicate that these effects are estimated from only three-fourths of the data. The
sums of squares for blocks and for the main effects are computed, in the usual
way, from all of the data. If contrasts are used to compute the sums of squares
for the unconfounded main effects the divisor of L^2 is 32.

The sums of squares for the confounded effects are computed from the data
in the replications in which the effects are not confounded. For AB we use the
data from replications II, III, and IV; for AC the data are in I, III, and IV; data
in I, II, and IV are used for BC; and for ABC we use I, II, and III. The divisor
of L^2 for these contrasts is 24. The error SS is obtained, as usual, by difference.

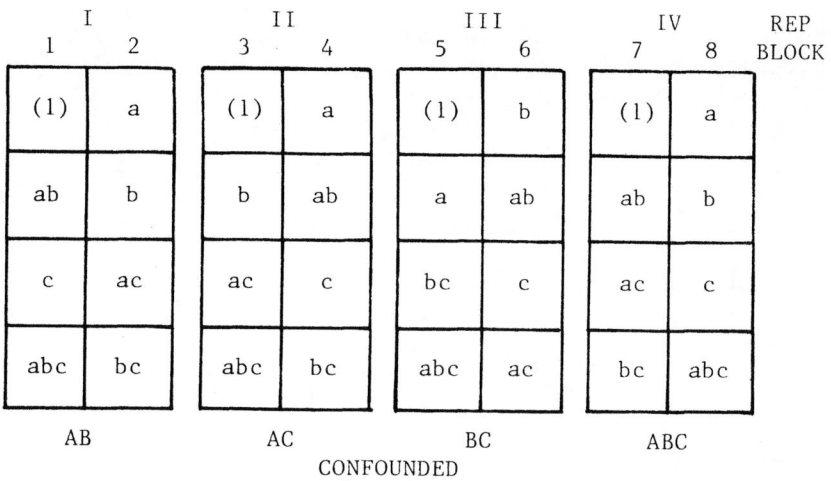

Fig. 9.2 Experimental plan for partially confounding a 2^3 factorial in blocks of four.

TABLE 9.5 ANOVA of 2^3 Factorial
Partially Confounded in Blocks of Four

Source	d.f.	Relative information
Total	31	
Block	7	
A	1	1
B	1	1
AB	1'	3/4
C	1	1
AC	1'	3/4
BC	1'	3/4
ABC	1'	3/4
Error	17	

Note: 1': These effects are estimated with only ¾ the precision of the main effect estimates.

9.5.2 Numerical Example: Partial Confounding

An experiment was conducted to determine the effects of three factors on the power required to operate a new type of lathe. The factors and their levels were:

A—Tool angle: $15°(-)$, $30°(+)$
B—Tool type: $1 (-1)$, $2 (+)$
C—Type of cut: continuous $(-)$, interrupted $(+)$

Because of the time required to set up a run only four runs could be made during a shift. It was decided to use a confounded factorial with a different interaction confounded in each pair of blocks of four runs in one shift. The experiment was replicated four times.

The design and power requirement, measured as a dynamometer deflection in millimeters, are presented in Fig. 9.3.

I

(1)	b
7	24
abc	ac
39	31
a	c
30	21
bc	ab
27	39

II

ab	bc
36	31
(1)	ac
19	36
abc	b
41	30
c	a
30	33

III

a	ac
28	31
c	b
24	19
ab	(1)
35	13
bc	abc
26	36

IV

abc	(1)
66	11
a	bc
31	29
c	ac
21	33
b	ab
25	43

Fig. 9.3 Experimental plan and power requirement for a partially confounded experiment on factors affecting performance of a new lathe.

We compute the sums of squares for the unconfounded main effects from Table 9.6 of contrasts using all the data.

The AB interaction is computed as a contrast of totals from the replications I, III, IV, in which it is not confounded:

	(1)	a	b	ab	c	ac	bc	abc		
Sum	31	89	68	117	66	95	82	141	L	S^2
AB	+	−	−	+	+	−	−	+	21	18.38

Similarly, AC is computed from replications I, II, and IV

	(1)	a	b	ab	c	ac	bc	abc		
Sum	37	94	79	118	72	100	87	146	L	S^2
AC	+	−	+	−	−	+	−	+	−9	3.38

BC is obtained from replications II, III, and IV

	(1)	a	b	ab	c	ac	bc	abc		
Sum	43	92	74	114	75	100	86	143	L	S^2
BC	+	+	−	−	−	−	+	+	1	.04

and ABC from replications I, II, and III

	(1)	a	b	ab	c	ac	bc	abc		
Sum	39	91	73	110	75	98	84	116	L	S^2
ABC	−	+	+	−	+	−	−	+	24	24.00

The block sum of squares is obtained in the usual way. If R_i is the total for the ith block then

$$\text{SS Block} = (1/4) \sum_{i=1}^{8} R_i^2 - (\sum_{i=1}^{8} R_i)^2/32$$

$$= 364.22, \quad \text{with 7 d.f.}$$

The error sum of squares, with 17 d.f., is obtained as the difference between the total sum of squares and the total of the sums of squares for blocks and treatments.

TABLE 9.6 Contrasts for Unconfounded Main Effects

Contrast	Treatment	(1)	a	b	ab	c	ac	bc	abc		
	Sum	50	122	98	153	96	131	113	182	L	S^2
A		−	+	−	+	−	+	−	+	231	1667.53
B		−	−	+	+	−	−	+	+	147	675.28
C		−	−	−	−	+	+	+	+	99	306.28

The analysis of variance of the data is presented in Table 9.7. Note that all of the main effects are significant. None of the interactions is significant.

Report of statistical analysis. The results of a study of factors affecting the use of energy by a new type of lathe may be summarized in three tables of means, Tables 9.8, 9.9, and 9.10.

It is clear that the factor which most affects energy use is tool angle. Increasing the angle from 15° to 30° increased the energy use by about 14.4 mm

TABLE 9.7 ANOVA of Power Requirement Data from a Partially Confounded 2^3 Experiment on Lathe Operation

Source	d.f.	SS	MS	F
Total	31	3605.97		
Block	7	364.22	52.03	1.62
A	1	1667.53	1667.53	51.84**
B	1	675.28	675.28	20.99**
AB	1'	18.38	18.38	.57
C	1	306.28	306.28	9.52**
AC	1'	3.38	3.38	.11
BC	1'	.04	.04	.00
ABC	1'	24.00	24.00	.75
Error	17	546.86	32.17	

**Significant at the 1% level.

Note: 1': These effects are estimated with only 3/4 the precision of the main effect estimates.

TABLE 9.8 Effect of Tool Angle on Mean Energy Use

Angle	15°	30°	Standard error
Mean	22.31	36.75	1.42

TABLE 9.9 Effect of Tool Type on Mean Energy Use

Type	1	2	Standard error
Mean	24.94	34.12	1.42

TABLE 9.10 Effect of Type of Cut on Mean Energy Use

Cut	Continuous	Interrupted	Standard error
Mean	26.44	32.62	1.42

on the dynamometer. The least effective factor appears to be type of cut. There was a difference of about 6.2 mm between an interrupted cut and a continuous cut. The effect of type of tool was intermediate to that of the other factors, type 1 requiring about 9.2 mm less energy than type 2.

Blocking on work shift apparently had little effect on increasing the efficiency of this trial. It is suggested that a less elaborate design might be used in future studies of this type.

9.6 CONFOUNDING IN THE 3^k SERIES

The principles of confounding used to reduce block size in the 2^k series can be applied to reduce block size in the 3^k series as well. We select a confounding effect in which we have no interest and with which we can partition the 3^k treatment combinations into three groups each with 3^{k-1} treatments. Two contrasts among these groups can be made equivalent to two comparisons among three blocks. To do this we assign those treatments for which the equation for the confounding effect takes the value 0 (mod 3) to one block, those for which the equation is 1 (mod 3) to a second block, and those with the value 2 (mod 3) to the third block.

9.6.1 Illustration

For example, suppose we want to run a 3^3 factorial in blocks of $3^{3-1} = 9$ experimental units. We choose a confounding contrast and assign treatments to the blocks based on the contrast. Suppose, for example, we select AB^2 as a confounding contrast. Then treatments would be assigned according to the value of $x_1 + 2x_2$ (mod 3). This assignment is shown in Table 9.11. If we replicated this design twice the ANOVA would be as shown in Table 9.12.

9.7 FURTHER SUBDIVISION

Just as was the case with the 2^k series we can choose a second defining effect to further subdivide the 3^k series. With one confounding effect we obtain three groups with 3^{k-1} units in each group. With two confounding effects we obtain

TABLE 9.11 Assignment of 3^3 Factorial to Blocks of 9 with AB^2 as the Confounding Contrast

	Block		
$x_1 + 2x_2$ = 0 mod 3	1 0 mod 3	2 1 mod 3	3 2 mod 3
	000	020	010
	001	021	011
	002	022	012
	110	100	120
	111	101	121
	112	102	122
	220	210	200
	221	211	201
	222	212	202

TABLE 9.12 Format for the ANOVA of a 3^3 Factorial Confounded in Blocks of 9

Source	d.f.
Total	53
Blocks	5
A	2
B	2
AB	2
C	2
AC	2
AC^2	2
BC	2
BC^2	2
ABC	2
ABC^2	2
AB^2C	2
AB^2C^2	2
Error	24

3^2 groups with 3^{k-2} units in each. As might be expected, when two effects, X and Y, are confounded their generalized interaction, XY and XY^2, is also confounded. Again, however, all of the unconfounded effects are estimated as if no confounding were present, and there is no pattern of aliases to be concerned with.

We can illustrate the subdivision of a 3^3 factorial into $3^2 = 9$ groups of $3^{3-2} = 3$ units each by choosing a second effect for the preceding example. We used AB^2 to obtain three groups of 9. Suppose we use AC^2 as a second defining effect. Then

$$XY = (AB^2)(AC^2) = A^2B^2C^2 = \underline{ABC}$$

and

$$XY^2 = (AB^2)(AC^2)^2 = A^3B^2C^4 = \underline{BC^2}$$

Thus ABC and BC^2 will also be confounded. Under AC^2 we partition the treatments into three groups for which $x_1 + 2x_3 = 0, 1, 2 \pmod 3$. These groups are shown in Table 9.13.

The final partition is obtained by considering the two equations jointly. We have the final partition as presented in Table 9.14.

If we ran two replications of this design the analysis of variance would be as given in Table 9.15.

9.7.1 Numerical Example: 3^2 Confounded in Blocks of 3

A nutritionist wanted to measure the effect of two feed additives, A and B each at three levels, on the weight gain in rats. He decided to use a 3×3 factorial set of treatments. He wanted to remove differences among litters from his treatment comparisons, but he could only obtain six litters with three rats per litter.

TABLE 9.13 Partition of 3^3 Factorial Treatments into Groups of 9 with AC^2 as a Confounding Contrast

$x_1 + 2x_3 = 0$	1	2
000	002	001
010	012	011
020	022	021
101	100	102
111	110	112
121	120	122
202	201	200
212	211	210
222	221	220

TABLE 9.14 Partition of 3^3 Factorial
Treatments into 9 Groups of 3 with AB^2
and AC^2 as Confounding Contrasts

		AB^2	
AC^2	0	1	2
0	000	020	010
	111	101	121
	222	212	202
1	002	022	012
	110	100	120
	221	211	201
2	001	021	011
	112	102	122
	220	210	200

TABLE 9.15 ANOVA of a 3^3 Factorial
Confounded in Blocks of Three Units

Source	d.f.
Total	53
Block	17
A	2
B	2
AB	2
C	2
AC	2
BC	2
ABC^2	2
AB^2C	2
AB^2C^2	2
Error	18

LITTER

1	2	3	4	5	6
11	21	20	22	02	12
18.8	24.9	22.3	27.1	20.5	24.0
00	10	01	11	21	20
18.7	20.8	19.2	23.5	24.2	22.9
22	02	12	00	10	01
25.6	20.8	22.0	17.5	20.5	21.3

Fig. 9.4 Experimental plan and 21-day gain (pounds) for a 3 × 3 feed additive experiment confounded in blocks of three.

TABLE 9.16 Block Totals

Block	1	2	3	4	5	6	Sum
Sum	63.1	66.5	63.5	68.1	65.2	68.2	394.6

Accordingly, he set up the experiment as a 3 × 3 factorial confounded in blocks of 3 with litters as blocks, replicated twice.

Using AB^2 as a confounding contrast, he obtained the following assignment of treatments to blocks (each pattern appears in two blocks)

$a + 2b =$

0	1	2 (mod 3)
00	02	01
11	10	12
22	21	20

The final layout and gain (grams per 21 days) are shown in Fig. 9.4. We begin the analysis by constructing two tables of totals, Table 9.16 and Table 9.17.

For the AB interaction we require the totals obtained from the combinations whose coordinates yield 0, 1, 2 (mod 3) for the equation $a + b$. These are:

1. $a + b = 0$: (00), (12) , (21)

   $$36.2 + 46.0 + 49.1 = 131.3$$

2. $a + b = 1$: (01), (10) , (22)

$$40.5 + 41.3 + 52.7 = 134.5$$

3. $a + b = 2$: (02), (11), (20)

$$41.3 + 42.3 + 45.2 = 128.8$$

The analysis of variance of the data is presented in Table 9.18. The computations for Table 9.18 are:

$$C = (394.6)^2/18 = 8650.51$$

$$\text{SS Block} = (1/3)(63.1^2 + 66.5^2 + 63.5^2 + 68.1^2 + 65.2^2 + 68.2^2) - C$$

$$\text{SSA} = (1/6)(118.0^2 + 129.6^2 + 147.0^2) - C$$

$$\text{SSB} = (1/6)(122.7^2 + 131.9^2 + 140.0^2) - C$$

$$\text{SSAB} = (1/6)(131.3^2 + 134.5^2 + 128.8^2) - C$$

$$\text{SS ERROR} = \text{Difference}$$

TABLE 9.17 Factor A × Factor B Totals

Factor B	Factor A			Sum
	0	1	2	
0	36.2	41.3	45.2	122.7
1	40.5	42.3	49.1	131.9
2	41.3	46.0	52.7	140.0
	118.0	129.6	147.0	394.6

TABLE 9.18 ANOVA of 21-Day Gain in Feed Additive Trial

Source	d.f.	SS	MS	F
Total	17	116.35		
Block	5	8.16	1.6320	
A	2	71.02	35.5100	22.47**
B	2	24.97	12.4850	7.90*
AB	2	2.72	1.3606	0.86 NS
Error	6	9.48	1.5800	

**Significant at the 1% level.
*Significant at the 5% level.
Note: Confounded with AB^2 interaction.
NS, Not significant.

Report of statistical analysis. The results of an experiment to measure the effect of two feed additives on weight gain in rats may be summarized in two tables of means:

TABLE 9.19 Mean Gain (g/21 days) at Different Levels of A

Level of A	0	1	2	Standard error
Mean gain	19.7	21.6	24.5	.51

TABLE 9.20 Mean Gain (g/21 days) at Different Levels of B

Level of B	0	1	2	Standard error
Mean gain	20.4	22.0	23.3	.51

Gain increased as both additives A and B were increased. There was no indication that response to change in the level of A depended on the level of B.

There was very little increase in precision from blocking. It is suggested that a less elaborate experimental design might serve in future trials of this type.

9.8 EXTENSION

Of course, we can use partial confounding with the 3^k series just as we did with the 2^k series.

We can also extend these principles to cover situations in which the factors are at more than three levels. For the more mathematically oriented reader Kempthorne (1973) gives an extensive coverage of both confounding and fractionation in the p^k factorial series where p is a prime number. The development is based on Galois field theory. On the more applied level Cochran and Cox (1957) present some designs and their analysis for factors at more than three levels.

9.9 MIXED LEVEL CONFOUNDING

Confounding of mixed level factorials is also possible. With some factors at two levels and other factors at more than two levels we can reduce the block size to half the total number of treatments by using an interaction involving only two-level factors as a confounding contrast. For example, suppose we want to confound a $2 \times 2 \times 3$ (12 treatment) factorial in blocks of six. We could use AB as a confounding contrast and put those combinations for which the equation $x_1 + x_2$ is 0 modulo 2 in one block and those for which the equation is 1 (mod 2) in the other. We would have the grouping shown in Table 9.21.

Similarly, to reduce block size to one-third of the total number of treatments we could use an interaction involving three-level factors as a confounding contrast. Suppose, for example, we want to confound a $3 \times 3 \times 2$ (18 treatment) factorial in blocks of six. We could use AB^2 as a confounding contrast and separate the treatments into three groups of six by the equation $x_1 + 2x_2 = 0$, 1, 2 (mod 3). The blocks in one replication would be as given in Table 9.22.

TABLE 9.21 Partition of a $2 \times 2 \times 3$ Factorial into Two Blocks of 6 Using AB as a Confounding Contrast

	Block
	1	2
$x_1 + x_2 = 0 \bmod 2$		1 mod 2
	000	010
	001	011
	002	012
	110	100
	111	101
	112	102

TABLE 9.22 Partition of a $3 \times 3 \times 2$ Factorial into Three Blocks of 6 Using AB^2 as a Confounding Contrast

	Block		
	1	2	3
$x_1 + 2x_2 = 0 \bmod 3$		1 mod 3	2 mod 3
	000	020	010
	001	021	011
	110	100	120
	111	101	121
	220	210	200
	221	211	201

Confounding schemes for factors at four levels may be constructed fairly easily by replacing the factor at four levels with two pseudo-factors, each at two levels. Confounding is then done just as for 2^k factorials. To illustrate the process suppose we have two factors: A at two levels and B at four levels in a 2×4 factorial. There are eight treatment combinations: a_0b_0, a_0b_1, a_0b_2, a_0b_3, a_1b_0, a_1b_1, a_1b_2, a_1b_3. Suppose we are restricted to a block size of four, but assume that we have sufficient experimental material for three replications. We can develop a partially confounded design as follows: Define two pseudo-factors, C and D each at two levels, to replace the four-level factor B. The transformation is $b_0 = c_0d_0$, $b_1 = c_0d_1$, $b_2 = c_1d_0$, $b_3 = c_1d_1$. Using 0,1 notation for 2^k factorials, the relationship between the original and the transformed notation is

Original AB	Transformed ACD
a_0b_0	000
a_0b_1	001
a_0b_2	010
a_0b_3	011
a_1b_0	100
a_1b_1	101
a_1b_2	110
a_1b_3	111

We can now partition the treatments into three pairs of groups of four according to whether the three equations, $x_1 + x_2$, $x_1 + x_3$, $x_1 + x_2 + x_3$, take the value 0 or 1 modulo 2. In detail, we have

$x_1 + x_2$		$x_1 + x_3$		$x_1 + x_2 + x_3$		
0	1	0	1	0	1	mod 2
000 010		000 001		000 001		
001 011		010 011		011 010		
110 100		101 100		101 100		
111 101		111 110		110 111		

Expressed in terms of the original symbols for the treatments, the final design plan is given in Fig. 9.5. With this design one degree of freedom of the AB interaction (with 3 d.f.) is confounded in each replication. Further, a different d.f. is used as a confounding contrast in each replication so that none of the three degrees of freedom is completely confounded.

In this brief presentation we have only tried to illustrate the procedure for confounding mixed level factorial experiments. A detailed presentation of the

I		II		III		REPLICATION
1	2	3	4	5	6	BLOCK
$a_0 b_0$	$a_0 b_2$	$a_0 b_0$	$a_0 b_1$	$a_0 b_0$	$a_0 b_1$	
$a_0 b_1$	$a_0 b_3$	$a_0 b_2$	$a_0 b_3$	$a_0 b_3$	$a_0 b_2$	
$a_1 b_2$	$a_1 b_0$	$a_1 b_1$	$a_1 b_0$	$a_1 b_1$	$a_1 b_0$	
$a_1 b_3$	$a_1 b_1$	$a_1 b_3$	$a_1 b_2$	$a_1 b_2$	$a_1 b_3$	

Fig. 9.5   Design for a 2×4 factorial confounded in blocks of four.

underlying principles is given by Kempthorne (1973), and detailed experimental plans are presented by Cochran and Cox (1957) and Connor and Young (1961).

9.10   BLOCKING IN FRACTIONAL REPLICATION

As the number of factors in a factorial experiment increases the number of treatments in the full factorial set increases rapidly. We have seen that we can reduce the size of such an experiment by using a fractional replication of the full set. Until now we have assumed that the experiment would be run using a completely randomized design. In some cases, however, there are too many treatments in even a fraction for a precise experiment. For this reason we might like to reduce block size in fractional factorials just as we did with full factorial experiments. We can do this by combining both fractionation and confounding in the same design.

Consider, for example, an experiment with six factors each at two levels for a total of $2^6 = 64$ treatments. We can reduce the number of treatments to 32 by choosing a defining contrast and using those treatments for which the equation for the contrast is either 0 (mod 2) or 1 (mod 2). Suppose we use the six-factor (*ABCDEF*) interaction as a defining contrast. The 32 treatments for which the equation $x_1 + x_2 + x_3 + x_4 + x_5 + x_6$ is equal to 0 (mod 2) for this contrast are

(1)	*ae*	*af*	*ef*
ab	*be*	*bf*	*abef*
ac	*ce*	*cf*	*acef*
bc	*abce*	*abcf*	*bcef*
ad	*de*	*df*	*adef*
bd	*abde*	*abdf*	*bdef*
cd	*acde*	*acdf*	*cdef*
abcd	*bcde*	*bcdf*	*abcdef*

At this point each main effect has a five-factor interaction as an alias; each first-order interaction has a four-factor interaction as an alias; and the three-factor interactions are aliases of each other. The ANOVA takes the form presented in Table 9.23.

Now, suppose we want to run the experiment in blocks of 16 units. We can choose any contrast except the defining one as a blocking contrast. The blocking (confounding) contrast and its generalized interaction with the defining contrast will be confounded with the block contrast, and cannot be estimated. *The alias pattern for the other contrasts is not changed.* All contrasts except the blocking contrast, the defining contrast, and their generalized interaction are estimated as before.

For example, suppose we choose *ACE* as a blocking contrast. Then *ACE* and $ACE \times ABCDEF = BDF$ will be confounded. We place those treatments in the 1/2 replication for which the equation $x_1 + x_3 + x_5$ takes the value 0 (mod 2) in one block and those for which the equation takes the value 1 (mod 2) in the other. We have the plan given in Fig. 9.6.

At this point the ANOVA key-out is as shown in Table 9.24.

TABLE 9.23 ANOVA of a 1/2
Replication of a 2^5 Factorial

Source	d.f.
Total	31
Main effects (five-factor)	6
Two-factor (four-factor)	15
Error (three-factor)	10

BLOCK 1 (0 MOD 2)		BLOCK 2 (1 MOD 2)	
(1)	bf	ab	af
ac	abcf	bd	cf
bd	df	ad	abdf
abcd	acdf	cd	bcdf
ae	abdf	be	ef
ce	bcef	abce	acef
abde	adef	de	bdef
bcde	cdef	acde	abcdef

Fig 9.6 One-half replication of a 2^5 factorial confounded in blocks of 16 units.

TABLE 9.24   ANOVA of 1/2
Replication of a 2^5 Factorial
Confounded in Blocks of 16

Source	d.f.
Total	31
Block ($ACE = BDF$)	1
Main effects (five-factor)	6
Two-factor (four-factor)	15
Error (three-factor)	9

Although this is probably as far as we should go with six factors, we can reduce the block size to eight by choosing another blocking contrast. This contrast and its generalized interaction with the defining contrast, the first blocking contrast, and their interaction are also confounded with block differences. It is not possible in this case (1/2 replication of a 2^6 factorial) to reduce the block size to eight units without confounding a two-factor interaction.

To illustrate the process, however, suppose we choose ABC as a second contrast to divide each of the blocks obtained using ACE as a blocking contrast. In this case

$$ABC \times ABCDEF = \underline{DEF}$$
$$ABC \times ACE \quad = \underline{BE}$$

and

$$ABC \times BDF \quad = \underline{ACDF}$$

will also be confounded with blocks and cannot be estimated. The remaining effects, on the other hand, are clear of confounding. At this point we have one defining contrast, $ABCDEF$, and six contrasts confounded with blocks:

$$ABC, ACE, ACDF, BE, BDF, DEF$$

none of these contrasts can be estimated. The four blocks defined by these contrasts are shown in Fig. 9.7.

With four blocks of eight units the analysis of variance is given in Table 9.25.

As with other topics in this chapter we have only presented the general principles of combining fractional replication and confounding. Detailed plans for designs of this type for from four to eight factors are given by Cochran and Cox (1957).

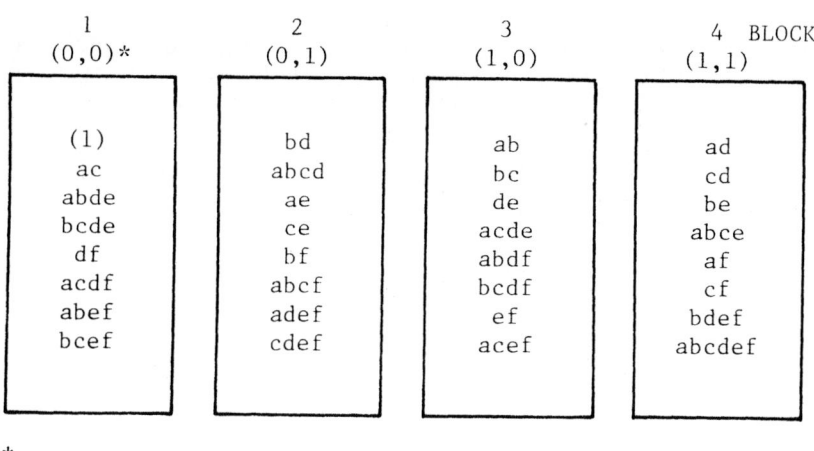

* $(x_1+x_3+x_5, \; x_1+x_2+x_3 \; \text{MOD } 2)$

Fig. 9.7 One-half replication of a 2^5 factorial confounded in blocks of 8 units.

TABLE 9.25 ANOVA of 1/2
Replication of a 2^5 Factorial
Confounded in Blocks of 8

Source	d.f.
Total	31
Block	3
Main effects	6
Two-factor (except *BE*)	14
Error	8

9.10.1 Numerical Example: Blocking in Fractional Replication

An agronomist wanted to introduce a new legume into the agricultural system of the area in which he worked. He was particularly concerned about the requirement for fertilizer. He decided to look at six fertilizers, each at two levels: absence and presence. The fertilizers were N, P, K, Mg, Mn, and S, designated *A, B, C, D, E,* and *F* in the experimental design. His resources were limited so he decided to use a half replication of the full 2^6 factorial set using *ABCDEF* as the defining contrast. Because of soil variability the maximum usable block size was eight plots. He decided to use *ACE* and *ABC* as confounding contrasts.

He realized that this would also confound, *BE, BDF, DEF*, and *ACDF*, but he thought that he could forgo information on the P × Mn interaction in the interests of experimental efficiency. His field plan, after randomization, was as illustrated in Fig. 9.8. Yields are also included in Fig. 9.8.

Using a modification of the Yates algorithm, we obtain the analysis given in Table 9.26.

$$\text{SSE} = (32.00 + 512.00 + 210.12 + 820.12 + 544.50 + 36.12 +$$
$$4.50 + 12.50)$$
$$= 2171.86$$
$$\text{MSE} = s^2 = \text{SSE}/8 = 2171.86/8 = 271.48$$

1	
abde	241
ac	178
(1)	202
acdf	258
df	256
bcde	251
bcef	178
abef	167

2	
ab	176
bc	174
abdf	244
de	219
acef	162
ef	178
bcdf	256
acde	256

3	
bdef	245
af	178
abcdef	232
cd	274
abce	163
cf	188
ad	240
be	176

4	
abcd	274
cdef	278
abcf	196
adef	235
bd	208
ae	162
bf	184
ce	159

Fig. 9.8 Field plan and yields in a one-half replication of a 2^6 fertilizer trial confounded in blocks of 8 plots.

TABLE 9.26 Analysis of a 1/2 Replication of a 2^6 Factorial Confounded in Blocks of 8 Using the Yates Method

Treatment	Yield	I	II	III	IV	V	SS	Effect
(1)	202	380	740	1,476	3,486	6,788	—	Defining
$a(f)$	178	360	736	2,010	3,302	− 64	128.00	A
$b(f)$	184	366	948	1,345	2	− 58	105.12	B
ab	176	370	1,062	1,957	− 66	106	351.12	AB
$c(f)$	188	496	683	− 20	− 62	166	861.12	C
ac	178	452	662	22	4	− 14	6.12	AC
bc	174	532	940	− 37	134	0	0.00	BC
$abc(f)$	196	530	1,017	− 29	− 28	− 4	0.50	Block
$d(f)$	256	340	− 32	− 16	110	1,146	41,041.12	D**
ad	240	343	12	− 46	56	50	78.12	AD
bd	208	321	20	23	26	− 72	162.00	BD
$abd(f)$	244	341	2	− 19	− 40	32	32.00	Error
cd	274	454	− 25	48	66	216	1,458.00	CD*
$acd(f)$	258	486	− 12	86	− 66	− 128	512.00	Error
$bcd(f)$	256	534	12	− 11	− 2	− 82	210.12	Error
$abcd$	274	483	− 41	− 17	− 2	14	6.12	EF
$e(f)$	178	− 24	− 20	− 4	534	− 184	1,058.00	E
ae	162	− 8	4	114	612	− 68	144.50	AE
be	176	− 10	− 44	− 21	42	66	136.12	Block
$abe(f)$	167	22	− 2	77	8	− 162	820.12	Error
ce	159	− 16	3	44	− 30	− 54	91.12	CE
$ace(f)$	162	36	20	− 18	− 42	− 66	136.12	Block
$bce(f)$	178	− 16	32	13	38	− 132	544.50	Error
$abce$	163	18	− 51	− 53	− 6	0	0.00	DF
de	219	− 16	16	24	118	78	190.12	DE
$ade(f)$	235	− 9	32	42	98	− 34	36.12	Error
$bde(f)$	245	3	52	17	− 62	− 12	4.50	Error
$abde$	241	− 15	34	− 83	− 66	− 44	60.50	CF
$cde(f)$	278	16	7	16	18	− 20	12.50	Error
$acde$	256	− 4	− 18	− 18	− 100	− 4	0.50	BF
$bcde$	251	− 22	− 20	− 25	− 34	− 118	435.12	AF
$abcde(f)$	232	− 19	3	23	48	82	210.12	F

Note: SS = $(V)^2/32$.

*Significant at 5%.

**Significant at 1%.

Treatment SS required for significance, Sig S_α^2

$$\text{Sig } S_\alpha^2 = \text{MSE}(F_{\alpha(1,8)})$$
$$\text{Sig } S_{.05}^2 = (271.48)(5.32) = 1444.27$$
$$\text{Sig } S_{.01}^2 = (271.48)(11.26) = 3056.86$$

Note that CD is significant at 5%, while D is significant at 1%. We can compute a table of means for C and D with the reverse Yates algorithm. This is shown in Table 9.27.

Report of statistical analysis. An experiment was conducted to determine the necessity for six fertilizer elements, N, P, K. Mg, Mn, and S, on a new forage crop. The pertinent results are summarized in Table 9.28.

Note that, in general, yield is increased by the addition of Mg. This increase is significantly greater when K is present than when K is absent. None of the other elements affected yield to any appreciable extent.

TABLE 9.27 Reverse Yates Algorithm to Compute Means

Effect	Total	(1)	(2)	Mean = (2)/32
CD	216	1362	8316	259.88
D	1146	6954	7552	236.00
C	166	930	5592	174.75
(1)	6788	6622	5692	177.88

Note: Standard error $= \sqrt{\text{MSE}/8} = \sqrt{271.48/8} = 5.82$.

TABLE 9.28 Mean Yield of Forage in the Absence and Presence of K and Mg

	K	
Mg	Absent	Present
Absent	177.88	174.75
Present	236.00	259.88

Note: Standard error = 5.82.

FURTHER READING*

Anderson, V. L., and R. A. McLean (1974), Ch. 10–12.
Cochran, W. G., and G. M. Cox (1957), Ch. 6.
Daniel, C. (1976), Ch. 10.
Finney, D. J. (1960), Ch. 5.
Kempthorne, O. (1973), Ch. 11–14.

REFERENCES

Cochran, W. G., and G. M. Cox (1957). *Experimental Designs,* 2nd ed., Wiley, New York, pp. 234–243.

Connor, W. S., and S. Young (1961). Fractional Factorial Designs for Experiments with Factors at Two and Three Levels, *Natl. Bur. Stand. Appl. Math. Ser., No. 58.*

Kempthorne, O. (1973). *The Design and Analysis of Experiments* (reprint), Krieger, Huntington, NY, Ch. 17.

---

*See Bibliography for complete citation.

10
Split-Plot Design: Variations

10.1 INTRODUCTION

We have seen that the basic split-plot design is one in which there are two sizes of experimental unit. Each block of the design contains a number of large units, the whole plots, to which the levels of one factor are applied. The whole plots are each divided into a number of smaller units, the subplots, to which the levels of the other factor are applied. Because we are dealing with two factors, one on the whole plots and one on the subplots within the whole plots, the treatments applied to the subplots constitute factorial treatment combinations. Further, if we consider that the whole plots consist of blocks of subplots we have the whole-plot factor main effect confounded with differences among these blocks.

We previously stated that three reasons are given for using split-plot designs:

1. They permit the efficient use of some treatments which require larger experimental units than others.
2. They provide greater precision in the estimation of some factor effects as compared to others.
3. They permit the introduction of new treatments into an experiment which is already in progress.

We have already considered the first reason (need for large units for some factors) for using split-plot designs. This is probably the best and most easily justified reason. It permits the conduct of some experiments, in a precise and efficient way, which otherwise could not be run.

The claim that it is desired to estimate some effects with greater precision than others is seldom a valid reason for confounding a main effect. A better procedure would be to treat the whole plots as blocks and confound a high-order interaction. Compare, for example, two designs for looking at a 2^3 factorial set of treatments replicated four times with a maximum whole-plot size of four units. Suppose it is claimed that increased precision is desired on factors A, B and their interaction with C at the expense of reduced precision on C. In this case we might run a split-plot experiment with levels of C on the whole plots. One replication of this design is shown in Fig. 10.1.

The ANOVA would take the form given in Table 10.1. Note that the main effect of factor C is tested against an error with only three degrees of freedom. The error for all the rest of the effects, on the other hand, has 18 degrees of freedom. If factor C were such that its levels required large units for their application in contrast to the requirement of small units for A and B this would be a good design. If this were not the case it would be a bad design.

Suppose, on the other hand, we consider that the whole plots of the previous design constitute two blocks of four plots within each replication. We could use ABC as a blocking contrast and confound this three-factor interaction with blocks

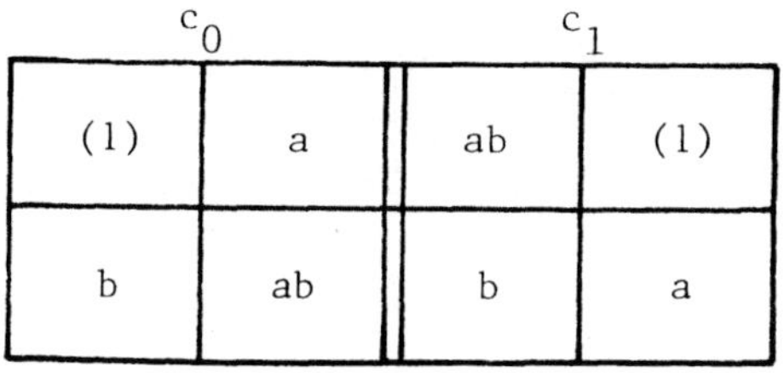

c_0		c_1	
(1)	a	ab	(1)
b	ab	b	a

Fig. 10.1 Diagram of one replication of a split-plot design with factor C on the whole plots, AB on the subplots.

TABLE 10.1 ANOVA of Split-Plot with C on the Whole Plots, AB on the Subplots

Source	d.f.
Total	31
Block	3
C	1
Error (C)	3
A	1
B	1
AB	1
AC	1
BC	1
ABC	1
Error (AB)	18

in each replication. In this case, one replication of the design would be as shown in Fig. 10.2. The ANOVA for the design in Fig. 10.2 is tabulated in Table 10.2.

With this design we sacrifice precision on the *ABC* interaction, which is hard to interpret anyway, in place of sacrificing precision on a main effect. Here the error for the high-order interaction has only three degrees of freedom. All of the main effects and first-order interactions are estimated with equally high precision. The error for these effects has 18 d.f. Certainly, if there is a choice, the second design is to be preferred over the first.

10.2 ALTERNATIVES

At first glance, splitting the plots of an established experiment to permit the introduction of new treatments would seem to be a reasonable thing to do. On second thought, however, there are seen to be alternatives which are as good as

(1)	ab	abc	c
ac	bc	b	a

Fig. 10.2 2^3 factorial in blocks of four with *ABC* as the confounding contrast.

TABLE 10.2 ANOVA for a
2^3 Factorial Confounded in
Blocks of 4 with *ABC* as
Confounding Contrast

Source		d.f.
Total		31
Block		7
Rep	3	
ABC	1	
Interblock error	3	
A		1
B		1
AB		1
C		1
AC		1
BC		1
Error		18

or better than creating a split-plot design. We might add the new factor in such a way as to convert the original design into a fractional replication of a larger design, or add the new factor to create a design in which high-order interactions are confounded with blocks.

To illustrate some of the alternatives, suppose that the original experiment consisted of two replications of a 2^2 factorial in a randomized block design. Figure 10.3 gives the design plan before randomization.

Suppose we now want to add factor C at two levels. One alternative would be to split the plots and apply the two levels of C at random to the halves of each plot. We would have the plan shown in Fig. 10.4. This would generate a split-plot design in which C, AC, BC, and ABC are estimated with greater precision than are A, B, and AB.

A second alternative would be to choose ABC as a confounding contrast and add the levels of C to the original plots as presented in Fig. 10.5. With this

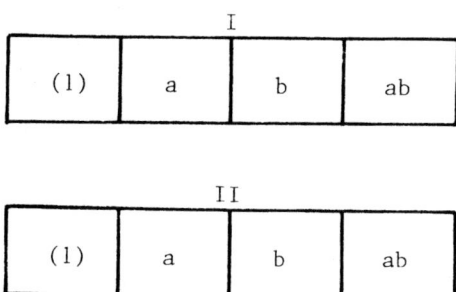

Fig. 10.3 Plan, before randomization, of a 2^2 factorial trial.

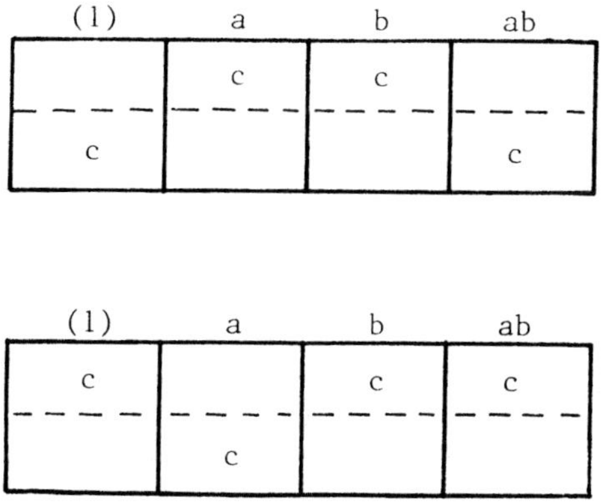

Fig. 10.4 2^2 factorial plan with plots split. Levels of C applied at random to the half plots.

(1)	a	b	ab
c			c

(1)	a	b	ab
	c	c	

Fig. 10.5   Factor C added to original plots to create a 2^3 factorial in blocks of four with ABC as a confounding contrast.

(1)	a	b	ab
cd	d	c	

(1)	a	b	ab
	c	d	cd

Fig. 10.6   Two replications of a 2^2 factorial converted to one-half replication of a 2^4 factorial confounded in blocks of four.

design ABC would not be estimable, but all main effects and first-order inter-actions would be estimated with equal precision.

Several interesting possibilities exist if we add two factors, C, D at two levels, to the original 2^2 set of treatments. One is to choose a suitable defining contrast so that the original design containing two replications of the 2^2 set is converted into a 1/2 replication of a 2^4 factorial. We would then choose another contrast to be confounded with the blocks of the original experiment. One such design is given in Fig. 10.6. Adding C and D as above creates a design in which $ABCD$ serves as a defining contrast to create a 1/2 replication of a 2^4 factorial set. Then, $AC = BD$ are confounded with blocks.

10.3 FURTHER SUBDIVISION

Split-plot design concepts are not restricted to a single hierarchy of division of the whole plots into subplots. The subplots themselves may be divided into sub-subplots. This would create a split-split-plot design in which the levels of a whole-plot factor would be applied at random to the whole plots within the blocks. The levels of another factor would then be applied at random to the subplots within the whole plots, using a separate randomization in each whole plot. Finally, the levels of a third factor would be applied at random to the sub-subplots, using a separate randomization in each subplot.

To illustrate, suppose we have four levels of the whole-plot factor A, three levels of the subplot factor B, and two levels of the sub-subplot factor C. Figure 10.7 illustrates one block of this design. Note that there are three sizes of experimental unit in this design, each with its own experimental error. These are indicated on Fig. 10.7 by the crosshatch pattern.

10.3.1 Analysis

To present the general format for the analysis of variance of the split-split-plot design suppose there are a levels of factor A on the whole plots, b levels of B on the subplots, and c levels of C on the sub-subplots. Assume that the whole plots are arranged in a randomized block design with r blocks. Table 10.3 shows the format for the analysis of variance of this design.

The addition of another factor and a second split to obtain the split-split-plot design greatly increases the number of types of comparisons which can be made and, along with these, their associated standard errors. The standard errors of the individual means are:

1. A factor means

    $$s_a = \text{MSEA}/rbc$$

    with $(r - 1)(a - 1)$ d.f.
2. B factor means

    $$s_b = \text{MSEAB}/rac$$

    $A \times B$ combination means

    $$s_{ab} = \text{MSEAB}/rc$$

    each with $a(r - 1)(b - 1)$ d.f.

a_2 a_4 a_1 a_3

	b_1		b_1		b_3	
b_1 c_2	c_2	c_1	c_1	c_2	c_2	c_1
b_1 c_2	b_3 c_1	c_2	b_2 c_1	c_2	b_2 c_1	c_2
b_3 c_1	b_2 c_1	c_2	b_3 c_2	c_1	b_1 c_2	c_1

EXPERIMENTAL UNIT FOR A

EXPERIMENTAL UNIT FOR B, AB

EXPERIMENTAL UNIT FOR C, AC, BC, ABC

Fig. 10.7   Plan for one block of a split-split-plot design.

3. *C* factor means

$$s_c = \sqrt{\text{MSEABC}/rab}$$

  A × *C* combination means

$$s_{ac} = \sqrt{\text{MSEABC}/rb}$$

  B × *C* combination means

$$s_{bc} = \sqrt{\text{MSEABC}/ra}$$

TABLE 10.3 ANOVA of a Split-Split-Plot Design

Source	d.f.	MS		F
Total	$rabc - 1$			
Block	$r - 1$	MSR	⭠⎤	F_R
A	$a - 1$	MSA	⭠	F_A
Error (A)	$(r - 1)(a - 1)$	MSEA	⎦	
B	$b - 1$	MSB	⭠⎤	F_B
AB	$(a - 1)(b - 1)$	MSAB	⭠	F_{AB}
Error (AB)	$a(r - 1)(b - 1)$	MSEAB	⎦	
C	$c - 1$	MSC	⭠⎤	F_C
AC	$(a - 1)(c - 1)$	MSAC	⭠	F_{AC}
BC	$(b - 1)(c - 1)$	MSBC	⭠	F_{BC}
ABC	$(a - 1)(b - 1)(c - 1)$	MSABC	⭠	F_{ABC}
Error (ABC)	$ab(r - 1)(c - 1)$	MSEABC	⎦	

$A \times B \times C$ combination means

$$s_{abc} = \sqrt{MSEABC/r}$$

each with $ab(r - 1)(c - 1)$ d.f.

10.3.2 Comparisons and Their Standard Errors

There are 12 ways to compare means in the split-split-plot design and, consequently, 12 different standard errors:

1. Two A means, $\bar{a}_1 - \bar{a}_0$

   $$s = \sqrt{2MSEA/rbc}, \qquad (r - 1)(a - 1) \text{ d.f.}$$

2. Two B means, $\bar{b}_1 - \bar{b}_0$

   $$s = \sqrt{2MSEAB/rac}, \qquad a(r - 1)(b - 1) \text{ d.f.}$$

3. Two B means at the same level of A, $\overline{a_1b_1} - \overline{a_1b_0}$

   $$s = \sqrt{2MSEAB/rc}, \qquad a(r - 1)(b - 1) \text{ d.f.}$$

4. Two A means at the same or different levels of B, $\overline{a_1b_1} - \overline{a_0b_1}$ or $\overline{a_1b_1} - \overline{a_0b_0}$

   $$s = \{2[(b - 1)MSEAB + MSEA]/rbc\}^{1/2}$$

   $$\text{d.f.} \cong \frac{a(r - 1)(a - 1)[b - 1)MSEAB + MSEA]^{2}*}{[(a - 1)(b - 1)(MSEAB)^{2} + a(MSEA)^{2}]}$$

*Satterthwaite's approximation (Satterthwaite, 1946).

5. Two C means, $\bar{c}_1 - \bar{c}_0$

$$s = \sqrt{2\text{MSEABC}/rab}, \qquad ab(r - 1)(c - 1) \text{ d.f.}$$

6. Two C means at the same level of A, $\overline{a_1c_1} - \overline{a_1c_0}$

$$s = \sqrt{2\text{MSEABC}/rb}, \qquad ab(r - 1)(c - 1) \text{ d.f.}$$

7. Two C means at the same level of B, $\overline{b_1c_1} - \overline{b_1c_0}$

$$s = \sqrt{2\text{MSEABC}/ra}, \qquad ab(r - 1)(c - 1) \text{ d.f.}$$

8. Two C means at the same level of A and B; $\overline{a_1b_1c_1} - \overline{a_1b_1c_0}$

$$s = \sqrt{2\text{MSEABC}/r}, \qquad ab(r - 1)(c - 1) \text{ d.f.}$$

9. Two B means at the same or different levels of C, $\overline{b_1c_1} - \overline{b_0c_1}$ or $\overline{b_1c_1} - \overline{b_0c_0}$

$$s = \{2[(c - 1)\text{MSEABC} + \text{MSEAB}]/rac\}^{1/2}$$

$$\text{d.f.} \cong \frac{ab(R - 1)(b - 1)[(c - 1)\text{MSEABC} + \text{MSEAB}]^{2}*}{[(b - 1)(c - 1)(\text{MSEABC})^{2} + b(\text{MSEAB})^{2}]}$$

10. Two B means at the same level of A and C, $\overline{a_1b_1c_1} - \overline{a_1b_0c_1}$

$$s = \{2[(c - 1)\text{MSEABC} + \text{MSEAB}]/rc\}^{1/2}$$

$$\text{d.f.} \cong \frac{ab(R - 1)(b - 1)[(c - 1)\text{MSEABC} + \text{MSEAB}]^{2}*}{[(b - 1)(c - 1)(\text{MSEABC})^{2} + b(\text{MSEAB})^{2}]}$$

11. Two A means at the same or different levels of C, $\overline{a_1c_1} - \overline{a_0c_1}$ or $\overline{a_1c_1} - \overline{a_0c_0}$

$$s = \{2[(c - 1)\text{MSEABC} + \text{MSEA}]/rbc\}^{1/2}$$

$$\text{d.f.} \cong \frac{ab(r - 1)(a - 1)[(c - 1)\text{MSEABC} + \text{MSEA}]^{2}*}{[(a - 1)(c - 1)(\text{MSEABC})^{2} + ab(\text{MSEA})^{2}]}$$

12. Two A means at the same or different levels of B and C, $\overline{a_1b_1c_1} - \overline{a_0b_1c_1}$ or $\overline{a_1b_1c_1} - \overline{a_0b_0c_0}$

$$s = \{2[b(c - 1)\text{MSEABC} + (b - 1)\text{MSEAB} + \text{MSEA}]/rbc\}^{1/2}$$

$$\text{d.f.} \cong \frac{a(r - 1)(a - 1)[b(c - 1)\text{MSEABC} + (b - 1)\text{MSEAB} + \text{MSEA}]^{2}}{b(a - 1)(c - 1)\text{MSEABC})^{2} + (a - 1)(b - 1)(\text{MSEAB})^{2} + a(\text{MSEA})^{2}}$$

As with other designs, these standard errors may be used in the construction of interval estimates and in computing the sample value of t for hypothesis tests.

---

*Satterthwaite's approximation (Satterthwaite, 1946).

10.4 FURTHER VARIATIONS

A number of additional variations are possible with split-plot type designs. As additional factors are included in the experiment further subdivision of the experimental units may be done. We will not consider further subdivision here, however. The basic principles are unchanged as the number of factors is increased.

10.4.1 Strip-Plot Designs

A particularly useful variant of the basic split-plot design is a type called the "split-block" design by Federer (1955) and called the "strip-plot" design by Petersen (1976). The basic idea behind these designs may be illustrated by considering an experiment involving two factors, a seedbed preparation factor and a rate of fertilization factor. Suppose we want to duplicate a grower's conditions and use farm-scale machinery both for seedbed preparation and for fertilizer application. It would be convenient to be able to use long, narrow strips for the levels of both factors in order to reduce the necessity for large turning areas within the blocks. This is possible if a strip-plot design is used.

In the strip-plot design the blocks are first divided into strips of plots running in one direction through the block. The levels of one factor are then randomly assigned to these strips, a different randomization being used in each block. The blocks are then divided into other strips of plots, with the new strips oriented perpendicular to the old. The levels of the second factor are then assigned at random to these strips, a separate randomization being used in each block.

To illustrate, suppose we have two blocks, three levels of factor A, and four levels of factor B. The field plan for a strip-plot design to meet these conditions is illustrated in Fig. 10.8. Note that there are three sizes of experimental unit in this design. These are delineated on Fig. 10.8 by the crosshatch patterns. As might be expected, with three types of experimental unit there will be three different experimental errors for this design.

10.4.2 Analysis

In general, suppose that there are a levels of factor A, b levels of factor B, and r blocks. The format for the analysis of variance for the strip-plot design is shown in Table 10.4.

Just as for the split-plot design, there are different standard errors associated with the various means for the strip-plot design. These are:

1. A factor means

$$s_a = \sqrt{\text{MSEA}/rb}, \qquad (r-1)(a-1) \text{ d.f.}$$

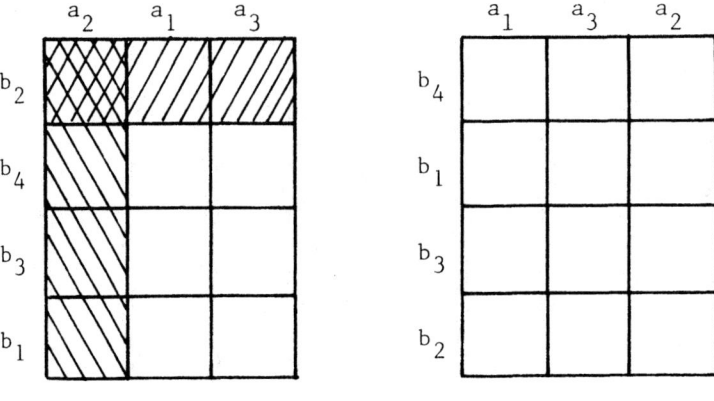

EXPERIMENTAL UNIT FOR A

EXPERIMENTAL UNIT FOR B

EXPERIMENTAL UNIT FOR AB

Fig. 10.8 Plan for a two-factor strip-plot design.

TABLE 10.4 ANOVA of a Strip-Plot Design

Source	d.f.	MS		F
Total	$rab - 1$			
Block	$r - 1$	MSR		
A	$a - 1$	MSA	←	F_A
Error (a)	$(r - 1)(a - 1)$	MSEA		
B	$b - 1$	MSB	←	F_B
Error (b)	$(r - 1)(b - 1)$	MSEB		
AB	$(a - 1)(b - 1)$	MSAB	←	F_{AB}
Error (ab)	$(r - 1)(a - 1)(b - 1)$	MSEAB		

2. B factor means

$$s_b = \sqrt{MSEB/ra}, \qquad (r-1)(b-1) \text{ d.f.}$$

3. $A \times B$ combination means

$$s_{ab} = \sqrt{MSEAB/r}, \qquad (r-1)(a-1)(b-1) \text{ d.f.}$$

Likewise, standard errors of the differences between means depend on which way the comparisons are made. We have

1. Two A means, $\bar{a}_1 - \bar{a}_0$

$$s = \sqrt{2MSEA/rb}, \qquad (r-1)(a-1) \text{ d.f.}$$

2. Two B means, $\bar{b}_1 - \bar{b}_0$

$$s = \sqrt{2MSEB/ra}, \qquad (r-1)(b-1) \text{ d.f.}$$

3. Two A means at the same level of B, $\overline{a_1 b_1} - \overline{a_0 b_1}$

$$s = \{2[(b-1)MSEAB + MSEA]/rb\}^{1/2}$$

$$\text{d.f.} \cong \frac{(r-1)(a-1)[(b-1)MSEAB + MSEA]^{2*}}{[(b-1)(MSEAB)^2 + (MSEA)^2]}$$

4. Two B means at the same level of A, $\overline{a_1 b_1} - \overline{a_1 b_0}$

$$s = \{2[(a-1)MSEAB + MSEB]/ra\}^{1/2}$$

$$\text{d.f.} \cong \frac{(r-1)(b-1)[(a-1)MSEAB + MSEB]^{2*}}{[(a-1)(MSEAB)^2 + (MSEB)^2]}$$

5. Two A means at different B, $\overline{a_1 b_1} - \overline{a_0 b_0}$

$$s = \{2[(ab-a-b)MSEAB + bMSEB + aMSEA]/rab\}^{1/2}$$

$$\text{d.f.} \cong \frac{(r-1)(a-1)(b-1)[(ab-a-b)MSEAB + bMSEB + aMSEA]^2}{[(ab-a-b)(MSEAB)]^2 + (a-1)[bMSEB]^2 + (b-1)[aMSEA]^2} *$$

10.4.3 Numerical Example: Strip-Plot Design

A sugar planter wanted to determine the effect of potassium and phosphorus fertilizers on the yield of sugarcane. He established an experiment with the following factors:

Factor A = potassium: K_1 = none, K_2 = 25 kg/ha, K_3 = 50 kg/ha
Factor B = phosphorus: P_1 = 25 kg/ha, P_2 = 50 kg/ha

---

*Satterthwaite's approximation (Satterthwaite, 1946).

He decided to use a design with three blocks. Because he wanted to use farm-scale equipment to apply the fertilizers, he assigned the potassium rates to strips of two plots within blocks. He then assigned phosphorus to strips of three plots at right angles to the potassium strips within blocks. The field plan and yields (kilograms per plot) of cane are shown in Fig. 10.9. The analysis of variance of the data is given in Table 10.5.

Three summary tables are required to compute the sums of squares in the analysis of variance in Table 10.5. These are shown as Table 10.6, Table 10.7, and Table 10.8.

The sums of squares for blocks and for potassium are computed from the border totals in Table 10.6. Error (A) is equivalent to the block $\times$ potassium interaction. Its sum of squares is computed from the entries in the body of Table 10.6.

The sum of squares for phosphorus is computed from the border totals in Table 10.7. Error (B) is equivalent to the block $\times$ phosphorus interaction. Its sum of squares is computed from the entries in the body of Table 10.7.

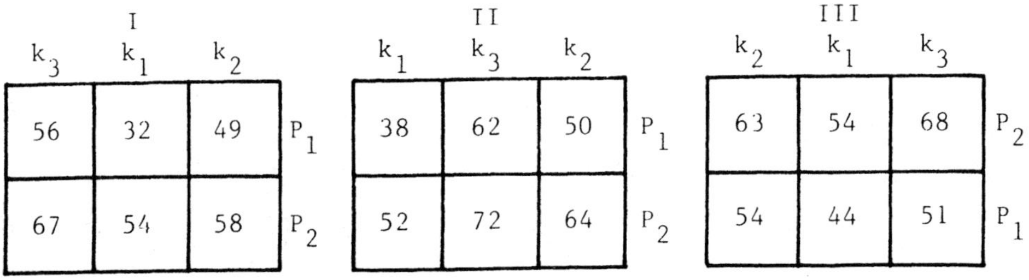

Fig. 10.9 Field plan and yields of sugarcane for a fertilizer experiment using a strip-plot design.

TABLE 10.5 ANOVA of Sugarcane Yields (kg/plot) from Strip-Plot Experiment with P and K Fertilizers

Source	d.f.	SS	MS	F
Total	17	1833.78		
Block	2	45.78	22.89	
Potassium	2	885.78	442.89	22.64**
Error (A)	4	78.22	19.56	
Phosphorus	1	747.56	747.56	479.21**
Error (B)	2	3.11	1.56	
K $\times$ P	2	16.44	8.22	.58
Error (AB)	4	56.89	14.22	

**Significant at the 1% level.

TABLE 10.6 Block × K Totals

Block	Potassium			
	K_1	K_2	K_3	Sum
I	86	107	123	316
II	90	114	134	338
III	98	117	119	334
Sum	274	338	376	988

TABLE 10.7 Block × P Totals

Block	Phosphorus		
	P_1	P_2	Sum
I	137	179	316
II	150	188	338
III	149	185	334
Sum	436	552	988

TABLE 10.8 P × K Totals

Phosphorus	Potassium			
	K_1	K_2	K_3	Sum
P_1	114	153	169	436
P_2	160	185	207	552
Sum	274	338	376	988

The sum of squares for the potassium × phosphorus interaction is computed from the entries in the body of Table 10.8. The sum of squares for error (*AB*), which is equivalent to the block × potassium × phosphorus interaction, is computed by difference.

Report of statistical analysis. The results of a trial designed to measure the effect of potassium and phosphorus fertilizers on the yield of sugarcane are summarized in Tables 10.9 and 10.10.

From Table 10.9 it is seen that over the range used in this trial sugarcane yield increased about 8.5 kg/plot for each 25-kg/ha increase in potassium fertilizer. This increase occurred regardless of the level of phosphorus.

TABLE 10.9 Mean Yields (kg/plot) of Sugarcane Under Different Levels of Potassium Fertilizer

K rate (kg/ha)	0	25	50	Standard error
Mean yield	45.67	56.33	62.67	1.81

TABLE 10.10 Mean Yields (kg/plot) of Sugarcane Under Different Levels of Phosphorus Fertilizer

P rate (kg/ha)	25	50	Standard error
Mean yield	48.44	61.33	0.42

In Table 10.10 it will be seen that increasing the phosphorus rate from 25 to 50 kg/ha resulted in an increase of about 13 kg/plot of sugarcane.

10.5 EXTENSION

The strip-plot designs can be expanded to include more than two factors. In general the same basic principles apply, but the possible permutations and combinations of groups and strips increase rapidly.

As the designs become more complex their analysis becomes more difficult. The following set of rules and algorithms developed by Federer (1975) is an excellent systematic procedure for attacking the analysis of complex designs:

Dependence of Split Plot and Whole Plot ANOVA's
(i) If the q split plot treatments are randomly allotted to the q experimental units within each whole plot, the experiment design for whole plots *does not* affect the split plot analysis.

(ii) If a complete block experiment design for split plot treatments is used within *each* whole plot treatment, the form of the split plot analysis is unaffected by the statistical design for whole plot treatments.

(iii) In a standard split block or two-way whole plot experiment design, the analysis of variance for one factor and for the two-factor interaction is unaffected by the experiment design utilized for the second factor.

Rules and Algorithms for Obtaining an ANOVA for a Complex Experiment Design
Eight rules and three algorithms are presented for obtaining an appropriate partitioning of degrees of freedom in the ANOVA and appropriate error variances for interval estimation and F-tests.

In the course of statistical consulting it becomes apparent that the experimenter usually does not know what type of design he has nor what type of confounding of effects is present in the experiment. Perhaps the only type of consulting the author receives is for completely and partially confounded experiments and surveys, but in nearly every case there is no simple textbook answer. Each investigation and its related statistical design must be approached as a unique situation and not one that appears on page X of textbook Y. It *may* be like that one on page X but more often than not there is something in the experiment for which no close analogy to a textbook example can be made. This leads to Rule I.

Rule I: Make no assumptions about the form of the statistical design; always determine the exact experimental procedure, not the stated one.

Quite often the investigator states that his statistical design was *D* when in fact it was *X*. One should always have the consultee describe the investigation in minute detail, and then one may come to a conclusion as to the statistical design.

Once one thinks he knows the statistical design, it is then possible to key-out the degrees of freedom for the ANOVA. However in doing this rules II, III, IV, and V have been found essential.

Rule II: Determine the experimental unit for levels of each category (factor, block, etc.); then determine any common experimental units for combinations of all possible pairs of categories, then for all possible triplets of categories, etc.

Rule III: Count the number of randomizations for each category (factor, block, etc.) in the experiment; then count randomizations for combinations of levels for all possible pairs of categories, then for all possible triplets, etc.

Rule IV: Determine which category levels are nested within another category level and determine which are cross classified.

Rule V: Ignore complexity of design in first key-out of degrees of freedom; relate key-out to nearest known design.

Application of rules I through V should enable one to key-out the degrees of freedom in an ANOVA as described in algorithm I.

Algorithm I: Keying-out degrees of freedom in the ANOVA.

1. At every step perform simplest key-out of degrees of freedom that is possible.
2. First determine total degrees of freedom and partition into one for the correction term and the remainder for the sum of squares corrected for the mean.
3. Key-out degrees of freedom for category or categories offering the least difficulty.
4. Key-out degrees of freedom for ANOVA's for all possible pairs of categories, then all possible triplets, etc., excluding any pairs, triplets, etc. not needed.
5. Isolate all sets of degrees of freedom in the ANOVA for which the partitioning is not understood.
6. Defer the partitioning of sets of degrees of freedom that are not completely understood.
7. Approach the partitioning in step 6 from different directions in order to reduce steps 5 and 6 to the null set. Note that partitioning may be impossible until more information becomes available.

In using the algorithm *always* approach the key-out of degrees of freedom from the direction which is simplest and easiest to understand. Keep picking away at the remainder degrees of freedom until one reaches the desired stage, which could of course be single degree of freedom contrasts for the total degrees of freedom. When one knows the total number of observations N, one knows the total degrees of freedom which is N. Then one can always partition these N degrees of freedom into one for the correction for the mean and $N - 1$ for the remainder. Then, if there are r blocks, one can always partition the $N - 1$ into a set of $r - 1$ and $N - r$ degrees of freedom. This procedure is continued until step 7 in the algorithm is reached.

Investigators and statisticians often start computing sums of squares prior to using algorithm I. This practice can result in misspent effort and hence Rule VI.

Rule VI: Do NO computing of sums of squares until the correctness of the degree of freedom key-out in the ANOVA has been ascertained and the appropriate error variances have been designated.

Before computing any sums of squares, it is well to recognize the difficulty encountered in keying-out degrees of freedom in certain types of experiments. It is wise to consider the following two rules whenever human or animal experiments are involved.

Rule VII: With almost probability one, experiments and surveys involving humans and animals will have effects completely or partially confounded and one will need to follow rules I through V in order to ascertain this.

Rule VIII: Be prepared to spend considerable time and effort unravelling the confounding schemes in any human or animal experiment as planned by the researcher (and perhaps even by a statistician).

When one is satisfied with the key-out of degrees of freedom for an investigation, then and only then should one consider computing totals, solutions for effects, and sums of squares. In connection with the last item algorithm II has been found useful.

Algorithm II: Computing sums of squares in the ANOVA

1. At every step compute the simplest ANOVA sums of squares, that is, sums of squares assuming nesting even though there was no nesting.
2. Compute sums of squares for degree of freedom key-outs in steps 2, 3, and 4 of algorithm I. For many investigations, this is a desk calculator job.
3. For partially confounded effects, it may be necessary to solve a set of normal equations prior to computing the sums of squares.
4. If steps 5 and 6 of algorithm I have not been reduced to the null set, nothing should be done about further partitioning of the sums of squares.

All too often computing specialists become imbued with a program or package for high speed computing and do not pay sufficient attention to simplifications.

The method of computing described indicates exactly what is being done whereas the use of a high speed nonorthogonal n-way classification program or a multiple regression program would not indicate the nature of quantities being computed. Likewise, rounding errors from high speed computer programs have always plagued this author, with complete nonsense resulting in several cases.

Once one computes the appropriate sums of squares in the ANOVA, then appropriate error variances need to be determined. Algorithm III is presented in this light.

Algorithm III: Determining appropriate error variances for F-tests

1. Factors with the same type of experimental unit *may* have the same error variance.
2. Factors with different experimental units almost always have different error variances.
3. In order to check the validity of an error variance, determine the appropriate error variance assuming other effects are absent from the experiment for single factors, for pairs of factors, etc.
4. Check to determine if partially confounded effects may be estimated from two sources and with two different error variances.

5. Check your decisions with known situations.*

A more naive approach which can be used as a first approximation is:

1. Identify all of the different types of experimental unit in the design.
2. Determine which of the effects are estimated on each of the types of unit.
3. Use the pooled interaction of effect × block for all of the effects estimated on a unit as the error for testing those effects.

A still different procedure was developed by Anderson (1970) and is presented by Anderson and McLean (1974). They have developed what they call "restriction errors" which are introduced into the design. Although these restriction errors have no degrees of freedom, and cannot be estimated in the analysis of variance, they serve to point out the places where randomization is restricted in a design.

The strip-plot design has essentially the same advantages, disadvantages, and uses as the split-plot design. The principal difference is that there is more restriction on randomization and more subdivision of the error variance in the strip-plot than in the split-plot design. From this standpoint the strip-plot is a more complex design and, given a choice, it would be preferable to use a split-plot design.

10.6   COMPACT DESIGNS FOR EXPLORATORY RESEARCH

In the initial stages of an investigation one aim is to identify those factors which have an effect and those which do not, and to see how the factors interact with each other. A further aim is to sample a broad spectrum of external conditions to determine which effects are stable and which effects vary with changing conditions. To meet these aims it is often necessary to conduct a large number of exploratory experiments. These experiments should be designed in such a way that they provide the required information but, at the same time, make efficient use of the limited resources of any research program. Such experiments should possess a number of characteristics:

1. They should be capable of examining a fairly large number of factors simultaneously.
2. They should be relatively small.

*Reproduced from The Misunderstood Split-Plot by W. T. Federer, in *Applied Statistics*, R. P. Gupta (ed.) (1975), with permission of the publisher, North Holland Publishing Co., Amsterdam.

3. They should be designed in such a way that the experimental units can be grouped into fairly small blocks.

4. They should provide reasonable precision, with minimum of about 10 degrees of freedom for error, for estimating the effects of interest.

5. In some cases they should permit the use of large experimental units for some factors while retaining the convenience of small units for others.

Designs which meet these criteria may be constructed using the 2^k factorial series as a base. The total design may be kept small by using the principle of fractional replication presented in Chapter 8. The experimental units may be grouped into blocks by using the technique of confounding discussed in Chapter 9. Experimental units of different size may be constructed by confounding a main effect and, in this way, creating a split-plot design as discussed in this chapter (Chapter 10). By using these techniques it is possible to construct compact designs for examining from four to eight factors, and which conform to the following restrictions:

1. No more than 64 experimental units
2. A maximum block size of 16 units
3. At least 13 degrees of freedom for the estimation of error
4. Except for split designs, all main effects and first-order interactions estimable

10.6.1   Example of a Compact Design

A representative compact design is illustrated in Fig. 10.10. In this plan there are seven factors which are examined at two levels in a total of 64 experimental units grouped into four blocks of 16 units each. The plan was constructed by taking a 1/2 replication of a 2^7 factorial set with *ABCDEFG* as a defining contrast. The design includes combinations with the minus $(-)$ sign under this contrast. The 64 combinations were split into two groups of 32 units by using the main effect of *G* as a confounding contrast. Finally, block size was reduced to four blocks of 16 units using *ABC* as a confounding contrast. This produces a grouping of treatments such that *ABCDEFG*, *G*, *ABCDEF*, *ABC*, *DEFG*, *ABCD*, and *DEF* cannot be estimated. However, all of the main effects except *G* and all of the first-order interactions are estimable. Note that none of the combinations in blocks 1 and 2 contain *G* at the high level, while the high level of *G* occurs in all of the combinations in blocks 3 and 4. Thus blocks 1 + 2 and 3 + 4 constitute, in effect, two whole plots of a split-plot design with levels of *G* assigned to them. This design has 33 d.f. for error.

Compact designs which conform to the limitations presented in Section 10.6 have been constructed for from four to eight factors. These have been presented

BLOCK

1	2	3	4
ef	abcdef	abeg	cdeg
ab	ad	bcdg	cdfg
ac	bdef	abdefg	adfg
de	ae	eg	cefg
df	cf	bcfg	bdfg
(1)	abcd	bcdg	abcefg
abdf	cd	aceg	abcdfg
bcdf	ce	acdg	cg
abef	adef	dg	abcdeg
bcef	cdef	defg	adeg
acdf	bf	bcdefg	aefg
bcde	abcf	abfg	abcg
bc	bd	acfg	bdeg
acef	be	fg	bg
acde	abce	acdefg	befg
abde	af	abdg	ag

(1)   g

Fig. 10.10 Plan of a compact design for seven factors with one split.

by Petersen (1976) for designs which do not involve the split-plot feature. Compact designs for five and six factors which allow one, two, or three splits have been described by Petersen (1980).

10.6.2 Analysis of Compact Designs

In the analysis of data from these designs it is assumed that second, and higher, order interactions are negligible. Some of these high-order interactions are used as defining contrasts to reduce the size of the experiment. Others are used as confounding contrasts to reduce block size. Sums of squares of the other high-order interactions are assumed to reflect only random variation. These are pooled to obtain an estimate of experimental error for testing main effects and low-order interactions. This procedure provides a conservative test of these effects. Any effects which are declared to be significant using the pooled interactions as error are certain to be significant if compared to the true error. This is because interactions, if they are real, would increase the mean square used as the denominator of the F ratio.

To introduce the split-plot feature into the compact designs main effects are used as confounding contrasts. If this is done the confounded main effects and, possibly, some low-order interactions are not estimable because they are made equivalent to block contrasts by the confounding. This has no effect on the estimation of the other main effects or their interaction with the confounded effects. These are estimated and tested as if the split-plot feature had not been introduced.

In the analysis of data from these compact designs the sums of squares, each with one degree of freedom, for main effects and first-order interactions are computed. This is most easily done using the modified Yates algorithm. The sums of squares for blocking contrasts are computed and removed from consideration. The remaining sums of squares are pooled as an estimate of error. The degrees of freedom for this error sum of squares are the number of sums of squares pooled to obtain the error.

FURTHER READING*

Cochran, W. G., and G. M. Cox (1957), Ch. 7.
Federer, W. T. (1955), Ch. 10.
Finney, D. J. (1960), Ch. 5.
Kempthorne, O. (1973), Ch. 19.

REFERENCES

Anderson, V. L. (1970). Restriction Errors for Linear Models, *Biometrics* 26:255–268.
Anderson, V. L., and R. A. McLean (1974). *Design of Experiments*, Marcel Dekker, New York, Ch. 5.
Federer, W. T. (1955). *Experimental Design*, Macmillan, New York, Ch. 10.
Federer, W. T. (1975). The Misunderstood Split-Plot, in *Applied Statistics*, R. P. Gupta (ed.), North Holland Publishing Co., Amsterdam.

---

*See Bibliography for complete citation.

Petersen, R. G. (1976). *Experimental Designs for Agricultural Research in Developing Areas*, O.S.U. Bookstores, Inc., Corvallis, OR.

Petersen, R. G. (1980). Experimental Designs for Off-Station Agronomy Trials, Discussion Paper No. 2, The International Center for Agricultural Research in the Dry Areas, Aleppo, Syria.

Satterthwaite, F. E. (1946). An Approximate Distribution of Estimates of Variance Components, *Biometrics Bull.* 2:110–114.

11

Response Surfaces

11.1 INTRODUCTION

In our analysis of the results of multifactor experiments we have, until now, been concerned with estimating means, and estimating and testing such factorial contrasts as main effects and interactions. We have not bothered, for the most part, about whether the factors were quantitative or qualitative. We now want to turn our attention to situations in which the factors are quantitative and yield is a measure of response to changes in the levels of these quantitative factors. That is, we are interested in characterizing the surface which describes the yield over a range of the quantitative factors.

We have already had a brief introduction to the ideas involved when we considered regression contrasts in Chapter 5 on the separation of means. There we were concerned with using the regression approach to describe yield response as a function of the level of a single quantitative treatment variable. We now want to expand consideration to include more than one variable, and to look for experimental designs which lead to the efficient characterization of response surfaces.

Because this chapter is concerned with multiple treatment variables which are quantitative in nature, and because we want to describe yield as a function of changes in these quantitative variables, the analytical methods of this chapter are the methods of multiple regression analysis. In multiple regression analysis it is convenient to use matrix notation and to express the analytical procedures using the methods of elementary matrix algebra. Our symbolic notation for the basic matrix algebra used in this chapter is presented in Section 11.9 at the end of the chapter.

11.2 BASIC IDEAS

Suppose we have a system which involves a response variable, y, which depends on the level of a number of input variables, $\xi_1, \xi_2, \ldots, \xi_k$. We assume that the levels of the ξ's can be controlled by the experimenter with negligible error. We conduct an experiment with design variables $x_1, x_2, \ldots, x_k$ which are usually simple transformations of the ξ_i. Each treatment can be represented by a point with coordinates $(x_{1j}, x_{2j}, \ldots, x_{kj})$ in a k-dimensional factor space. At each point we observe a value of y_j.

Since we are free to choose the levels of the ξ's it is convenient to choose them to be equally spaced. The relationship between the ξ_{ij} and the x_{ij} can be expressed as

$$x_{ij} = (\xi_{ij} - \bar{\xi_i})/\Delta$$

where

$$\bar{\xi_i} = \sum_i \xi_{ij}/n = \text{mean of the } \xi_{ij}$$

and

$$\Delta = \xi_{ij} - \xi_{i(j-1)} = \text{the difference in levels of the } \xi_{ij}$$

In general, response is a function of the input variables

$$y = f(\xi_1, \xi_2, \ldots, \xi_k)$$

Often $f(\xi_i)$ is unknown and, perhaps, complicated. The basic response surface procedure is to approximate $f(\xi_i)$ with a low-order polynomial, and to use sample data to fit the least squares estimates of the coefficients of the polynomial.

11.2.1 Basic Assumptions

The basic assumptions are:

1. A structure, $y = f(\xi_1, \xi_2, \ldots, \xi_k)$, exists and is either complicated or unknown.
2. The variables, ξ_i, are quantitative and continuous.
3. The true function, $f(\xi_i)$, can be approximated in the region of interest by a low-order polynomial.

4. The design variables, $x_1, x_2, \ldots, x_k$, are controlled and measured without error.

The goal of most response surface research is twofold:

1. To find a suitable approximating function for the purpose of predicting future response
2. To find levels of the input variables for which, in some sense, the response is optimized

It should be noted that the usual response surface procedures are not generally designed to increase understanding of the mechanism of the underlying system. Rather, the aim is to determine optimum operating conditions or to define a region in the space of the input variables where certain operating specifications are met.

Suppose we conduct an experiment with design variables $x_1, x_2, \ldots, x_k$ representing the input variables ξ_i. We can obtain an approximate representation of $f(\xi_i)$ from the low-order terms of a Taylor series expansion around the point $x_1 = x_2 = \cdots = x_k = 0$. If we use only the first-order terms we have as our model the linear polynomial

$$E(y) = \beta_0 + \sum_{i=1}^{k} \beta_i x_i$$

where

$E(y)$ = expected response

β_0 = intercept

β_i = linear coefficient relating response to the level of the ith input variable

x_i = level of the ith input variable

The estimating equation for this first-order model will be

$$\hat{y} = b_0 + \sum_{i=1}^{k} b_i x_i$$

where

b_0, b_i are sample estimates of β_0, β_i

This estimating equation can be conveniently written in matrix form as

$$\hat{y} = b_0 + \mathbf{x}'\mathbf{b}$$

in which

$$\mathbf{x} = \begin{bmatrix} x_1 \\ x_2 \\ \vdots \\ x_k \end{bmatrix}, \qquad \mathbf{b} = \begin{bmatrix} b_1 \\ b_2 \\ \vdots \\ b_k \end{bmatrix}$$

The linear polynomial model will often provide a reasonable approximation of the true response surface over a limited range of values of x_i.

Over a wider range of x_i a satisfactory approximation can often be obtained by taking both first- and second-order terms of the expansion to obtain the quadratic polynomial model

$$E(y) = \beta_0 + \sum_i \beta_i x_i + \sum_i \beta_{ii} x_i^2 + \sum_{i<i'} \beta_{ii'} x_i x_{i'}$$

where

β_0, β_i, and x_i are as for the first-order model

β_{ii} = quadratic coefficient for the ith variable

$\beta_{ii'}$ = interaction coefficient for the interaction of variables i and i'

The estimating equation for this model can be written, in matrix form, as

$$\hat{y} = b_0 + \mathbf{x}'\mathbf{b} + \mathbf{x}'\mathbf{B}\mathbf{x}$$

in which

$$\mathbf{x} = \begin{bmatrix} x_1 \\ x_2 \\ \vdots \\ x_k \end{bmatrix}, \qquad \mathbf{b} = \begin{bmatrix} b_1 \\ b_2 \\ \vdots \\ b_k \end{bmatrix}, \qquad \mathbf{B} = \begin{bmatrix} b_{11} & (b_{12})/2 & \cdots & (b_{1k})/2 \\ & b_{22} & \cdots & (b_{2k})/2 \\ & \text{Symm} & \cdots & \vdots \\ & & & b_{kk} \end{bmatrix}$$

11.2.2 Analysis

Regardless of the design used to obtain the data, the analysis usually follows a fixed pattern. To illustrate, suppose we are interested in the yield, y, of a chemical reaction as a function of temperature, ξ_1, and pressure, ξ_2. Suppose we choose to do two replications of a 3×3 factorial experiment with three equally spaced levels of temperature and pressure. Figure 11.1 illustrates the configuration of points in the ξ_1, ξ_2 factor space.

Using the transformation $x_{ij} = (\xi_{ij} - \bar{\xi})/\Delta$ we obtain the points:

PRESSURE

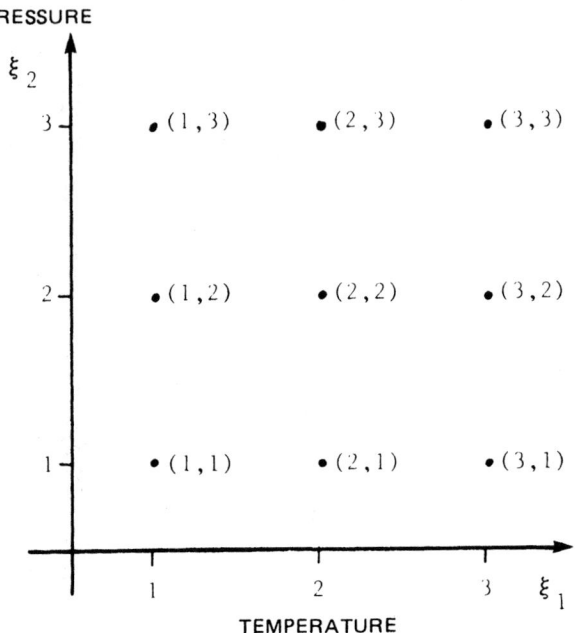

Fig. 11.1 Configuration of design points for a 3×3 factorial design on temperature and pressure.

$$
\begin{array}{lll}
(-1, \ 1) & (0, \ 1) & (1, \ 1) \\
(-1, \ 0) & (0, \ 0) & (1, \ 0) \\
(-1,-1) & (0,-1) & (1,-1)
\end{array}
$$

as coordinates of points in the x_1, x_2 design space corresponding to the ξ_1, ξ_2 points in the temperature, pressure factor space.

The analysis proceeds in the following stages:

1. Fit the usual analysis of variance model for a randomized block design

    $$E(y) = \mu + \rho_i + \tau_j$$

    With two blocks and nine treatments the format for the analysis of variance is as shown in Table 11.1.

2. Now, partition the treatment sum of squares into two parts:
    a. A part, with 5 d.f., associated with the regression of yield on x_1, x_2 described by the model

TABLE 11.1 ANOVA for
Standard Analysis of Two Blocks
of a 3 × 3 Factorial

Source	d.f.
Total	17
Block	1
Treatment	8
Error	8

TABLE 11.2 Final ANOVA of
3 × 3 Factorial After Regression
Partition of Treatment Source of
Variation

Source	d.f.
Total	17
Block	1
Fitted surface	5
Lack of fit	3
Error	8

$$E(y) = \beta_0 + (\beta_1 x_1 + \beta_2 x_2) + (\beta_{11} x_1^2 + \beta_{12} x_1 x_2 + \beta_{22} x_2^2)$$

$$= \beta_0 + [x_1 \ x_2] \begin{bmatrix} \beta_1 \\ \beta_2 \end{bmatrix} + [x_1 \ x_2] \begin{bmatrix} \beta_{11} & 1/2\beta_{12} \\ 1/2\beta_{12} & \beta_{22} \end{bmatrix} \begin{bmatrix} x_1 \\ x_2 \end{bmatrix}$$

$$= \beta_0 + \mathbf{x}'\boldsymbol{\beta} + \mathbf{x}'\mathbf{B}\mathbf{x}$$

b. A part, with 3 d.f., associated with "lack of fit." This sum of squares is computed as the difference between the treatment SS computed in stage 1 and the regression SS computed in stage 2. This sum of squares for lack of fit is a measure of the failure of the treatment means to conform to the quadratic model in x_1, x_2.

The final form of the analysis of variance is shown in Table 11.2. At this point we can use the error mean square in an F test for lack of fit and for the goodness of fit of the quadratic approximation. If lack of fit is not significant we also, at this point, have estimates of the parameters in the predicting equation

$$\hat{y} = b_0 + [x_1 \ x_2] \begin{bmatrix} b_1 \\ b_2 \end{bmatrix} + [x_1 \ x_2] \begin{bmatrix} b_{11} & 1/2b_{12} \\ 1/2b_{12} & b_{22} \end{bmatrix} \begin{bmatrix} x_1 \\ x_2 \end{bmatrix}$$

$$= b_0 + \mathbf{xb} + \mathbf{x'Bx}$$

11.2.3 Examining the Fitted Surface

Having found and tested a quadratic polynomial which approximates the response function, our next step is to look for the essential features of the fitted surface. There are a number of questions we might want to ask:

1. For what values of x_1, x_2, . . . , x_k is y a minimum or maximum?
2. What is the value of y at this point?
3. What is the shape of the surface within the range of the input variables covered in the experiment?

Using calculus we can find the point for which y is a maximum or minimum. To do this we differentiate $\hat{y}$ with respect to each x_i in turn, equate the result to zero, and solve for $\mathbf{x}_0$, the coordinates of a point in x_1, x_2, . . . , x_k space. This point may be a minimum, maximum, or saddle point for y. It is called a stationary point because it is the point at which the derivative (rate of change) is zero. Given the general quadratic estimating equation

$$\hat{y} = b_0 + \mathbf{x'b} + \mathbf{x'Bx}$$

the coordinates, $\mathbf{x}_0$, of the stationary point are found to be

$$\mathbf{x}_0 = -\mathbf{B}^{-1}\mathbf{b}/2$$

where $\mathbf{b}$ and $\mathbf{B}$ are obtained from the estimating equation.

At $\mathbf{x}_0$ the value of y might be a maximum, a minimum, or a minimax (minimum for some variables and maximum for others). The nature of y at the stationary point is determined by the signs of the b_{ii} in $\hat{y}$. This is illustrated for two input variables in Fig. 11.2.

The value of y, y_0, at the stationary point may be found by substituting $\mathbf{x}_0$ into the estimating equation, or else the following direct solution may be used:

$$y_0 = b_0 + (\mathbf{x}_0' \ \mathbf{b})/2$$

This is normally as far as we need to go. At this point we know the coordinates of the stationary point and the yield at the stationary point. We can get some idea of the shape of the surface at the stationary point from the signs on the quadratic (b_{ii}) coefficients. We also have an equation into which we can substitute values of the input variables and solve for the expected yield.

We can gain some further information on the nature of the stationary point and the shape of the surface by performing what is called a canonical analysis

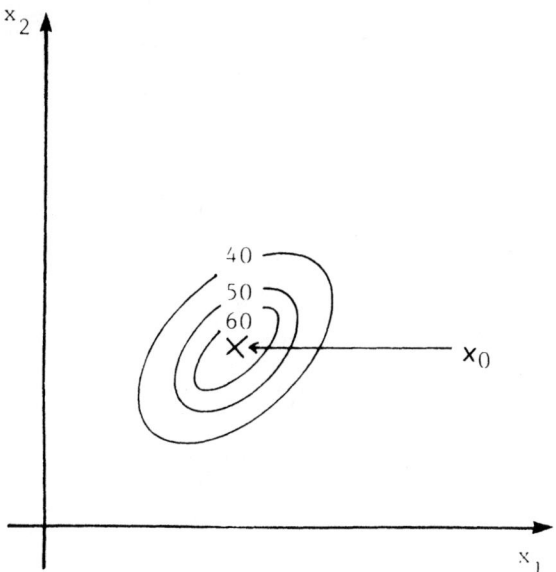

Fig. 11.2a Sign of both b_{11} and b_{22} is minus. y_0 is a maximum.

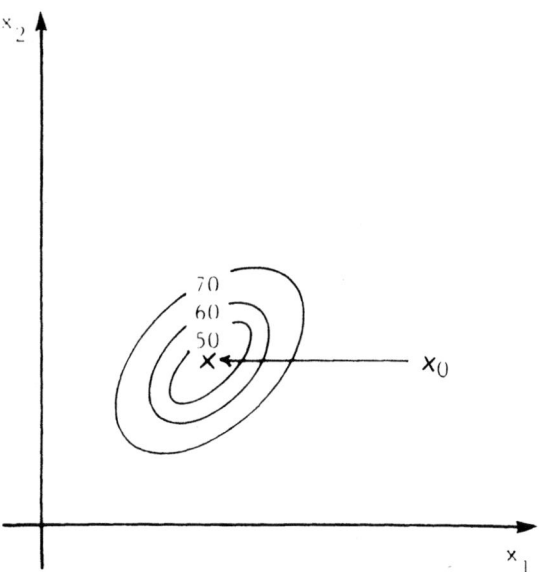

Fig. 11.2b Sign of both b_{11} and b_{22} is plus. y_0 is a minimum.

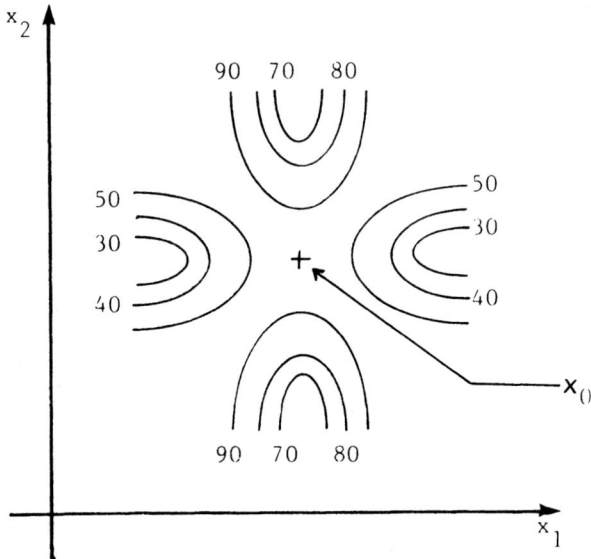

Fig. 11.2c Sign b_{11} is minus, sign of b_{22} is plus. y_0 is a saddle point (maximum for x_1, minimum for x_2).

of the surface. We will not go into the details here. They are covered in detail by Myers (1971). The basic idea is to shift the center of the coordinate system from $(x_1, x_2, \ldots, x_k = 0)$ to $\mathbf{x}_0$, then rotate the axes so that they coincide with the principal axes of the surface.

11.2.4 Numerical Example: Response Surface Analysis

A biochemist conducted an experiment to study the effect of temperature (T_p) and reaction time (T_m) on the yield of a process used to produce an antibiotic. The experiment was run using a randomized block design with three blocks. Trials completed in each of 3 days constituted the blocks. The treatments consisted of the following 3×3 factorial combinations of T_p (°C) and T_m (min), where $T_p = \xi_1$ and $T_m = \xi_2$.

Tmt.	T_p	T_m	Tmt.	T_p	T_m	Tmt.	T_p	T_m
1.	20	20	4.	50	20	7.	80	20
2.	20	50	5.	50	50	8.	80	50
3.	20	80	6.	50	80	9.	80	80

The design variables, x_{1j} for T_p and x_{2j} for T_m, were obtained by the following transformation:

$$x_{ij} = (\xi_{ij} - 50)/30$$

The resulting design points and yields (grams per liter of solution) are given in Table 11.3.

A preliminary analysis of variance of the yield data is presented in Table 11.4.

Since the two treatment factors are quantitative variables we will see if we can fit a response surface using a quadratic polynomial as an approximate equation to describe the surface:

$$\hat{y} = b_0 + b_1 x_1 + b_2 x_2 + b_{11} x_1^2 + b_{12} x_1 x_2 + b_{22} x_2^2$$

The basic design matrix $(\mathbf{X})$ for this experiment is shown in Table 11.5. The pattern in Table 11.5 is repeated three times, once for each block.

The normal equations, $\mathbf{X'Xb} = \mathbf{X'y}$, for this experiment are given in Table 11.6, and Table 11.7 gives the solution, $\mathbf{b} = (\mathbf{X'X})^{-1}\mathbf{X'y}$, of these normal equations.

TABLE 11.3 Design Points and Yields (g/liter solution) for a 3 × 3 Experiment on Antibiotic Production

Treatment	Design points x_1	x_2	Block I	II	III
1	−1	−1	13.66	13.16	15.05
2	−1	0	12.23	12.84	11.85
3	−1	1	7.97	10.19	8.38
4	0	−1	16.98	13.42	14.29
5	0	0	13.80	15.55	13.92
6	0	1	12.86	18.27	15.00
7	1	−1	10.00	10.01	11.10
8	1	0	10.88	12.18	11.13
9	1	1	13.28	10.06	11.96

TABLE 11.4 ANOVA (Preliminary) of Antibiotic Yields

Source	d.f.	SS	MS	F
Total	26	154.1606		
Block	2	.9704	.4852	
Treatment	8	118.3145	14.7893	6.78**
Error	16	34.8757	2.1797	

**Significant at the 1% level.

TABLE 11.5 Basic Design Matrix (**X**)
for the 3 × 3 Experiment on Antibiotic
Production

$$
\mathbf{X} = \begin{bmatrix}
1 & -1 & -1 & 1 & 1 & 1 \\
1 & -1 & 0 & 1 & 0 & 0 \\
1 & -1 & 1 & 1 & -1 & 1 \\
1 & 0 & -1 & 0 & 0 & 1 \\
1 & 0 & 0 & 0 & 0 & 0 \\
1 & 0 & 1 & 0 & 0 & 1 \\
1 & 1 & -1 & 1 & -1 & 1 \\
1 & 1 & 0 & 1 & 0 & 0 \\
1 & 1 & 1 & 1 & 1 & 1
\end{bmatrix}
$$

TABLE 11.6 Normal Equations for Fitting a Quadratic Response
Surface to the Data on Antibiotic Yield

X'Xb = X'y

$$
\begin{bmatrix}
27 & 0 & 0 & 18 & 0 & 18 \\
0 & 18 & 0 & 0 & 0 & 0 \\
0 & 0 & 18 & 0 & 0 & 0 \\
18 & 0 & 0 & 18 & 0 & 12 \\
0 & 0 & 0 & 0 & 12 & 0 \\
18 & 0 & 0 & 12 & 0 & 18
\end{bmatrix}
\begin{bmatrix}
b_0 \\
b_1 \\
b_2 \\
b_{11} \\
b_{12} \\
b_{22}
\end{bmatrix}
=
\begin{bmatrix}
340.02 \\
-4.73 \\
-9.70 \\
205.93 \\
19.52 \\
222.64
\end{bmatrix}
$$

The sum of squares, SSR, for the fitted surface (corrected for the mean) is

$$\text{SSR} = \mathbf{b'X'y} - (\Sigma y)^2/27 = 4394.7113 - 4281.9852 = 112.7261$$

The sum of squares for lack of fit, SSLF, is

$$\text{SSLF} = \text{SST} - \text{SSR} = 118.3145 - 112.7261 = 5.5884$$

The final analysis of variance, with the treatment sum of squares partitioned
into a component associated with the fitted surface and a component associated
with lack of fit, is shown in Table 11.8.

TABLE 11.7 Solution of the Normal Equations in Table 11.6

$(X'X)^{-1}X'y = b$

$$
\begin{bmatrix}
5/27 & 0 & 0 & -1/9 & 0 & -1/9 \\
0 & 1/18 & 0 & 0 & 0 & 0 \\
0 & 0 & 1/18 & 0 & 0 & 0 \\
-1/9 & 0 & 0 & 1/6 & 0 & 0 \\
0 & 0 & 0 & 0 & 1/12 & 0 \\
-1/9 & 0 & 0 & 0 & 0 & 1/6
\end{bmatrix}
\begin{bmatrix}
340.02 \\
-4.73 \\
-9.70 \\
205.93 \\
19.52 \\
222.64
\end{bmatrix}
=
\begin{bmatrix}
15.3478 \\
-.2628 \\
-.5389 \\
-3.4583 \\
1.6267 \\
-.6733
\end{bmatrix}
$$

TABLE 11.8 Final ANOVA of Yields (g/liter solution) of Antibiotic

Source	d.f.	SS	MS	F
Total	26	154.1606		
Block	2	.9704	.4852	
Fitted surface	5	112.7261	22.5452	10.34**
Lack of fit	3	5.5884	1.8628	.85 NS
Error	16	34.8757	2.1797	

**Significant at the 1% level.
NS, not significant.

On considering the ANOVA in Table 11.8 it is seen that the fitted surface accounts for a significant amount of the variation in antibiotic yield, while the lack of fit is nonsignificant. The proportion of the total variation in yield which is accounted for by the fitted surface is

$$R^2 = \text{SSR/SSTot} = 112.7261/154.1606 = .73$$

The equation which describes the surface in terms of the design variables is

$$
\hat{y} = 15.35 - .26x_1 - .54x_2 - 3.46x_1^2 + 1.63x_1x_2 - .67x_2^2
$$

$$
= 15.35 + [x_1\ x_2]
\begin{bmatrix} -.26 \\ -.54 \end{bmatrix}
+ [x_1\ x_2]
\begin{bmatrix} -3.46 & .81 \\ .81 & -.67 \end{bmatrix}
\begin{bmatrix} x_1 \\ x_2 \end{bmatrix}
$$

$$
= b_0 + x'b + x'Bx
$$

To examine the surface further we can find the coordinates, x_0, of the stationary point and the yield, y_0, at this point. We have

$$\mathbf{x}_0 = (-\mathbf{B}^{-1}\,\mathbf{b})(1/2)$$

$$= -\begin{bmatrix} -.4040 & -.4880 \\ -.4880 & -2.0747 \end{bmatrix}\begin{bmatrix} -.26 \\ -.54 \end{bmatrix}(1/2) = \begin{bmatrix} -.18 \\ -.625 \end{bmatrix}$$

and

$$y_0 = b_0 + (\mathbf{x}_0'\,\mathbf{b})(1/2)$$

$$= 15.35 + [-.18 \ -.625]\begin{bmatrix} -.26 \\ -.54 \end{bmatrix}(1/2) = 15.54$$

Since both b_{11} and b_{22} are negative the stationary point is a maximum.

Report of statistical analysis.   An experiment was run to examine the effect of temperature and reaction time on the yield of an antibiotic produced by a new process. A quadratic equation was then fitted to the resulting data in an attempt to describe the yield surface as a function of temperature (T_p) and reaction time (T_m). The fitted equation was

$$\hat{y} = 9.7239 + 2852T_p - .0335T_m - .0038T_p^2 + .0018(T_p)(T_m) - .0007T_m^2$$

where

$\hat{y}$ = expected antibiotic yield (g/liter solution)

T_p = temperature (°C)

T_m = reaction time (min)

This equation provides an adequate description of the yield response surface over the range of temperature and reaction time used in this study. About 73% (R^2 = .7312) of the variation in yield is accounted for by the fitted equation.

Further analysis of the fitted surface reveals that a maximum yield of about 15.5 g/liter solution would be obtained at a temperature of 44.6°C and a reaction time of 31.2 min. The nature of the surface can be seen by examining the diagram presented in Fig. 11.3.

11.3 DESIGN SELECTION

In selecting an experimental design for response surface experiments we want to be able to:

1. Estimate the coefficients of the equation of the surface
2. Assess the reliability of the estimates
3. Minimize variance
4. Minimize bias
5. Measure lack of fit

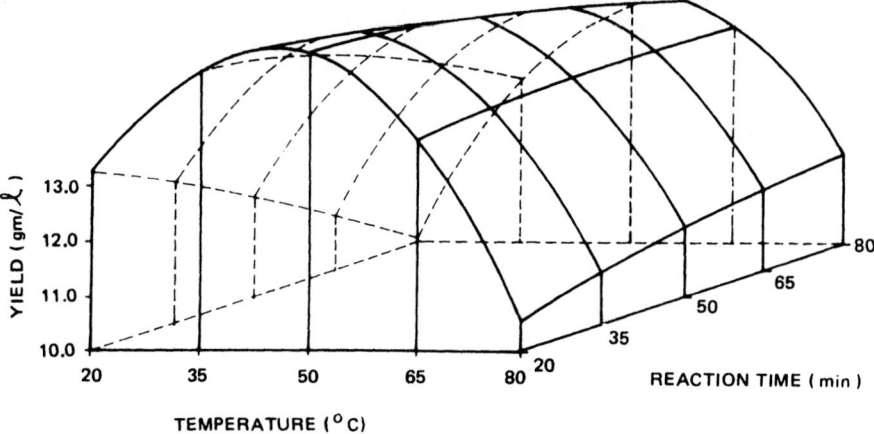

Fig. 11.3 Yield of antibiotic at different temperatures and reaction times.

   6.   Add a minimum number of points, if necessary, to estimate higher order coefficients

The nature of the design will depend, of course, on the complexity of the underlying surface. We will begin with designs for the least complex surface and work our way up.

11.3.1 First-Order Surfaces

The simplest kind of surface which could describe the response, y, as a function of k factors (input variables) is a first-order (linear) surface. This surface is a hyperplane in a space of $k + 1$ dimensions. The equation for this type of surface is

$$y = b_0 + \sum_{i=1}^{k} b_i x_i = \mathbf{x'b} = [1 \; x_1 \; x_2 \; \ldots \; x_k] \begin{bmatrix} b_0 \\ b_1 \\ \vdots \\ b_k \end{bmatrix}$$

Suppose we want to use $n > (k + 1)$ points to find the sample values, $\mathbf{b}$, of the linear coefficients in this equation. In general, the equation for this set of data will be

$$\mathbf{y} = \mathbf{Xb} + \mathbf{e}$$

in which the $n \times (k + 1)$ matrix $\mathbf{X}$ will take the form

$$\mathbf{X} = [\mathbf{1}\ \mathbf{D}]$$

where

$$\mathbf{1} = \begin{bmatrix} 1 \\ 1 \\ \vdots \\ 1 \end{bmatrix} \quad \text{and} \quad \mathbf{D} = \begin{bmatrix} x_{11} & x_{21} & \cdots & x_{k1} \\ x_{12} & x_{22} & \cdots & x_{k2} \\ \vdots & \vdots & & \vdots \\ x_{1n} & x_{2n} & \cdots & x_{kn} \end{bmatrix}$$

The rows of $\mathbf{D}$ are the coordinates of the n experimental points in the k-dimensional factor space. $\mathbf{D}$ is called the design matrix.

If ξ_{ij} is the level of input for the ith factor at the jth experimental point we will choose the design levels, x_{ij}, to meet the following restrictions:

$$\sum_{j=1}^{n} x_{ij}^2 = n$$
$$\qquad\qquad i = 1, 2, \ldots, k$$
$$\sum_{j=1}^{n} x_{ij} = 0$$

We meet these restrictions as follows: if ξ_{ij} is the jth level of factor i, and $\bar{\xi}_i$ is the mean of the ith variable, then

$$x_{ij} = (\xi_{ij} - \bar{\xi}_i)/S_i$$

where

$$S_i^2 = [\sum_{j=1}^{n} (\xi_{ij} - \bar{\xi}_i)^2]/n$$

This procedure results in an orthogonal design under the proper choice of the n design points. An orthogonal design has the property that

$$\mathbf{X}'\mathbf{X} = \begin{bmatrix} n & 0 & \cdots & 0 \\ 0 & n & \cdots & 0 \\ \vdots & \vdots & \ddots & \vdots \\ 0 & & & n \end{bmatrix} = n\mathbf{I}$$

These orthogonal designs have the important property that they are minimum variance designs for estimating the coefficients, b_i, of a first-order equation.

There are a number of classes of orthogonal designs for fitting first-order surfaces. The most often used are the 2^k factorials and fractional factorials.

Suppose we consider, for example, a problem which involves three input variables. We might use a 2^3 factorial combination of treatments to estimate the coefficients of the model

$$E(y) = \beta_0 + \beta_1 x_1 + \beta_2 x_2 + \beta_3 x_3$$

With one replication of this design the data equation $\mathbf{y} = \mathbf{Xb} + \mathbf{e}$ takes the form shown in Table 11.9.

Note that with this design $n = 8$, and

$$\mathbf{X'X} = \begin{bmatrix} 8 & 0 & 0 & 0 \\ 0 & 8 & 0 & 0 \\ 0 & 0 & 8 & 0 \\ 0 & 0 & 0 & 8 \end{bmatrix}, \quad (\mathbf{X'X})^{-1} = \begin{bmatrix} 1/8 & 0 & 0 & 0 \\ 0 & 1/8 & 0 & 0 \\ 0 & 0 & 1/8 & 0 \\ 0 & 0 & 0 & 1/8 \end{bmatrix}$$

Hence $\mathbf{b} = (\mathbf{X'X})^{-1}\mathbf{X'y} = (1/8)\mathbf{X'y}$, and the variance of any b_i, $V(b_i)$, is $V(b_i) = \sigma^2/8$.

11.4 BIAS

Suppose that the true model for a surface is of higher order than the approximating equation used to fit the data. The coefficients of the fitted equation will then be biased because the higher order coefficients have not been included. In this case we might be interested in determining the biases of the estimated coefficients.

TABLE 11.9 Data Equation for a 2^3 Factorial Set of Treatments

$$\mathbf{y} = \mathbf{XB} + \mathbf{e}$$

$$\begin{bmatrix} (1) \\ a \\ b \\ ab \\ c \\ ac \\ bc \\ abc \end{bmatrix} = \begin{bmatrix} 1 & -1 & -1 & -1 \\ 1 & 1 & -1 & -1 \\ 1 & -1 & 1 & -1 \\ 1 & 1 & 1 & -1 \\ 1 & -1 & -1 & 1 \\ 1 & 1 & -1 & 1 \\ 1 & -1 & 1 & 1 \\ 1 & 1 & 1 & 1 \end{bmatrix} \begin{bmatrix} b_0 \\ b_1 \\ b_2 \\ b_3 \end{bmatrix} + \begin{bmatrix} e_1 \\ e_2 \\ e_3 \\ e_4 \\ e_5 \\ e_6 \\ e_7 \\ e_8 \end{bmatrix}$$

$$\mathbf{y} = [\mathbf{1} \quad \mathbf{D}]\mathbf{b} + \mathbf{e}$$

In general we postulate the model

$$\mathbf{y} = \mathbf{X}_1\boldsymbol{\beta}_1 + \boldsymbol{\epsilon}$$

where

> $\mathbf{y}$ has dimension $n \times 1$, $\mathbf{X}_1$ has dimension $n \times p_1$, and $\boldsymbol{\beta}_1$ has dimension $p_1 \times 1$

Suppose that the true response function involves these p_1 parameters plus an additional p_2 parameter giving a true model of the form

$$\mathbf{y} = \mathbf{X}_1\boldsymbol{\beta}_1 + \mathbf{X}_2\boldsymbol{\beta}_2 + \boldsymbol{\epsilon}$$

in which $\mathbf{X}_2$ is $n \times p_2$ and $\boldsymbol{\beta}_2$ is $p_2 \times 1$. We fit the equation

$$\mathbf{y} = \mathbf{x}_1'\mathbf{b}_1$$

and our goal is to find the biases of

$$\mathbf{b}_1 = (\mathbf{X}_1'\mathbf{X}_1)^{-1}\mathbf{X}_1'\mathbf{y}$$

in the presence of the true model.

If we find the expected value of $\mathbf{b}$ under the conditions of the true model we have

$$E(\mathbf{b}_1) = E\{(\mathbf{X}_1'\mathbf{X}_1)^{-1}\mathbf{X}_1'[\mathbf{X}_1\boldsymbol{\beta}_1 + \mathbf{X}_2\boldsymbol{\beta}_2 + \boldsymbol{\epsilon}]\}$$

$$= [(\mathbf{X}_1'\mathbf{X}_1)^{-1}(\mathbf{X}_1'\mathbf{X}_1)\boldsymbol{\beta}_1 + (\mathbf{X}_1'\mathbf{X}_1)^{-1}(\mathbf{X}_1'\mathbf{X}_2)\boldsymbol{\beta}_2 + (\mathbf{X}_1'\mathbf{X}_1)^{-1}\mathbf{X}_1'\boldsymbol{\epsilon}]$$

Since

$$E(\boldsymbol{\epsilon}) = \mathbf{0}$$

then

$$E(\mathbf{b}_1) = \boldsymbol{\beta}_1 + (\mathbf{X}_1'\mathbf{X}_1)_1^{-1}(\mathbf{X}_1'\mathbf{X}_2)\boldsymbol{\beta}_2$$

$$= \boldsymbol{\beta}_1 + \mathbf{A}\boldsymbol{\beta}_2$$

in which $\mathbf{A} = (\mathbf{X}_1'\mathbf{X}_1)^{-1}(\mathbf{X}_1'\mathbf{X}_2)$ is called the alias matrix. Each coefficient in b_1 has as its bias a linear function of the parameters in $\boldsymbol{\beta}_2$. This linear function is determined by the alias matrix, $\mathbf{A}$.

To illustrate, suppose we use a 1/2 replication of a 2^3 factorial with *ABC* as the defining contrast. We will postulate that the surface is described by the first-order model

$$E(y) = \beta_0 + \beta_1 x_1 + \beta_2 x_2 + \beta_3 x_3$$

The data equation in this instance would be

$$\mathbf{y} = \mathbf{X}_1\mathbf{b}_1 + \mathbf{e}$$

which is

$$
\overset{\mathbf{y}}{\begin{bmatrix} (1) \\ ab \\ ac \\ bc \end{bmatrix}} = \overset{\mathbf{X}_1}{\begin{bmatrix} 1 & -1 & -1 & -1 \\ 1 & 1 & 1 & -1 \\ 1 & 1 & -1 & 1 \\ 1 & -1 & 1 & 1 \end{bmatrix}} \overset{\mathbf{b}_1}{\begin{bmatrix} b_0 \\ b_1 \\ b_2 \\ b_3 \end{bmatrix}} = \mathbf{e}
$$

Note: b_1, b_2, b_3 estimate the main effects of A, B, and C.

Now, assume that the true model contains all of the first-order interaction terms b_{12}, b_{13}, b_{23}, as well as the main effects. Then the true model would be

$$E(y) = \beta_0 + \beta_1 x_1 + \beta_2 x_2 + \beta_3 x_3 + \beta_{12} x_1 x_2 + \beta_{13} x_1 x_3 + \beta_{23} x_2 x_3$$

For this data set we have

$$E(\mathbf{y}) = \mathbf{X}_1\boldsymbol{\beta}_1 + \mathbf{X}_2\boldsymbol{\beta}_2$$

or

$$
E\begin{bmatrix} (1) \\ ab \\ ac \\ bc \end{bmatrix} = \begin{bmatrix} 1 & -1 & -1 & -1 \\ 1 & 1 & 1 & -1 \\ 1 & 1 & -1 & 1 \\ 1 & -1 & 1 & 1 \end{bmatrix} \begin{bmatrix} \beta_0 \\ \beta_1 \\ \beta_2 \\ \beta_3 \end{bmatrix} + \begin{bmatrix} 1 & 1 & 1 \\ 1 & -1 & -1 \\ -1 & 1 & -1 \\ -1 & -1 & 1 \end{bmatrix} \begin{bmatrix} \beta_{12} \\ \beta_{13} \\ \beta_{23} \end{bmatrix}
$$

Since the design is orthogonal

$$\mathbf{X}_1'\mathbf{X}_1 = 4\mathbf{I} \qquad \text{and} \qquad (\mathbf{X}_1'\mathbf{X}_1)^{-1} = (1/4)\mathbf{I}$$

then

$$
\mathbf{X}_1'\mathbf{X}_2 = \begin{bmatrix} 1 & 1 & 1 & 1 \\ -1 & 1 & 1 & -1 \\ -1 & 1 & -1 & 1 \\ -1 & -1 & 1 & 1 \end{bmatrix} \begin{bmatrix} 1 & 1 & 1 \\ 1 & -1 & -1 \\ -1 & 1 & -1 \\ -1 & -1 & 1 \end{bmatrix} = \begin{bmatrix} 0 & 0 & 0 \\ 0 & 0 & -4 \\ 0 & -4 & 0 \\ -4 & 0 & 0 \end{bmatrix}
$$

and

$$\mathbf{A} = (\mathbf{X}_1'\mathbf{X}_1)^{-1}(\mathbf{X}_1'\mathbf{X}_2)$$

$$= (1/4)\mathbf{I} \begin{bmatrix} 0 & 0 & 0 \\ 0 & 0 & -4 \\ 0 & -4 & 0 \\ -4 & 0 & 0 \end{bmatrix} = \begin{bmatrix} 0 & 0 & 0 \\ 0 & 0 & -1 \\ 0 & -1 & 0 \\ -1 & 0 & 0 \end{bmatrix} = \mathbf{A}$$

Hence

$$E(\mathbf{b}_1) = \boldsymbol{\beta}_0 + \mathbf{A}\boldsymbol{\beta}_2$$

$$= \begin{bmatrix} \beta_1 \\ \beta_1 \\ \beta_2 \\ \beta_3 \end{bmatrix} + \begin{bmatrix} 0 & 0 & 0 \\ 0 & 0 & -1 \\ 0 & -1 & 0 \\ -1 & 0 & 0 \end{bmatrix} \begin{bmatrix} \beta_{12} \\ \beta_{13} \\ \beta_{23} \end{bmatrix} = \begin{bmatrix} \beta_0 \\ \beta_1 - \beta_{23} \\ \beta_2 - \beta_{13} \\ \beta_3 - \beta_{12} \end{bmatrix}$$

Thus, b_0 is an unbiased estimator for β_0, but the linear coefficients (main effects) have first-order interactions as biases.

As a further illustration consider the use of a full replication of a 2^2 factorial experiment to fit the first-order model

$$E(y) = \beta_0 + \beta_1 x_1 + \beta_2 x_2$$

when the true surface is a quadratic described by the model

$$E(y) = \beta_0 + \beta_1 x_1 + \beta_2 x_2 + \beta_{11} x_1^2 + \beta_{12} x_1 x_2 + \beta_{22} x_2^2$$

The X matrices needed to compute the biases are

$$\begin{matrix} & x_1 \quad x_2 \\ \mathbf{X}_1 = & \begin{bmatrix} 1 & -1 & -1 \\ 1 & 1 & -1 \\ 1 & -1 & 1 \\ 1 & 1 & 1 \end{bmatrix} \end{matrix}, \qquad \begin{matrix} x_1^2 \quad x_1 x_2 \quad x_2^2 \\ \mathbf{X}_2 = \begin{bmatrix} 1 & 1 & 1 \\ 1 & -1 & 1 \\ 1 & -1 & 1 \\ 1 & 1 & 1 \end{bmatrix} \end{matrix}$$

Then, since the design is orthogonal

$$\mathbf{X}_1'\mathbf{X}_1 = 4\mathbf{I} \qquad \text{and} \qquad (\mathbf{X}_1'\mathbf{X}_1)^{-1} = (1/4)\mathbf{I}$$

also

$$\mathbf{X}_1'\mathbf{X}_2 = \begin{bmatrix} 4 & 0 & 4 \\ 0 & 0 & 0 \\ 0 & 0 & 0 \end{bmatrix}, \quad \text{and} \quad \mathbf{A} = (\mathbf{X}_1'\mathbf{X}_1)^{-1}(\mathbf{X}_1'\mathbf{X}_2) = \begin{bmatrix} 1 & 0 & 1 \\ 0 & 0 & 0 \\ 0 & 0 & 0 \end{bmatrix}$$

Hence

$$E(\mathbf{b}_1) = \boldsymbol{\beta}_1 + \mathbf{A}\boldsymbol{\beta}_2$$

$$E\begin{bmatrix} b_0 \\ b_1 \\ b_2 \end{bmatrix} = \begin{bmatrix} \beta_0 \\ \beta_1 \\ \beta_2 \end{bmatrix} + \begin{bmatrix} 1 & 0 & 1 \\ 0 & 0 & 0 \\ 0 & 0 & 0 \end{bmatrix}\begin{bmatrix} \beta_{11} \\ \beta_{12} \\ \beta_{22} \end{bmatrix} = \begin{bmatrix} \beta_0 + \beta_{11} + \beta_{22} \\ \beta_1 \\ \beta_2 \end{bmatrix}$$

We see that b_0 is biased by $\beta_{11} + \beta_{22}$ while b_1 and b_2 are unbiased.

In general, if a full 2^k factorial is used to fit a first-order model in the presence of a second-order system b_0 is biased by $\sum_{i=1}^{k} \beta_{ii}$ while $b_1, b_2, \ldots, b_k$ remain unbiased. A first-order model is usually used because the experimenter believes that the true surface is nearly planar in the region of interest. It is well to consider, however, the possibility of bias from high-order terms if they exist. For this reason the full replication of a 2^k factorial set of treatments should be used if at all possible.

11.5 AUGMENTED 2^k FACTORIALS

Although there are many advantages to the 2^k factorial experiments they also have some disadvantages for use as response surface designs. Unless they are replicated, for example, they provide no measure of error variance. Consider a single replicaton of a 2^3 factorial experiment. The analysis of variance from fitting a first-order model to the eight observations is shown in Table 11.10.

The remainder contains not only pure error but also a contribution from the failure of the first-order equation to describe the true surface (lack of fit). Estimates of the cross-product terms can be obtained but no measure of the quadratic terms is available with factors at only two levels.

If we replicate the 2^k design we can obtain an estimate of pure error. However, complete replication is not always desirable because of the large number of experimental units involved. Further, even with replication the quadratic terms cannot be estimated because the factors are at only two levels.

TABLE 11.10 ANOVA of First-Order Response Surface on a 2^3 Factorial

Source	d.f.
Total	8
β_0	1
β_1	1
β_2	1
β_3	1
Remainder	4

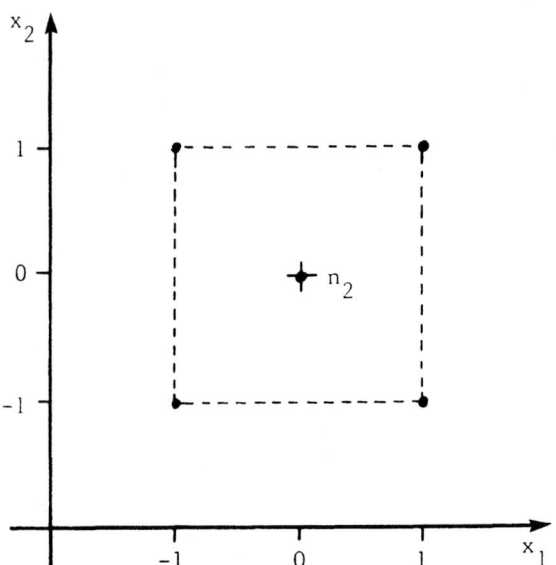

Fig. 11.4 Configuration of design points in an augmented 2^2 design.

11.5.1 Design Characteristics

Suppose we start with a 2^k design containing $2^k = n_1$ points, and suppose we add n_2 points at the center of the design, at the point whose coordinates in the $x_1, x_2, \ldots, x_k$ factor space are 0, 0, . . , 0. For a two-factor design the configuration of points in the factor space would be as shown in Fig. 11.4.

The result is a design which gives

1. $n_2 - 1$ degrees of freedom for estimating pure error
2. One degree of freedom for measuring the existence of pure quadratic terms

If we call the mean of the n_1 observations in the factorial part of the design $\bar{y}_1$, and the mean of the n_2 observations at the center of the design $\bar{y}_2$, then

$$E(\bar{y}_1 - \bar{y}_2) = \sum_{i=1}^{k} \beta_{ii}$$

A single-degree-of-freedom sums of squares, S_Q^2, for testing

$$H_0: \sum_{i=1}^{k} \beta_{ii} = 0$$

against

$$H_a: \sum_{i=1}^{k} \beta_{ii} \neq 0$$

is given by

$$S_Q^2 = \frac{n_1 n_2 (\bar{y}_1 - \bar{y}_2)^2}{n_1 + n_2}$$

If $y_{01}, y_{02}, \ldots, y_{0n}$ represent the yields of the center points, then the pure error sum of squares, S_{PE}^2, is computed as

$$S_{PE}^2 = \sum_{j=1}^{n_2} y_{0j}^2 - (\sum_{j=1}^{n_2} y_{0j})^2 / n_2$$

This sum of squares has $n_2 - 1$ degrees of freedom.

11.5.2 Analysis

To illustrate the computations involved in the analysis of data from an augmented 2^k factorial experiment consider a 2^3 design with four center points. The data equation is

$$\mathbf{y} = \mathbf{Xb} + \mathbf{e}$$

Table 11.11 gives the details of the matrices and vectors in this data equation. The analysis of variance is shown in Table 11.12. The sums of squares in Table 11.12 are computed as

$$\text{SSTotU} = \mathbf{y}'\mathbf{y} = \sum_{j=1}^{12} y_j^2$$

$$S_0^2 = (\sum_{j=1}^{12} y_j)^2 / 12$$

Then the sums of squares, S_1^2, S_2^2, S_3^2, for the linear coefficients are computed according to the contrasts in Table 11.13.

TABLE 11.11 Data Equation for a 2^3 Factorial Design Augmented with Four Center Points

$$
\begin{bmatrix}
(1) \\
a \\
b \\
ab \\
c \\
ac \\
bc \\
abc \\
y_{01} \\
y_{02} \\
y_{03} \\
y_{04}
\end{bmatrix}
=
\begin{bmatrix}
1 & -1 & -1 & -1 \\
1 & 1 & -1 & -1 \\
1 & -1 & 1 & -1 \\
1 & 1 & 1 & -1 \\
1 & -1 & -1 & 1 \\
1 & 1 & -1 & 1 \\
1 & -1 & 1 & 1 \\
1 & 1 & 1 & 1 \\
1 & 0 & 0 & 0 \\
1 & 0 & 0 & 0 \\
1 & 0 & 0 & 0 \\
1 & 0 & 0 & 0
\end{bmatrix}
\begin{bmatrix}
b_0 \\
b_1 \\
b_2 \\
b_3
\end{bmatrix}
+ \mathbf{e}
$$

TABLE 11.12 ANOVA of Augmented 2^3 Factorial

Source	d.f.	SS
Total (uncorr.)	12	SSTotU
β_0	1	S_0^2
β_1	1	S_1^2
β_2	1	S_2^2
β_3	1	S_3^2
Remainder	8	
Cross products	4	S_{CP}^2
Quadratic	1	S_Q^2
Pure error	3	S_{PE}^2

TABLE 11.13 Contrasts for the Linear Coefficients

Effect	(1)	a	b	ab	c	ac	bc	abc	L_i
				Treatment					
β_1	$-$	$+$	$-$	$+$	$-$	$+$	$-$	$+$	L_1
β_2	$-$	$-$	$+$	$+$	$-$	$-$	$+$	$+$	L_2
β_3	$-$	$-$	$-$	$-$	$+$	$+$	$+$	$+$	L_3

Note: $L_i = \sum_{j=1}^{8} k_{ij} y_j$

$S_i^2 = L_i^2/8$

Pure quadratic:

$$S_Q^2 = \frac{n_1 n_2 (\bar{y}_1 - \bar{y}_2)^2}{n_1 + n_2} = \frac{32(\bar{y}_1 - \bar{y}_2)^2}{12}$$

Pure error:

$$S_{PE}^2 = (y_{01}^2 + y_{02}^2 + y_{03}^2 + y_{04}^2) - (y_{01} + y_{02} + y_{03} + y_{04})^2/4$$

Cross products:

$$S_{CP}^2 = \text{SSTotU} - S_0^2 - S_1^2 - S_2^2 - S_3^2 - S_Q^2 - S_{PE}^2$$

Significance tests are the usual F tests with $\text{MSE} = S_{PE}^2/3$ as the denominator. The fitted first-order equation is

$$\hat{y} = b_0 + b_1 x_1 + b_2 x_2 + b_3 x_3$$

where

$$b_0 = (\sum_{j=1}^{12} y_j)/12$$

$$b_i = L_i/8$$

11.5.3 Augmented Two-Level Factorial: Example

A chemical company wanted to examine the effect of three factors, temperature, concentration, and pressure, on the yield of a process under pilot plant conditions. Because the plant was in operation the number of experimental runs had to be limited. The project manager believed that the response over the range of conditions being considered was described by a planar surface, but he was not sure and he wanted to check this point out. The company statistician proposed a 2^3

factorial combination of the three factors, augmented by four points at the center of the factor space. The chosen levels of the three factors were:

x_{ij}	Temperature (a)	Concentration (b)	Pressure (c)
-1	160°C	20%	1.0 atm
0	170	30	1.5
1	180	40	2.0

The treatment combinations and yields are presented in Table 11.14. The analysis of variance of the data is given in Table 11.15. The computations for this experiment are:

$$\bar{y}_1 = (60 + 72 + \cdots + 80)/8 = 64.25$$

$$\bar{y}_2 = (65 + 59 + 68 + 60)/4 = 63.00$$

$$\bar{\bar{y}} = (60 + 72 + \cdots + 60)/12 = 63.83$$

$$\text{SSTotU} = 60^2 + 72^2 + \cdots + 60^2 = 50{,}272.00$$

$$S_0^2 = (60 + 72 + \cdots 60)^2/12 = 48{,}896.33$$

The coefficient contrasts and their sums of squares are computed in Table 11.16.

$$S_Q^2 = [n_1 n_2 (\bar{y}_1 - \bar{y}_2)^2]/(n_1 + n_2) = [(8)(4)(64.25 - 63.00)^2]/(8 + 4)$$
$$= 4.17$$

TABLE 11.14 Treatment Combinations and Yields in an Augmented Factorial Experiment Run Under Pilot Plan Conditions

Treatment	Design points			Yield
	x_{1j}	x_{2j}	x_{3j}	
(1)	-1	-1	-1	60
a	1	-1	-1	72
b	-1	1	-1	54
ab	1	1	-1	68
c	-1	-1	1	52
ac	1	-1	1	83
bc	-1	1	1	45
abc	1	1	1	80
y_{01}	0	0	0	65
y_{02}	0	0	0	59
y_{03}	0	0	0	68
y_{04}	0	0	0	60

TABLE 11.15 ANOVA of Yields from Pilot Plant Experiment

Source	d.f.	SS	MS	F
Total	12	50,272.00		
β_0	1	48,896.33		
β_1	1	1,058.00	1,058.00	58.78**
β_2	1	50.00	50.00	2.78
β_3	1	4.50	4.50	.25
Cross products	4	205.00	51.25	2.85
Pure quadratic	1	4.17	4.17	
Pure error	3	54.00	18.00	

**Significant at the 1% level.

TABLE 11.16 Regression Coefficient Contrasts and Their Sums of Squares

Yield	60	72	54	68	52	83	45	80	
Effect	(1)	a	b	ab	c	ac	bc	abc	L_i
β_1	−	+	−	+	−	+	−	+	92
β_2	−	−	+	+	−	−	+	+	−20
β_3	−	−	−	−	+	+	+	+	6

$S_1^2 = (92)^2/8 = 1,058.00$

$S_2^2 = (-20)^2/8 = 50.00$

$S_3^2 = (6)^2/8 = 4.50$

$$S_{PE}^2 = (65^2 + 59^2 + 68^2 + 60^2) - (1/4)(65 + 59 + 68 + 60)^2 = 54.00$$

$$S_{CP}^2 = 50,272.00 - 48,896.33 - 1,058.00 - 50.00 - 4.50 - 4.17 - 54.00 = 205.00$$

Coefficients of the fitted surface:

$$b_0 = \bar{\bar{y}} = 63.83$$

$$b_1 = L_1/8 = 92/8 = 11.50$$

$$b_2 = L_2/8 = -20/8 = -2.50$$

$$b_3 = L_3/8 = 6/8 = .75$$

$$\hat{y} = b_0 + b_1x_1 + b_2x_2 + b_3x_3 = 63.83 + 11.50x_1 - 2.50x_2 + .75x_3$$

Report of statistical analysis.   Examination of data from an experiment to measure the effect of temperature, concentration, and pressure on the yield of a chemical process under pilot plant conditions revealed that the only significant factor was temperature. Over the range studied the yield is described by the equation

$$\hat{y} = 1.15T - .25C + 1.50P - 126.42$$

where

$$\hat{y} = \text{predicted yield}$$
$$T = \text{temperature (°C)}$$
$$C = \text{concentration (\%)}$$
$$P = \text{pressure (atm)}$$

There is no indication that a more complex equation is required over the range of variables studied here.

11.6   SECOND-ORDER SURFACES

11.6.1   Introduction

A second-order surface for k input variables is described by the quadratic model

$$E(y) = \beta_0 + \sum_{i=1}^{k} \beta_i x_i + \sum_{i=1}^{k} \beta_{ii} x_i^2 + \sum_{i<i'} \beta_{ii'} x_i x_{i'}$$

in which

$$\beta_0 = \text{the intercept}$$
$$\beta_i = \text{the linear coefficient for the } i\text{th input}$$
$$\beta_{ii} = \text{the quadratic coefficient for the } i\text{th input}$$
$$\beta_{ii'} = \text{the coefficient for the cross-product of the } i\text{th and } i'\text{th inputs}$$
$$x_i = \text{the level of the } i\text{th input}$$

For example, with two inputs $k = 2$ and

$$E(y) = \beta_0 + \beta_1 x_1 + \beta_2 x_2 + \beta_{11} x_1^2 + \beta_{22} x_2^2 + \beta_{12} x_1 x_2$$

Experimental designs for obtaining the data to estimate the coefficients of a second-order model must have at least three levels for each variable. With fewer than three levels the quadratic coefficients cannot be estimated. This suggests that perhaps a 3^k factorial design might be useful here.

11.6.2 Illustration

Suppose we have a problem involving $k = 2$ input variables. A single replication of a 3^2 design, under the coding procedure we have adopted, would have the design matrix, **D**, given in Table 11.17. A graph of the design in the two-dimensional space of factors x_1 and x_2 is shown in Fig. 11.5.

Ordinarily we would fit the equation

$$y = b_0 + b_1 x_1 + b_2 x_2 + b_{11} x_1^2 + b_{22} x_2^2 + b_{12} x_1 x_2$$

For this regression equation the data equation, in matrix form, for the 3^2 factorial set of treatments is ($\mathbf{y} = \mathbf{Xb} + \mathbf{e}$), which is presented in Table 11.18. With the set of data equations in Table 11.18 the normal equations would be $\mathbf{X'Xb} = \mathbf{X'y}$, as detailed in Table 11.19. With the set of normal equations in Table 11.19 the sample coefficients are computed as

$$b_0 = (1/3)\mathbf{x}_0'\mathbf{y} - (1/3)\mathbf{x}_{11}'\mathbf{y} - (1/3)\mathbf{x}_{22}'\mathbf{y}$$

$$b_1 = (1/6)\mathbf{x}_1'\mathbf{y}$$

TABLE 11.17 Design Matrix for One Replication of a 3^2 Design

$$\mathbf{D} = \begin{bmatrix} -1 & -1 \\ 0 & -1 \\ 1 & -1 \\ -1 & 0 \\ 0 & 0 \\ 1 & 0 \\ -1 & 1 \\ 0 & 1 \\ 1 & 1 \end{bmatrix}$$

$\qquad\qquad x_1 \qquad x_2$

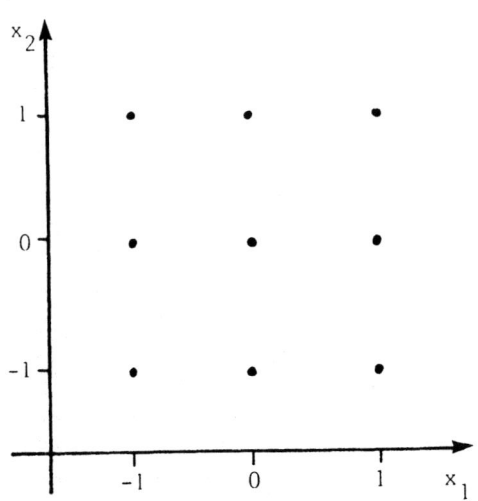

Fig. 11.5   Graph of the design points for a 3^2 design.

TABLE 11.18   Data Equation for a 3^2 Design

$$
\begin{bmatrix}
y_{00} \\
y_{10} \\
y_{20} \\
y_{01} \\
y_{11} \\
y_{21} \\
y_{02} \\
y_{12} \\
y_{22}
\end{bmatrix}
=
\begin{bmatrix}
1 & -1 & -1 & 1 & 1 & 1 \\
1 & 0 & -1 & 0 & 1 & 0 \\
1 & 1 & -1 & 1 & 1 & -1 \\
1 & -1 & 0 & 1 & 0 & 0 \\
1 & 0 & 0 & 0 & 0 & 0 \\
1 & 1 & 0 & 1 & 0 & 0 \\
1 & -1 & 1 & 1 & 1 & -1 \\
1 & 0 & 1 & 0 & 1 & 0 \\
1 & 1 & 1 & 1 & 1 & 1
\end{bmatrix}
\begin{bmatrix}
b_0 \\
b_1 \\
b_2 \\
b_{11} \\
b_{22} \\
b_{12}
\end{bmatrix}
+ \mathbf{e}
$$

$\quad\quad$ **y** $\quad\quad\quad\quad\quad$ **X** $\quad\quad\quad\quad\quad\quad\quad\quad$ **b**

TABLE 11.19 Normal Equations for a 3^2 Design

$$
\begin{bmatrix}
9 & 0 & 0 & 6 & 6 & 0 \\
0 & 6 & 0 & 0 & 0 & 0 \\
0 & 0 & 6 & 0 & 0 & 0 \\
6 & 0 & 0 & 6 & 4 & 0 \\
6 & 0 & 0 & 4 & 6 & 0 \\
0 & 0 & 0 & 0 & 0 & 4
\end{bmatrix}
\begin{bmatrix}
b_0 \\
b_1 \\
b_2 \\
b_{11} \\
b_{22} \\
b_{12}
\end{bmatrix}
= \mathbf{X'y}
$$

TABLE 11.20 Normal Equations for a 3^2 Model with Transformed x_1^2, x_2^2

$$
\begin{bmatrix}
9 & 0 & 0 & 0 & 0 & 0 \\
0 & 6 & 0 & 0 & 0 & 0 \\
0 & 0 & 6 & 0 & 0 & 0 \\
0 & 0 & 0 & 2 & 0 & 0 \\
0 & 0 & 0 & 0 & 2 & 0 \\
0 & 0 & 0 & 0 & 0 & 4
\end{bmatrix}
\begin{bmatrix}
b_0^* \\
b_1 \\
b_2 \\
b_{11} \\
b_{22} \\
b_{12}
\end{bmatrix}
= \mathbf{X'y}
$$

$$b_2 = (1/6)\mathbf{x}_2'\mathbf{y}$$

$$b_{11} = (-1/3)\mathbf{x}_0'\mathbf{y} + (1/2)\mathbf{x}_{11}'\mathbf{y}$$

$$b_{22} = (-1/3)\mathbf{x}_0'\mathbf{y} + (1/2)\mathbf{x}_{22}'\mathbf{y}$$

$$b_{12} = (1/4)\mathbf{x}_{12}'\mathbf{y}$$

We can simplify the computations, however, by transforming x_1^2 to $(x_1^2 - \overline{x_1^2})$ and x_2^2 to $(x_2^2 - \overline{x_2^2})$, where $\overline{x_1^2}$ and $\overline{x_2^2}$ are means of the squares of x_1 and x_2. The equation to be fitted is now written

$$y = b_0^* + b_1 x_1 + b_2 x_2 + b_{11}(x_1^2 - \overline{x_1^2}) + b_{22}(x_2^2 - \overline{x_2^2}) + b_{12} x_1 x_2$$

The normal equations are now as given in Table 11.20.

The transformation results in an orthogonal (diagonal) $\mathbf{X'X}$ matrix, and a new intercept is defined which is orthogonal to b_{11} and b_{22}. The only coefficient affected by this transformation is b_0^*.

The use of a 3^k factorial design is limited to situations where the number of factors is small, say two or three. With this design we are using 3^k points (observations) to fit an equation with $1 + 2k + C_2^k$ coefficients. With five factors, for example, we are using 125 observations to fit an equation with only 21 coefficients. This is a highly inefficient use of experimental resources.

11.6.3   Central Composite Designs

A much more efficient class of designs for fitting second-order response surface equations has been developed by Box and Wilson (1951) and their successors. These are the so-called central composite designs (CCD) and their variants. The basic central composite design for k variables consists of a 2^k factorial design with each factor at two levels: -1, $+1$. This is augmented by points placed at the coordinates shown in Table 11.21.

The value of α in Table 11.21 is chosen by the experimenter to achieve certain design properties. Similarly, $n_2 > 1$ center points $(0, 0, \ldots, 0)$ are used to attain certain properties. In general, the design contains $2^k + 2k + n_2$ points.

To illustrate, consider a central composite design for three variables. Suppose we use $n_2 = 4$ center points. The design will have $2^k + 2k + n_2 = 2^3 + (2)(3) + 4 = 18$ observations. The design matrix, $\mathbf{D}$, for this design is as shown in Table 11.22.

The configuration of points in the three-dimensional factor space is diagrammed in Fig. 11.6. Note that each factor is present at five levels: $-\alpha$, -1, 0, $+1$, $+\alpha$. The four center points are located at the point symbolized by $\odot$.

TABLE 11.21   Coordinates of Points Added to a 2^k Factorial to Create a Central Composite Design

	x_1	x_2	$\cdots$	x_k
1. Center points	0	0	$\cdots$	0
2. Axial points	$-\alpha$	0	$\cdots$	0
	α	0	$\cdots$	0
	0	$-\alpha$	$\cdots$	0
	0	α	$\cdots$	0
	$\vdots$	$\vdots$		$\vdots$
	0	0	$\cdots$	$-\alpha$
	0	0	$\cdots$	α

TABLE 11.22 Design Matrix for a Three-Variable Central Composite Design with Four Center Points

$$
\mathbf{D} = \begin{bmatrix}
-1 & -1 & -1 \\
1 & -1 & -1 \\
-1 & 1 & -1 \\
1 & 1 & -1 \\
-1 & -1 & 1 \\
1 & -1 & 1 \\
-1 & 1 & 1 \\
1 & 1 & 1 \\
0 & 0 & 0 \\
0 & 0 & 0 \\
0 & 0 & 0 \\
0 & 0 & 0 \\
-\alpha & 0 & 0 \\
\alpha & 0 & 0 \\
0 & -\alpha & 0 \\
0 & \alpha & 0 \\
0 & 0 & -\alpha \\
0 & 0 & \alpha
\end{bmatrix}
$$

The first eight points are the points in an ordinary 2^3 factorial design used to fit a first-order surface. The four center points give us an augmented 2^3 factorial, while the six axial points enable us to fit the second-order surface.

We can do the experimentation sequentially. First run the augmented 2^3 factorial. If the quadratic coefficients are not significant we can stop and fit a first-order equation. If they are significant we can then run the axial points and then fit a second-order surface.

11.6.4 General Configuration

The general expression for a central composite design is described by the model

$$
E(y) = \beta_0 + \sum_{i=1}^{k} \beta_i x_i + \sum_{i=1}^{k} \beta_{ii} x_i^2 + \sum_{i<i'} \beta_{ii'} x_i x_{i'}
$$

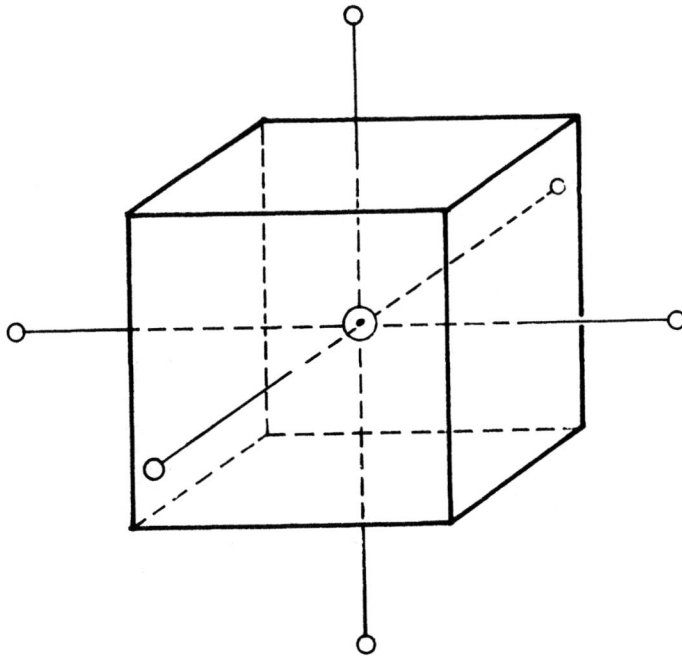

Fig. 11.6 Configuration of points in a three-factor central composite design.

Let

F = number of points in the factorial part ($= 2^k$ if a full factorial is used)

T = $2k + n_2$ = number of axial plus center points

Then, the **X** matrix for the data equations is given in Table 11.23.

That is, the pure quadratic terms in the **X** matrix have been corrected for their mean.

The **X'X** matrix is shown in Table 11.24. The elements in the submatrix, **Q**, associated with the pure quadratic terms of **X'X** are presented in Table 11.24. Except for **Q**, $(\mathbf{X'X})^{-1}$ is diagonal with elements equal to the reciprocal of the diagonal elements of **X'X**.

The least squares estimates, **b**, of the parameters, **β**, of the model are obtained in the usual way

$$\mathbf{b} = (\mathbf{X'X})^{-1}\mathbf{X'y}$$

The variances of the several classes of estimates are:

$$V(b_0) = \sigma^2/(F + T)$$
$$V(b_i) = \sigma^2(F + 2\alpha^2)$$

TABLE 11.23 General Representation of the **X** Matrix for a Central Composite Design

$$
\mathbf{X} =
\begin{array}{c}
\begin{array}{ccccc|cccc|cccc}
x_0 & x_1 & x_2 & \cdots & x_k & x_{11} & x_{22} & \cdots & x_{kk} & x_{12} & x_{13} & \cdots & x_{k-1,k}
\end{array}\\
\left[
\begin{array}{ccccc|cccc|cccc}
1 & -1 & -1 & \cdots & -1 & 1-c & 1-c & \cdots & 1-c & 1 & 1 & \cdots & 1 \\
1 & 1 & -1 & \cdots & -1 & 1-c & 1-c & \cdots & 1-c & -1 & -1 & \cdots & 1 \\
1 & -1 & 1 & \cdots & -1 & 1-c & 1-c & \cdots & 1-c & -1 & 1 & \cdots & 1 \\
\vdots & \vdots & \vdots & & \vdots & \vdots & \vdots & & \vdots & \vdots & \vdots & & \vdots \\
1 & 0 & 0 & \cdots & 0 & -c & -c & \cdots & -c & 0 & 0 & \cdots & 0 \\
1 & -\alpha & 0 & \cdots & 0 & (\alpha^2-c) & -c & \cdots & -c & 0 & 0 & \cdots & 0 \\
1 & \alpha & 0 & \cdots & 0 & (\alpha^2-c) & -c & \cdots & -c & 0 & 0 & \cdots & 0 \\
1 & 0 & -\alpha & \cdots & 0 & -c & (\alpha^2-c) & \cdots & -c & 0 & 0 & \cdots & 0 \\
1 & 0 & \alpha & \cdots & 0 & -c & (\alpha^2-c) & \cdots & -c & 0 & 0 & \cdots & 0 \\
\vdots & \vdots & \vdots & & \vdots & \vdots & \vdots & & \vdots & \vdots & \vdots & & \vdots \\
1 & 0 & 0 & \cdots & -\alpha & -c & -c & \cdots & (\alpha^2-c) & 0 & 0 & \cdots & 0 \\
1 & 0 & 0 & \cdots & \alpha & -c & -c & \cdots & (\alpha^2-c) & 0 & 0 & \cdots & 0
\end{array}
\right]
\end{array}
$$

$$\mathbf{D} = \text{design matrix}$$

Note: $\quad c = \dfrac{F + 2\alpha^2}{F + T}$

$$V(b_{ij}) = \sigma^2/F$$

$$V(b_{ii}) = e\sigma^2$$

$$\text{Cov}(b_{ii}, b_{i'i'}) = f\sigma^2$$

We see that α, the radius of the axial points, appears both in the estimators of the coefficients and in their variance. We can, then, change some of the characteristics of the design by our choice of α.

11.6.5 Choice of α and n_2 for a CCD

There are a number of properties we might want to consider in selecting a central composite design:

1. Orthogonality
2. Rotatability
3. Uniformity of precision

TABLE 11.24   Configuration of the **X'X** Matrix for a Central Composite Design

$$
\begin{bmatrix}
F + T & \mathbf{O'} & \mathbf{O'} & \mathbf{O'} \\[2mm]
\mathbf{O} & \begin{matrix} (F + 2\alpha^2) & 0 & \cdots & 0 \\ & (F + 2\alpha^2) & \cdots & 0 \\ \vdots & \vdots & & \vdots \\ 0 & 0 & \cdots & (F + 2\alpha^2) \end{matrix} & \mathbf{O} & \mathbf{O} \\[2mm]
\mathbf{O} & \mathbf{O} & \begin{matrix} p & q & \cdots & q \\ q & p & \cdots & q \\ \vdots & \vdots & & \vdots \\ q & q & \cdots & p \end{matrix} & \mathbf{O} \\[2mm]
\mathbf{O} & \mathbf{O} & \mathbf{O} & \begin{matrix} F & 0 & \cdots & 0 \\ 0 & F & \cdots & 0 \\ \vdots & \vdots & & \vdots \\ 0 & 0 & \cdots & F \end{matrix}
\end{bmatrix}
$$

Note: $p = [FT - 4F\alpha^2 - 4\alpha^4 + 2(F + T)\alpha^4]/(F + T)$

 $q = (FT - 4F\alpha^2 - 4\alpha^4)/(F + T)$

TABLE 11.25   Submatrix, **Q**, of Quadratic Terms Abstracted from **X'X**

$$
\mathbf{Q}^{-1} = \begin{bmatrix} p & q & \cdots & q \\ q & p & \cdots & q \\ \vdots & \vdots & & \vdots \\ q & q & \cdots & p \end{bmatrix}^{-1} = \begin{bmatrix} e & f & \cdots & f \\ f & e & \cdots & f \\ \vdots & \vdots & & \vdots \\ f & f & \cdots & e \end{bmatrix}
$$

Note: $e = [p + (k - 2)q]/[(p - q)(p + kq - q)]$

 $f = q/(q - p)[p + (k - 1)q]$

By proper choice of α we can make the design orthogonal. When the design is orthogonal the entire $X'X$ matrix is diagonal, the computation of the parameter estimates is simplified, and these estimates are uncorrelated with each other. On considering the general form of $X'X$ given in Section 11.6.4 we see that in order to diagonalize $X'X$ we must find α such that

$$q = FT - 4F\alpha^2 - 4\alpha^4 = 0$$

If a full 2^k factorial is used with only one center point, values of α for which $X'X$ is diagonal are

k	2	3	4	5	6	7	8
α	1.000	1.216	1.414	1.596	1.761	1.910	2.045

Thus, for $k = 2$ the orthogonal CCD with one center point is a 3^2 factorial design.

11.6.6 Rotatability

When using certain designs to estimate a response surface it will be found that $\hat{y}$ will be estimated with greater precision at some points than at others. This will depend on the variance of $\hat{y}$, which will, in turn, depend on the orientation of the true response surface with respect to the design points in the factor space. There is a class of designs which may be used to avoid this problem. These are the so-called rotatable designs.

A design is said to be rotatable if the variance of $\hat{y}_j$, the estimated response, is a function only of the distance of the point $x_{1j}, x_{2j}, \ldots, x_{kj}$ from the center of the design. It is not dependent on the orientation of the design with respect to the true response surface. With a rotatable design all points at a distance of radius ρ from the center of the design are measured with equal precision.

A CCD may be made into a rotatable design by a suitable choice of α, the distance of the axial points from the center of the design. With a full factorial set of points in the hypercube α must be $2^{k/4}$ for a design with k factors to be rotatable. If a one-half replication of the factorial set is used in the hyperbcube then α must be $2^{(k-1)/4}$ for rotatability. Thus, values of α for rotatability with the full factorial set are:

k	2	3	4	5	6	7	8
α	1.4142	1.6818	2.0000	2.3784	2.8284	3.3636	4.0000

We should point out that even though it is rotatable a CCD may not be satisfactory with only one point at the center. First, with only one center point the design provides no estimate of experimental error unless it is replicated. Second, when we select α to obtain a rotatable design we no longer have an orthogonal one. Third, consider a graph of the variance, $V(\hat{y})$, of the estimated response, $\hat{y}$, as a function of the distance, ρ, from the center of a rotatable CCD with only one center point. This graph is shown in Fig. 11.7. We see that $V(\hat{y})$ is not constant. It varies with ρ, reaching a minimum at $\rho = 1$.

11.6.7 Uniform Precision

Box and Hunter (1957) have suggested the concept of a rotatable CCD which has uniform precision over the factor space spanned by $\rho = 0$ to $\rho = 1$. Their reasoning is that the experimenter is most interested in the center of the design where, hopefully, a stationary point on the surface is located. Uniform precision can be accomplished by adding points at the center of a rotatable CCD. Similarly, a rotatable CCD may be made orthogonal by proper choice of the number of center points.

Design parameters for uniform precision CCDs are given in Table 11.26. Design parameters for orthogonal CCDs are given in Table 11.27. In all cases the designs are rotatable.

11.6.8 Estimators and Variances

The sample estimators of the coefficients in the second-order model for the various designs, as well as the sample variances of the estimates, are summarized for uniform precision designs in Table 11.28 and for orthogonal designs in Table 11.29. In these tables the following symbols are used:

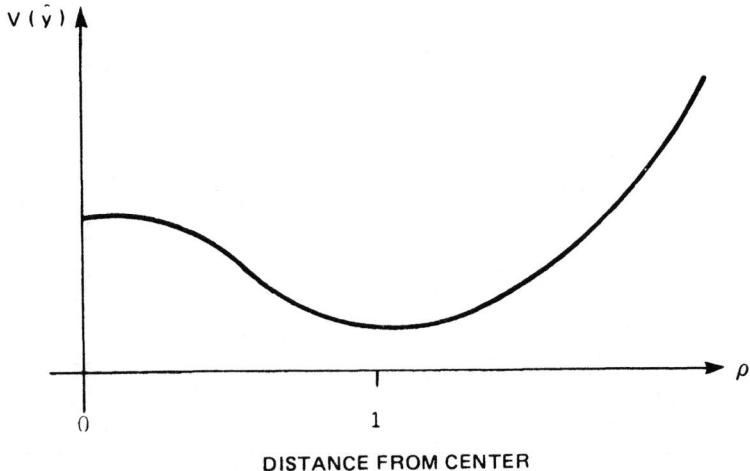

DISTANCE FROM CENTER

Fig. 11.7 Graph of $V(\hat{y})$ as a function of radius, ρ.

TABLE 11.26 Design Parameters for Uniform Precision Designs

Parameter	$k*$				
	2	3	4	5	6

I. Parameters for uniform precision, rotatable CCD with full factorial in the hypercube

	2	3	4	5	6
n_2	5	6	7	10	15
F	4	8	16	32	64
N	13	20	31	52	91
α	1.4142	1.6818	ˋ.0000	2.3784	2.8284
c	.6154	.6829	.7742	.8330	.8791
$F + 2\alpha^2$	8.0000	13.6570	24.0000	43.3136	80.0000
e	.1438	.0694	.0350	.0171	.0084
f	.0188	.0069	.0037	.0015	.0005

	$k*$			
	5	6	7	8

II. Parameters for uniform precision, rotatable CCD with 1/2 replication of a factorial in the hypercube

	5	6	7	8
n_2	6	9	14	20
F	16	32	64	128
N	32	53	92	164
α	2.0000	2.3784	2.8284	3.3636
c	.7500	.8172	.8696	.9185
$F + 2\alpha^2$	24.0000	43.3136	80.0000	150.6276
e	.0341	.0168	.0083	.0041
f	.0028	.0012	.0005	.0002

$k* =$ number of factors; $n_2 =$ number of center points; $F =$ points in hypercube; $N =$ total points $= F + 2k + n_2$; $\alpha =$ distance to axial points; $c = (F + 2\alpha^2)/(F + T)$; e, f are defined in Table 11.25.

$b_0 =$ intercept

$0y = \mathbf{x}_0'\mathbf{y}$

$b_i =$ linear coefficient for the ith input variable

$iy = \mathbf{x}_i'\mathbf{y}$

$b_{ii} =$ quadratic coefficient for the ith variable

$iiy = \mathbf{x}_{ii}'\mathbf{y}$

$b_{ij} =$ coefficient for the interaction (cross product) of the ith and jth variables

TABLE 11.27   Design Parameters for Orthogonal Designs

Parameter	$k*$				
	2	3	4	5	6

I. Parameters for orthogonal, rotatable CCD with full factorial in the hypercube

n_2	8	9	12	17	24
F	4	8	16	32	64
N	16	23	36	59	100
α	1.4142	1.6818	2.0000	2.3784	21.8284
c	.5000	.5938	.6667	.7341	.8000
$F + 2\alpha^2$	8.0000	13.6570	24.0000	43.3136	80.0000
e	.1250	.0625	.0313	.0156	.0078

$k*$			
5	6	7	8

II. Parameters for orthogonal, rotatable CCD with 1/2 replication of a factorial in the hypercube

n_2	10	15	22	33
F	16	32	64	128
N	36	59	100	177
α	2.0000	2.3784	2.8284	3.3636
c	.6667	.7341	.8000	.8510
$F + 2\alpha^2$	24.0000	43.3136	80.0000	150.6276
e	.0313	.0156	.0078	.0039
$f = 0$ for all orthogonal designs				

Note: $k*$ = number of factors; n_2 = number of center points; F = points in hypercube;
N = total points = $F + 2k + n_2$; α = distance to axial points; $c = (F + 2\alpha^2)/(F + T)$; e, f
are defined in Table 11.25.

$$ijy = \mathbf{x}'_{ij}\mathbf{y}$$

$$V(b) = \text{sample variance of the coefficient}$$

$$s^2 = \text{MSE} = \text{error mean square, an estimate of } \sigma^2$$

11.7 NUMERICAL EXAMPLE

An experiment was designed to study the interrelationships among manganese (Mn), molybdenum (Mo), and iron (Fe) in a nutrient solution used for raising tomatoes in a greenhouse (Kirsch et al., 1960). Each of these elements was supplied in five concentrations as shown in Table 11.30.

TABLE 11.28 Estimators and Variances for Uniform Precision Designs

			k		
	2	3	4	5	6

I. Uniform precision, rotatable CCD, full factorial

b_0	.0769($0y$)	.0500($0y$)	.0323($0y$)	.0192($0y$)	.0110($0y$)
$V(b_0)$	.0769 s^2	.0500 s^2	.0323 s^2	.0192 s^2	.0110 s^2
b_i	.1250(iy)	.0732(iy)	.0417(iy)	.0231(iy)	.0125(iy)
$V(b_i)$	.1250 s^2	.0732 s^2	.0417 s^2	.0231 s^2	.0125 s^2
b_{ii}	.1250(iiy) +.0188$\Sigma(iiy)$	.0625(iiy) +.0069$\Sigma(iiy)$	.0313(iiy) +.0037$\Sigma(iiy)$	.0156(iiy) +.0015$\Sigma(iiy)$	.0079(iiy) +.0005$\Sigma(iiy)$
$V(b_{ii})$	.1438 s^2	.0694 s^2	.0350 s^2	.0171 s^2	.0084 s^2
b_{ij}	.2500(ijy)	.1250(ijy)	.0625(ijy)	.0313(ijy)	.0156(ijy)
$V(b_{ij})$	.2500 s^2	.1250 s^2	.0625 s^2	.0313 s^2	.0156 s^2

			k	
	5	6	7	8

II. Uniform precision, rotatable CCD, 1/2 factorial

b_0	.0313($0y$)	.0189($0y$)	.0109($0y$)	.0061($0y$)
$V(b_0)$	.0313 s^2	.0189 s^2	.0109 s^2	.0061 s^2
b_i	.0417(iy)	.0231(iy)	.0125(iy)	.0066(iy)
$V(b_i)$	.0417 s^2	.0231 s^2	.0125 s^2	.0066 s^2
b_{ii}	.0313(iiy) +.0028$\Sigma(iiy)$	.0156(iiy) +.0012$\Sigma(iiy)$	.0078(iiy) +.0005$\Sigma(iiy)$	.0039(iiy) +.0002$\Sigma(iiy)$
$V(b_{ii})$	.0341 s^2	.0168 s^2	.0083 s^2	.0041 s^2
b_{ij}	.0625(ijy)	.0313(ijy)	.0156(ijy)	.0078(ijy)
$V(b_{ij})$	.0625 s^2	.0313 s^2	.0156 s^2	.0078 s^2

TABLE 11.29 Estimators and Variances for Orthogonal Designs

	2	3	4	5	6
			k		
I. Orthogonal, rotatable CCD, full factorial					
b_0	.0625($0y$)	.0436($0y$)	.0278($0y$)	.0169($0y$)	.0100($0y$)
$V(b_0)$	.0625 s^2	.0435 s^2	.0278 s^2	.0169 s^2	.0100 s^2
b_i	.1250(iy)	.0732(iy)	.0417(iy)	.0231(iy)	.0125(iy)
$V(b_i)$	.1250 s^2	.0732 s^2	.0417 s^2	.0231 s^2	.0125 s^2
b_{ii}	.1250(iiy)	.0625(iiy)	.0313(iiy)	.0156(iiy)	.0078(iiy)
$V(b_{ii})$	.1250 s^2	.0625 s^2	.0313 s^2	.0156 s^2	.0078 s^2
b_{ij}	.2500(ijy)	.1250(ijy)	.0625(ijy)	.0313(ijy)	.0156(ijy)
$V(b_{ij})$	.2500 s^2	.1250 s^2	.0625 s^2	.0313 s^2	.0156 s^2

	5	6	7	8
		k		
II. Orthogonal, rotatable CCD, 1/2 factorial				
b_0	.0278($0y$)	.0169($0y$)	.0100($0y$)	.0056($0y$)
$V(b_0)$	.0278 s^2	.0169 s^2	.0100 s^2	.0056 s^2
b_i	.0417 (iy)	.0231(iy)	.0125(iy)	.0066(iy)
$V(b_i)$	.0417 s^2	.0231 s^2	.0125 s^2	.0066 s^2
b_{ii}	.0313(iiy)	.0156(iiy)	.0078(iiy)	.0039(iiy)
$V(b_{ii})$	.0313 s^2	.0156 s^2	.0078 s^2	.0039 s^2
b_{ij}	.0625(ijy)	.0313(ijy)	.0156(ijy)	.0078(ijy)
$V(b_{ij})$	.0625 s^2	.0313 s^2	.0156 s^2	.0078 s^2

The experimental design used in the study was a uniform precision, rotatable central composite design. Twenty opaque plastic 10-liter containers were filled with clean sand and five tomato seedlings were planted in each container. Each container was supplied with a basic nutrient solution, which was changed once a week, at which time the required amounts of Mn, Mo, and Fe were also supplied.

The plants were allowed to grow for 36 days, at which time they were harvested. The plant material was washed briefly with deionized water to remove surface contaminants. It was then dried in a forced-air oven at 65°C and weighed.

The **X** matrix and vector, **y**, of yields (grams per pot) are given in Table 11.31.

TABLE 11.30 Nutrient Concentrations and Corresponding Coded Design Variables

Coded design variables x_i	Final concentration (ppm)		
	Mn	Mo	Fe
-1.68	0.001	0.000065	0.001
-1.00	0.005	0.000500	0.005
0.00	0.050	0.010000	0.050
1.00	0.500	0.200000	0.500
1.68	2.390	1.530000	2.390

11.7.1 Analysis

We first compute

$$\mathbf{X'y} = \begin{bmatrix} 0y \\ 1y \\ 2y \\ 3y \\ 11y \\ 22y \\ 33y \\ 12y \\ 13y \\ 23y \end{bmatrix} = \begin{bmatrix} 214.6700 \\ -4.2695 \\ 45.0780 \\ -4.7014 \\ -59.5920 \\ -33.8809 \\ 26.4510 \\ 35.7200 \\ 14.3200 \\ 1.1800 \end{bmatrix}$$

Then, using

$$b_0 = (0y)/(F + T) = (0y)/20$$
$$b_0 = (iy)/(F + 2\alpha^2) = (iy)/13.6570$$
$$b_{ii} = (iiy)e + [\Sigma(iyy)]f = (.0625)(iiy) - .4625$$
$$b_{ij} = (ijy)/F = (ijy)/8$$

we have

TABLE 11.31 **X** Matrix and Yield Vector, **y**, for Culture Solution Experiment with Tomatoes

	Mn	Mo	Fe							
x_0	x_1	x_2	x_3	x_{11}	x_{22}	x_{33}	x_{12}	x_{13}	x_{23}	
1	−1	−1	−1	.3171	.3171	.3171	1	1	1	11.94
1	1	−1	−1	↓	↓	↓	−1	−1	1	.48
1	−1	1	−1				−1	1	−1	9.59
1	1	1	−1				1	−1	−1	14.42
1	−1	−1	1				1	−1	−1	7.67
1	1	−1	1				−1	1	−1	1.70
1	−1	1	1				−1	−1	1	4.34
1	1	1	1	↓	↓	↓	1	1	1	17.90
1	0	0	0	−.6829	−.6829	−.6829	0	0	0	14.18
1	0	0	0	↓	↓	↓	0	0	0	14.60
1	0	0	0				0	0	0	15.17
1	0	0	0				0	0	0	13.25
1	0	0	0				0	0	0	13.73
1	0	0	0	↓	↓	↓	0	0	0	16.08
1	−1.6818	0	0	2.1456	−.6829	−.6829	0	0	0	4.86
1	1.6818	0	0	2.1456	−.6829	−.6829	0	0	0	1.81
1	0	−1.6818	0	−.6829	2.1456	−.6829	0	0	0	1.78
1	0	1.6818	0	−.6829	2.1456	−.6829	0	0	0	13.98
1	0	0	−1.6818	−.6829	−.6829	2.1456	0	0	0	18.48
1	0	0	1.6818	−.6829	−.6829	2.1456	0	0	0	18.61
					X					**y**

$$
\mathbf{b} = \begin{bmatrix} b_0 \\ b_1 \\ b_2 \\ b_3 \\ b_{11} \\ b_{22} \\ b_{33} \\ b_{12} \\ b_{13} \\ b_{23} \end{bmatrix} = \begin{bmatrix} 10.7335 \\ -.3125 \\ 3.2997 \\ -.3441 \\ -4.1870 \\ -2.5800 \\ 1.1907 \\ 4.4650 \\ 1.7900 \\ .1475 \end{bmatrix}
$$

The estimated yield, $\hat{y}$, is given by the equation

$$
\hat{y} = \mathbf{x'b} = 10.73 - .31x_1 + 3.30x_2 - .34x_3 - 4.19x_1^2 - 2.58x_2^2 \\
+ 1.19x_3^2 + 4.46x_1x_2 + 1.79x_1x_3 + .15x_2x_3
$$

The analysis of variance of these data is presented in Table 11.32. The sums of squares in Table 11.32 are computed as

1. Total SS = $\mathbf{y'y} - n\bar{y}^2 = 11.94^2 + .48^2 + \cdots + 18.62^2 -$
$(20)(10.74)^2 = 721.7807$

2. Regression SS = $\mathbf{b'X'y} - n\bar{y}^2$
$= (10.7335)(214.6700) + (-4.2695)(-.3125)$
$+ \cdots + (1.1800)(.1475) - (20)(10.74)^2$
$= 705.4123$

3. Pure Error SS is computed from the center points:
Pure Error SS = $(14.18^2 + 14.60^2 + \cdots + 16.08^2)$
$- (1.6)(14.18 + \cdots + 16.08)^2$
$= 5.2131$

4. Lack of Fit SS = Total SS − Regression SS − Pure Error SS
$= 721.7807 - 705.4123 - 5.2131 = 11.1553$

An F statistic for testing the significance of the individual regression coefficients is computed as

$$
F = b^2/V(b)
$$

We have

$$
F_1 = b_1^2/(.0732)s^2 = (-.3125)^2/(.0732)(1.0426) = 1.28
$$

TABLE 11.32   ANOVA of Culture Solution Yield Data

Source	d.f.	SS	MS	F
Total	19	721.7807		
Regression	9	705.4121	78.3791	75.18**
Lack of fit	5	11.1555	2.2311	2.14
Pure error	5	5.2131	1.0426	

**Significant at the 1% level.

$$F_2 = b_2^2 /(.0732)s^2 = (3.2997)^2/(.0732)(1.0426) = 142.66**$$

$$F_3 = b_3^2 /(.0732)s^2 = (-.3441)^2/(.0732)(1.0426) = 1.55$$

$$F_{11} = b_{11}^2/(.0694)s^2 = (-4.1870)^2/(.0694)(1.0426) = 242.29**$$

$$F_{22} = b_{22}^2/(.0694)s^2 = (-2.5800)^2/(.0694)(1.0426) = 91.99**$$

$$F_{33} = b_{33}^2/(.0694)s^2 = (1.1907)^2/(.0694)(1.0426) = 19.59**$$

$$F_{12} = b_{12}^2/(12.50)s^2 = (4.4650)^2/(.1250)(1.0426) = 152.97**$$

$$F_{13} = b_{13}^2/(.1250)s^2 = (1.7900)^2/(.1250)(1.0426) = 24.58**$$

$$F_{23} = b_{23}^2/(.1250)s^2 = (.1475)^2/(.1250)(1.0426) = .17$$

**Significant at the 1% level.

11.7.2   Report of Statistical Analysis

Analysis of data from an experiment to measure the interrelationship of Mn (x_1), Mo (x_2), and Fe (x_3) on the yield of tomatoes indicates that yield can be described by a quadratic function of the logarithm of the concentration of these elements. The equation for this relationship is

$$\hat{y} = 10.74 - .31x_1 + 3.30x_2 - .34x_3 - 4.19x_1^2 - 2.58x_2^2$$
$$+ 1.19x_3^2 + 4.46x_1x_2 + 1.79x_1x_3 + .15x_2x_3$$

where

$\hat{y}$ = expected yield

x_1, x_2, x_3 = coded levels of Mn, Mo, and Fe, respectively

On considering the magnitudes of the coefficients it is seen that the concentrations of manganese and molybdenum have a greater effect on yield than does that of iron. If we hold the level of iron constant at some concentration, say 0.050 ppm $(x_3 = 0)$, we can illustrate the relationship between expected yield, $\hat{y}$, and concentration of Mn and Mo. At $x_3 = 0$ we have

$$\hat{y} = 14.54 - .31x_1 + 3.30x_2 - 4.19x_1^2 - 2.58x_2^2 + 4.46x_1x_2$$

$$= 14.54 + [x_1 \; x_2] \begin{bmatrix} -.31 \\ 3.30 \end{bmatrix} + [x_1 \; x_2] \begin{bmatrix} -4.19 & 2.23 \\ 2.23 & -2.58 \end{bmatrix} \begin{bmatrix} x_1 \\ x_2 \end{bmatrix}$$

$$= b_0 + \mathbf{x'b} + \mathbf{x'Bx}$$

Note: $14.54 = 10.73 + (0.68) \sum_{i=1}^{3} b_{ii}$.

The stationary point, $\mathbf{x}_0$, when $x_3 = 0$ is

$$\mathbf{x}_0 = -\mathbf{B}^{-1}\mathbf{b}/2 = \begin{bmatrix} .4420 & .3820 \\ .3820 & .7178 \end{bmatrix} \begin{bmatrix} -.1550 \\ 1.6500 \end{bmatrix} = \begin{bmatrix} .5618 \\ 1.1252 \end{bmatrix}$$

The yield at the stationary point, y_0, may be obtained from

$$y_0 = b_0 + (\mathbf{x}_0'\mathbf{b})/2$$

$$= 14.54 + [.5618 \; 1.1252] \begin{bmatrix} -.31 \\ 3.30 \end{bmatrix} (1/2)$$

$$= 14.54 + 1.77 = 16.31$$

Since both b_{11} and b_{22} are negative the value of y_0 is a maximum.

Two ways to illustrate the relationship between expected yield, $\hat{y}$, and concentration of Mn and Mo at $x_3 = 0$ are:

1. A plot of the contours of constant yield as the concentrations of Mn and Mo are changed. This is shown in Fig. 11.8.
2. A sketch of the yield surface over the range of concentrations of Mn and Mo. This is shown in Fig. 11.9.

11.8 SEARCHING FOR OPTIMA

One of the problems facing an experimenter is that of determining the level at which each of a number of input factors should be set to obtain a response that is optimum in some sense. The goal is to find stationary points (maxima, minima, or saddle points) on a response surface, and to determine the nature of the surface in the neighborhood of the stationary points. A number of strategies have been proposed for this purpose. These include:

1. Single-factor methods (Friedman and Savage, 1947)
2. Method of steepest ascent (Box and Wilson, 1951)
3. Single large experiment
4. Random selection of test points (Anderson, 1953)

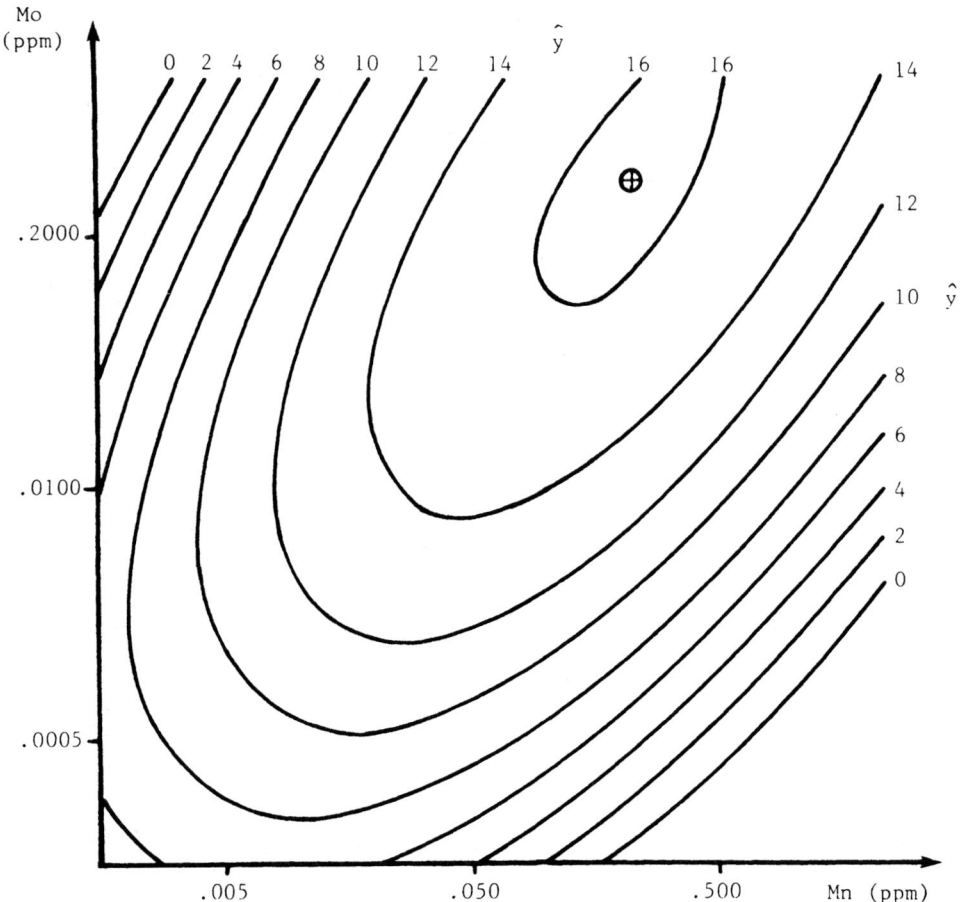

Fig. 11.8 Contours of constant yield, $\hat{y}$, as a function of concentration of Mn and Mo.

Each of these methods involves conducting a coordinated series of tests with a decision at the conclusion of each test. Each method has its advantages and disadvantages, but under many conditions the most generally useful is the *method of steepest ascent*.

To use the method of steepest ascent it is assumed that some preliminary work has been done to:

1. Identify the important factors
2. Establish the range of levels of these factors

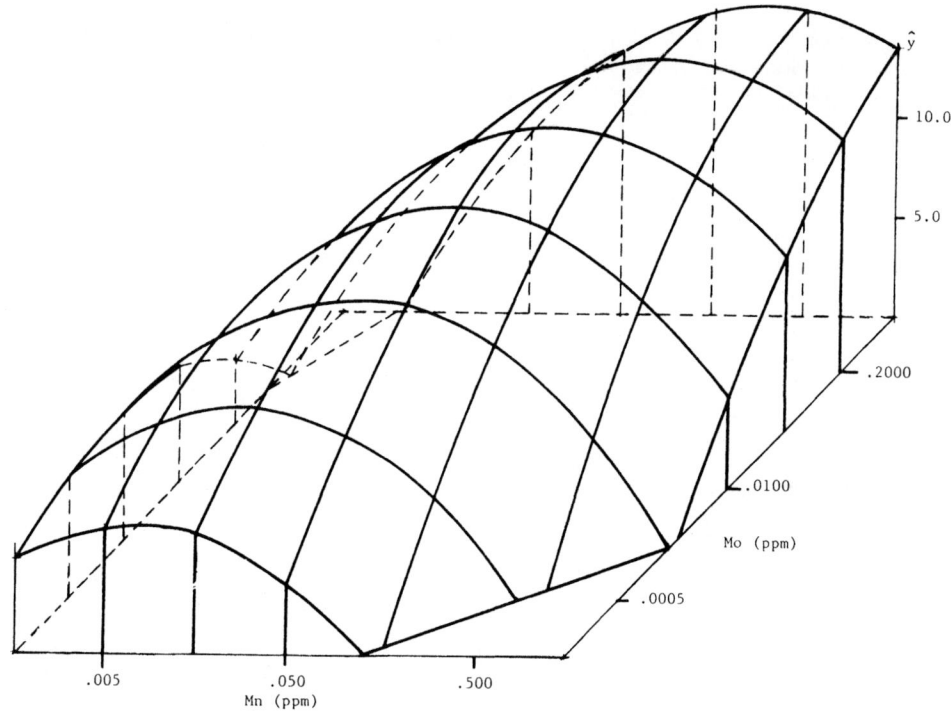

Fig. 11.9 Relationship between yield, $\hat{y}$, and concentration of Mn and Mo.

3. Make a guess as to where the optimum is likely to occur within the range

Since the method involves sequential experimentation it is most useful when results can be obtained quickly.

The process proceeds as follows:

Step 1. Run a 2^k factorial centered on the expected stationary point, $\mathbf{x}_0$. Augment the factorial with points at the center to obtain an estimate of error.

Step 2. Fit the first-order equation

$$\hat{y} = b_0 + b_1 x_1 + b_2 x_2 + \cdots + b_k x_k \tag{1}$$

then test for lack of fit.

Decision

a. If lack of fit is not significant go to step 5.

b. If lack of fit is significant go to step 3.

Step 3. Augment the 2^k factorial with points required to construct a second-order design (central composite, for example).

Step 4. Fit the second-order model, compute $\mathbf{x}_0$ and $\hat{y}_0$, then *Stop!*

Step 5. Taking the center of the design as the origin, move along the path of steepest ascent and conduct a new trial.

    a.   For each unit change in x_1 change x_i by b_i/b_1 units, where b_1 and b_i are coefficients in (1).

Step 6. Compute $\hat{y}$ at the new point using (1) and compare $\hat{y}$ with the observed y, y_{obs}.

    Decision

    a.   If $\hat{y}$ is close to y_{obs} go to step 5.

    b.   If $\hat{y}$ is not close to y_{obs} let the last point be the expected $\mathbf{x}_0$ and return to step 1.

11.9 LIST OF SYMBOLS

11.9.1 General

1. Greek letters denote parameters; English letters denote sample estimates.
2. Letters at the beginning of the alphabet stand for constants, letters at the end of the alphabet stand for variables.
3. Matrices are denoted by boldface uppercase letters: $\mathbf{A}$, $\mathbf{B}$, $\mathbf{X}$, $\mathbf{Y}$.
4. Vectors are designated by boldface lowercase letters: $\mathbf{a}$, $\mathbf{b}$, $\mathbf{x}$, $\mathbf{y}$.

    a.   Unless otherwise noted, all vectors used in this chapter are column vectors.

11.9.2 Special Symbols

1. $\mathbf{1}$ = a vector whose elements are all one.
2. $\mathbf{I}$ = an identity matrix with ones on the principal diagonal and zeros elsewhere.

11.9.3 Basic Matrix Operations

1. *Transpose:* Given a matrix $\mathbf{A}$, the transpose of $\mathbf{A}$, symbolized $\mathbf{A}'$, is obtained by interchanging the rows and columns of $\mathbf{A}$.
2. *Addition:* If two matrices $\mathbf{A}$, $\mathbf{B}$ have the same dimension, then $\mathbf{A} + \mathbf{B} = [a_{ij} + b_{ij}]$ for all i,j.
3. *Multiplication:* The product of two matrices $\mathbf{A}_{mn}$ and $\mathbf{B}_{np}$ is the matrix $\mathbf{C}_{mp}$ whose elements are obtained as

$$c_{ij} = \sum_{k=1}^{n} a_{ik}b_{kj}$$

Multiplication is defined only if the column dimension of the first matrix is equal to the row dimension of the second. The product matrix has the row dimension of the first matrix and the column dimension of the second.

4. The product of a scalar, k, and a matrix, $\mathbf{A}$, is the matrix

$$\mathbf{C} = k\mathbf{A}$$

such that

$$c_{ij} = ka_{ij} \quad \text{for all} \quad i,j$$

5. *Inverse:* Given a square matrix $\mathbf{A}$, the inverse of $\mathbf{A}$, symbolized $\mathbf{A}^{-1}$, is the matrix such that

$$\mathbf{A}^{-1}\mathbf{A} = \mathbf{A}\mathbf{A}^{-1} = \mathbf{I}$$

FURTHER READING*

Box, G. E. P., W. G. Hunter, and J. S. Hunter (1978), Ch. 14 and 15.
Cochran, W. G., and G. M. Cox (1957), Ch. 8A.
Davies, O. L. (ed.) (1963), Ch. 11.
Mead, R., and D. J. Pike (1975).
Myers, R. H. (1971).

REFERENCES

Anderson, R. L. (1953). Recent Advances in Finding Best Operating Conditions, *J. Amer. Stat. Assoc.* 48:789–798.

Box, G. E. P., and J. S. Hunter (1957). Multifactor Experimental Designs for Exploring Response Surfaces, *Ann. Math. Stat.* 28:195–241.

Box, G. E. P., and K. B. Wilson (1951). On the Experimental Attainment of Optimum Conditions, *J. R. Stat. Soc. B* 13:1–45.

Friedman, M., and L. J. Savage (1947). Planning Experiments Seeking Maxima, in *Techniques of Statistical Analysis*, McGraw-Hill, New York.

Kirsch, R. K., M. E. Harward, and R. G. Petersen (1960). Interrelationships Among Iron, Manganese, and Molybdenum in the Growth and Nutrition of Tomatoes Grown in Culture Solution, *Plant and Soil* XII:259–275.

Myers, R. H. (1971). *Response Surface Methodology*, Allyn & Bacon, Boston.

---

*See Bibliography for complete citation.

12

Change-Over Trials

12.1 INTRODUCTION

In some experimental situations it is desirable to apply two or more treatments consecutively to the same experimental unit. In dairy cattle feeding trials, for example, two or more feeds may be given, one after the other, to the same animal. Similarly, two or more drugs may be given sequentially to the same individual in a clinical trial. Experiments which involve the sequential application of several treatments to the same experimental unit are called change-over trials.

The primary advantage to conducting an experiment in this way is that, since treatments are compared on the same unit, the between-unit variability does not enter into experimental error. Because treatments are applied in sequence, however, change-over trials must provide control on time trends. These time trends are often called "period" effects.

12.2 CARRY-OVER (RESIDUAL) EFFECTS

When two or more treatments are applied in sequence to the same experimental unit it is logical to suppose that response to the second and subsequent treatments may be conditioned by the preceding treatments in the sequence. Such conditioning effects are called "carry-over" or "residual" effects. Depending on the nature of the treatments, carry-over effects may or may not exist. If they do exist we will make the following simplifying assumptions:

1. Carry-over effects in any period are due entirely to the immediately preceding treatment. That is, treatments have no effect beyond the period following the one in which they are applied.
2. The carry-over effect of a given treatment is independent of the effect of the following treatment in the sequence.

We would like to be able to estimate and test carry-over effects if they exist. Further, in some cases we may be interested in the sum of the direct effect and the carry-over effect. This sum is a measure of the so-called permanent effect, or equilibrium effect, of the treatment. These effects are more easily measured with some types of change-over design than with others, We shall comment on this as we consider the various designs.

We will consider a number of classes of change-over design, beginning with the simple designs involving only a few treatments and a small number of periods, then progressing to the more complex designs. For the simple designs we will present the basic pattern and discuss something of their use. For the most part carry-over effects are not very well estimated with these designs so we will not consider the analysis for residual effects in these designs. On the other hand, we will examine in detail one or two of the more useful designs for the estimation of carry-over effects. Numerical examples will be given for each of the change-over designs included here.

12.3 CROSS-OVER, OR SIMPLE REVERSAL, DESIGN

The most basic change-over design is the cross-over, or simple reversal. In this design two treatments, a and b, are studied in a trial lasting for two periods. Each unit receives the treatments in either the sequence $a \rightarrow b$ or else the sequence $b \rightarrow a$. The basic pattern of this design is

| | Sequence group | |
Period	1	2
1	a	b
2	b	a

Periods should be of equal length. Units are assigned at random to the sequence groups. Ordinarily the same number of units are assigned to each group since this provides maximum information per unit. Carry-over effects are difficult to estimate. They should be minimized, if necessary, by taking observations only at the end of the period, or by using a uniform treatment on both groups between experimental periods.

The cross-over is not a very useful design in practice. It does, however, illustrate the change-over concept. It may be useful for preliminary trials involving two extreme treatments.

12.3.1 Analysis

Let y_{ijk} = performance of the jth unit in the ith sequence group during the kth period. Assume that there are n_1 units in group 1 and n_2 units in group 2. The data are tabulated as shown in Table 12.1. The analysis of variance contains

TABLE 12.1   Data Table for a Cross-Over Design

Period	Treatment	Group 1				Sum
1	a	y_{111}	y_{121}	$\cdots$	y_{1n_11}	P_{11}
2	b	y_{112}	y_{122}	$\cdots$	y_{1n_12}	P_{12}
Difference		D_{11}	D_{12}	$\cdots$	D_{1n_1}	G_1
		Group 2				
1	b	y_{211}	y_{221}	$\cdots$	y_{2n_21}	P_{21}
2	a	y_{212}	y_{222}	$\cdots$	y_{2n_22}	P_{22}
Difference		D_{21}	D_{22}	$\cdots$	D_{2n_2}	G_2

Note: $D_{ij} = y_{ij1} - y_{ij2}$, $P_{ik} = \sum_j y_{ijk}$, $G_i = P_{i1} - P_{i2}$.

TABLE 12.2   Format for the ANOVA of a Cross-Over Design

Source	d.f.	SS	MS
Treatment	1	SST	MST
Error	$n_1 + n_2 - 2$	SSE	MSE

only two lines. The format is given in Table 12.2. The entries in Table 12.2 are computed as:

$$SST = (n_2 G_1 - n_1 G_2)^2 / [2n_1 n_2 (n_1 + n_2)]$$
$$SSE = (1/2)(\sum_i \sum_j D_{ij}^2) - (1/2n_1)G_1^2 - (1/2n_2)G_2^2$$

The treatment means are computed as:

$$\bar{y}_a = \bar{\bar{y}} + t, \qquad \bar{y}_b = \bar{\bar{y}} - t$$

where

$$\bar{\bar{y}} = (\sum_i \sum_k P_{ik}) / [2(n_1 + n_2)]$$

$$t = (n_2 G_1 - n_1 G_2)/4n_1 n_2$$

The sample variance of a mean, $V(\bar{y}_i)$, is $V(\bar{y}_i) = (n_1 + n_2)MSE/4n_1 n_2$ and the variance of the difference between means, $V(\bar{y}_a - \bar{y}_b)$, is

$$V(\bar{y}_a - \bar{y}_b) = 2[V(\bar{y}_i)] = (n_1 + n_2)(MSE)/2n_1 n_2$$

Note: If $n_1 = n_2 = n$ then

$$SST = (G_1 - G_2)^2/4n$$

$$SSE = (1/2) (\sum_i \sum_j D_{ij}^2) - (1/2n)(G_1^2 + G_2^2)$$

$$\bar{\bar{y}} = (\sum_i \sum_k P_{ik})/4n$$

$$t = (G_1 - G_2)/4n$$

$$V(\bar{y}_i) = MSE/2n$$

$$V(\bar{y}_a - \bar{y}_b) = MSE/n$$

12.3.2 Numerical Example: Cross-Over Design

A clinical psychologist wanted to test two drugs, *a* and *b*, which are intended to increase the reaction time to a certain stimulus. He was concerned about the inherent differences in reaction time among individuals so he decided to use a cross-over design in an attempt to eliminate individual differences. He obtained eight individuals and randomly assigned four to each of the two sequence groups. To conduct the trial he administered a drug to an individual, waited 15 min for absorption, applied the stimulus, and then measured the reaction time. These results are given in Table 12.3. The analysis of variance of the reaction time data is presented in Table 12.4.

The entries in Table 12.4 are computed as follows:

$$SST = (G_1 - G_2)^2/4n = (17 - 7)^2/(4)(4) = 6.2500$$

$$SSE = (1/2)(\sum_i \sum_j D_{ij}^2) - (1/2n)(G_1^2 + G_2^2)$$

$$= (1/2)(100.00) - [1/(2)(4)](338) = 7.7500$$

TABLE 12.3 Reaction Time (tenths of a second)

Period Group I	Drug	Subject				
		1	2	3	4	Sum
1	a	30	54	42	56	182
2	b	28	50	38	49	165
Difference		2	4	4	7	17
Group II		1	2	3	4	Sum
1	b	22	44	18	28	112
2	a	21	41	17	26	105
Difference		1	3	1	2	7

TABLE 12.4 ANOVA of Reaction Time

Source	d.f.	SS	MS	F
Treatment	1	6.2500	6.2500	4.84 NS
Error	6	7.7500	1.2917	

NS, not significant.

TABLE 12.5 Mean Reaction Time (tenth of a second) Under Two Drugs

Drug	a	b	Standard error
Reaction time	35.87	34.63	.40

The means are computed in the following way:

$$\bar{\bar{y}} = (\sum_i \sum_k P_{ik})/4n = 564/16 = 35.25$$

$$t = (G_1 - G_2)/4n = 10/16 = 0.62$$

$$\bar{y}_a = 35.25 + .62 = 35.87$$

$$\bar{y}_b = 35.25 - .62 = 34.63$$

The standard errors for this trial are

$$s_{\bar{y}} = \sqrt{\text{MSE}/2n} = \sqrt{1.2917/8} = .40$$

$$s_{(\bar{y}_a - \bar{y}_b)} = \sqrt{\text{MSE}/n} = \sqrt{1.2917/4} = .57$$

Report of statistical analysis. A cross-over trial was conducted to evaluate two drugs for their ability to increase the reaction time to a certain stimulus. The results are shown in Table 12.5.

The difference between the two drugs was not statistically significant.

12.4 SWITCH-BACK, OR DOUBLE REVERSAL, DESIGN

The switch-back (double reversal) is a simple extension of the cross-over design. In the switch-back design two treatments, *a* and *b*, are studied in a trial lasting for three periods. Each experimental unit receives both treatments, one treatment once and the other treatment twice in either the sequence $a \rightarrow b \rightarrow a$ or else the sequence $b \rightarrow a \rightarrow b$. The basic pattern for the design is

	Sequence group	
Period	1	2
1	*a*	*b*
2	*b*	*a*
3	*a*	*b*

Periods should be of equal length. Units are assigned at random to the sequence groups. Ordinarily, but not necessarily, the same number of units is assigned to each group. If information is missing for a unit for any period the data from that unit for the other periods should be discarded. Carry-over effects are difficult to estimate. They may be minimized by using a uniform treatment between comparison periods.

The switch-back design may be useful in two situations:

1. For preliminary work with two extreme treatments and a small number of units
2. For final work with small differences between treatments and a large number of units

12.4.1 Analysis

Let y_{ijk} = performance of the jth unit of the ith group during the kth period. The data may be tabulated as in Table 12.6. The analysis of variance has only two lines, as shown by the format in Table 12.7. The sums of squares in Table 12.7 are computed as:

$$SST = (n_2 G_1 - n_1 G_2)^2 / [6n_1 n_2 (n_1 + n_2)]$$

$$SSE = (1/6) \sum_i \sum_j D_{ij}^2 - (G_1^2 / 6n_1) - (G_2^2 / 6n^2)$$

To obtain the sample means, first compute

$$\bar{\bar{y}} = [\sum_i \sum_k P_{ik} - (n_1 - n_2)(n_2 G_1 - n_1 G_2)/8n_1 n_2]/[3(n_1 + n_2)]$$

and

$$t = (n_2 G_1 - n_1 G_2)/8n_1 n_2$$

Then

$$\bar{y}_a = \bar{\bar{y}} + t \quad \text{and} \quad \bar{y}_b = \bar{\bar{y}} - t$$

The sample variances of the means, $V(\bar{y}_i)$, and differences between means, $V(\bar{y}_a - \bar{y}_b)$, are:

$$V(y_i) = 3(n_1 + n_2)(MSE)/16n_1 n_2$$

$$V(\bar{y}_a - \bar{y}_b) = 2[V(y_i)] = 3(n_1 + n_2)(MSE)/8n_1 n_2$$

TABLE 12.6 Data Table for a Switch-Back Design

Period	Treatment	Group 1				Sum
1	a	y_{111}	y_{121}	$\cdots$	y_{1n_11}	P_{11}
2	b	y_{112}	y_{122}	$\cdots$	y_{1n_12}	P_{12}
3	a	y_{113}	y_{123}	$\cdots$	y_{1n_13}	P_{13}
Difference		D_{11}	D_{12}	$\cdots$	D_{1n_1}	G_1
		Group 2				
1	b	y_{211}	y_{221}	$\cdots$	y_{2n_21}	P_{21}
2	a	y_{212}	y_{222}	$\cdots$	y_{2n_22}	P_{22}
3	b	y_{213}	y_{223}	$\cdots$	y_{2n_23}	P_{23}
Difference		D_{21}	D_{22}	$\cdots$	D_{2n_2}	G_2

Note: $D_{ij} = y_{ij1} - 2y_{ij2} + y_{ij3}$, $\quad P_{ik} = \sum_j y_{ijk}$, $\quad G_i = P_{i1} - 2P_{i2} + P_{i3}$.

TABLE 12.7 ANOVA for a Switch-Back Design

Source	d.f.	SS	MS
Treatment	1	SST	MST
Error	$n_1 + n_2 - 2$	SSE	MSE

Note: If $n_1 = n_2 = n$

$$SST = (G_1 - G_2)^2/12n$$

$$SSE = (1/6) \sum_i \sum_j D_{ij}^2 - (1/6n)(G_1^2 + G_2^2)$$

$$\bar{\bar{y}} = (\sum_i \sum_k P_{ij})/6n$$

$$t = (G_1 - G_2)/8n$$

$$V(\bar{y}_i) = 3MSE/8n$$

$$V(\bar{y}_a - \bar{y}_b) = 3MSE/4n$$

12.4.2 Numerical Example: Switch-Back Design

The following data are from a study of the effect of concentrate type on the daily production of fat-corrected milk (FCM). Two concentrates were used: *a*, high fat; and *b*, low fat. Five animals were assigned to each of two sequence groups in a switch-back design. During the trial one animal in the first group developed mastitis and was removed from the trial. The results, average daily FCM (pounds), are summarized in Table 12.8.

The analysis of variance of the data in Table 12.8 is shown in Table 12.9. The entries in Table 12.9 are computed as follows:

$$\text{SST} = \frac{(n_2 G_1 - n_1 G_2)^2}{6 n_1 n_2 (n_1 + n_2)} = \frac{[(5)(8.7) - (4)(-11.4)]^2}{(6)(4)(5)(4 + 5)}$$

$$= \frac{(89.1)^2}{1080} = 7.3508$$

$$\text{SSE} = (1/6)(\sum_i \sum_j D_{ij}^2) - (G_1^2/6n_1) - (G_2^2/6n_2)$$

$$= (1/6)(64.27) - (8.7^2/24) - (-11.4^2/30) = 3.2259$$

TABLE 12.8 Average Daily FCM (pounds) from Animals in a Switch-Back Trial

Group 1		Cow				
Period	Treatment	1	2	3	4	Sum
1	*a*	40.8	21.5	48.4	50.3	161.0
2	*b*	35.2	18.4	44.4	45.7	143.7
3	*a*	30.8	17.8	42.7	43.8	135.1
Difference		1.2	2.5	2.3	2.7	8.7

Group 2		Cow					
		1	2	3	4	5	
1	*b*	43.3	27.6	57.8	49.4	36.6	214.7
2	*a*	40.9	30.2	53.2	48.5	35.9	208.7
3	*b*	37.6	27.4	45.5	45.5	35.3	191.3
Difference		−0.9	−5.4	−3.1	−2.1	0.1	−11.4

TABLE 12.9 ANOVA of Average Daily FCM Yields

Source	d.f.	SS	MS	F
Treatment	1	7.3508	7.3508	15.95**
Error	7	3.2259	.4608	

**Significant at the 1% level.

TABLE 12.10 Mean Daily FCM Production (pounds)
with Two Types of Concentrate

Concentrate	High fat	Low fat	Standard error
FCM (lb)	39.64	38.52	.20

Means are computed in the following way:

$$\bar{\bar{y}} = \frac{[\sum_i \sum_k P_{ik} - (n_1 - n_2)(n_2 G_1 - n_1 G_2)/8n_1 n_2]}{3(n_1 + n_2)}$$

$$= [1054.5 - (-1)[(5)(8.7) - (4)(-11.4)]/(8)(20)]/27 = 39.08$$

$$t = \frac{n_2 G_1 - n_1 G_2}{8n_1 n_2} = \frac{(5)(8.7) - (4)(-11.4)}{(8)(4)(5)} = \frac{89.1}{160} = .56$$

Then

$$\bar{y}_a = \bar{\bar{y}} + t = 39.08 + .56 = 39.64$$

$$\bar{y}_b = \bar{\bar{y}} - t = 38.52$$

Standard errors for this trial are

$$s_{\bar{y}_i} = [3(n_1 + n_2)MSE/16n_1 n_2]^{1/2}$$

$$= [(3)(9)(.4608)/(16)(4)(5)]^{1/2} = .1972$$

$$s_{(\bar{y}_a - \bar{y}_b)} = (\sqrt{2})(s_{\bar{y}_i}) = (1.4142)(.1972) = .2789$$

Report of statistical analysis. A study was conducted, using a switch-back design, to compare the effects of two types of concentrate on the production of fat-corrected milk (FCM) by dairy cattle. The results are summarized in Table 12.10.

Difference in daily FCM production was 1.12 lb (standard error = .28) in favor of the high-fat concentrate.

12.5 INCOMPLETE BLOCK SWITCH-BACK DESIGN

The basic switch-back design suffers from the limitation that only two treatments may be compared. However, a number of these basic designs may be combined, using the principles of incomplete block designs, to obtain a single design for comparing any number of treatments in three periods.

Basically, the design consists of blocks of size two, and requires one block for each possible couplet of treatments. For p treatments the design requires

$p(p - 1)$ units in $p(p - 1)/2$ blocks. For example, three treatments, a, b, c, would require six units in three blocks of two. The design would be

Block	I		II		III	
Sequence	1	2	3	4	5	6
Period 1	a	b	a	c	b	c
Period 2	b	a	c	a	c	b
Period 3	a	b	a	c	b	c

Experimental units are assigned at random to the sequences. The number of units on each sequence must be the same for all sequences. Further, the period length must be constant. Carry-over effects are difficult to estimate, and should be minimized by a standard treatment between periods.

The incomplete block switch-back design is useful when several treatments are to be compared and the number of periods must be kept small.

12.5.1 Analysis

In constructing this design it is helpful to form blocks. It is not necessary, however, to consider blocks in the analysis. Let y_{ij} = the response of the ith unit in the jth period. The data may be tabulated as shown in Table 12.11. Before proceeding with the ANOVA we compute Q_k, the adjusted treatment sums. These are summarized in Table 12.12. The entries in Table 12.12 are computed as follows:

$\Sigma D_k F$ = sum of the differences for sequences with the kth treatment in periods 1 and 3

TABLE 12.11 Data Table for an Incomplete Block Switch-Back Trial

Period	Sequence					
	1	2	3	$\cdots$	$p(p - 1)$	Sum
1	y_{11}	y_{21}	y_{31}	$\cdots$	y_{n_1}	P_1
2	y_{12}	y_{22}	y_{32}	$\cdots$	y_{n_2}	P_2
3	y_{13}	y_{23}	y_{33}	$\cdots$	y_{n_3}	P_3
Difference	D_1	D_2	D_3	$\cdots$	D_n	G

Note: D_i $y_{i1} - 2y_{i2} + y_{i3}$, $P_j = \sum_i y_{ij}$,

$G = \sum_i D_i = P_1 - 2P_2 + P_3$.

TABLE 12.12 Adjusted Treatment Sums

Treatment	Sum of D's for sequence with treatment in period 1 and 3	2	Q
1	$\Sigma D_1 F$	$\Sigma D_1 M$	Q_1
2	$\Sigma D_2 F$	$\Sigma D_2 M$	Q_2
3	$\Sigma D_3 F$	$\Sigma D_3 M$	Q_3
$\vdots$	$\vdots$	$\vdots$	$\vdots$
p	$\Sigma D_p F$	$\Sigma D_p M$	Q_p
Sum	G	G	0

TABLE 12.13 ANOVA of an Incomplete Block Switch-Back Trial

Source	d.f.	SS	MS
Total	$p(p - 1) - 1$	SSTot	
Treatment	$p - 1$	SST	MST
Error	$p^2 - 2p$	SSE	MSE

$\Sigma D_k M$ = sum of the differences for sequences with the kth treatment in period 2

$$Q_k = \Sigma D_k F - \Sigma D_k M$$

The analysis of variance takes the form shown in Table 12.13. The sums of squares in Table 12.13 are computed as:

$$SSTot = (1/6)(\sum_i D_i^2) - G^2/6p(p - 1)$$

$$SST = (1/12p)(\sum_k Q_k^2)$$

$$SSE = SSTot - SST$$

The sample means are $\bar{y}_k = \bar{\bar{y}} + Q_k/4p$, where $\bar{\bar{y}} = (\sum_j P_j)/3p(p - 1)$. The sample variance of the means, $V(\bar{y}_k)$, and differences, $V(\bar{y}_k - \bar{y}_{k'})$, are

$$V(\bar{y}_k) = 3(MSE)/4p$$

$$V(\bar{y}_k - \bar{y}_{k'}) = 3(MSE)/2p$$

12.5.2 Numerical Example: Incomplete Block Switch-Back Design

The purchasing agent for the home office of a large insurance company wanted to buy a quantity of word processing machines for use by the secretaries in the stenographic pool. Of the many models on the market he found three which appeared to meet his specifications. He arranged to conduct a test of these three, which he designated as models *a*, *b*, and *c*, by determining the time required by a number of secretaries to process a test document using the three models. He was concerned about differences in basic speed among the secretaries, and he also wanted to adjust for a learning effect if it existed. He decided to use an incomplete block switch-back design in which each of six secretaries would process the test document twice on one model and once on another. The allocation of machines to the secretaries and the processing sequence are displayed in Table 12.14.

The data obtained in the trial are given in Table 12.15, and the adjusted treatment totals are summarized in Table 12.16. The analysis of variance of the processing time data is given in Table 12.17.

TABLE 12.14 Allocation of Word Processing Machines to Secretaries During a Three-Period Sequence in an Incomplete Block Switch-Back Trial

Period	Secretary					
	1	2	3	4	5	6
1	*a*	*b*	*a*	*c*	*b*	*c*
2	*b*	*a*	*c*	*a*	*c*	*b*
3	*a*	*b*	*a*	*c*	*b*	*c*

TABLE 12.15 Time (min) Required to Process a Test Document

Period	Secretary						Sum
	1	2	3	4	5	6	
1	38.7	21.8	48.9	29.9	25.7	22.4	187.4
2	37.4	23.9	43.9	35.1	23.1	26.0	189.4
3	34.4	21.7	42.0	24.5	23.4	20.9	166.9
Difference	−1.7	−4.3	3.1	−15.8	2.9	−8.7	−24.5

TABLE 12.16   Adjusted Treatment Totals

Treatment	$\Sigma D_k F$	$\Sigma D_k M$	Q
a	1.4	−20.1	21.5
b	−1.4	−10.4	9.0
c	−24.5	6.0	−30.5
Sum	−24.5	−24.5	0.0

TABLE 12.17   ANOVA of Document Processing Time

Source	d.f.	SS	MS	F
Total	5	44.1147		
Treatment	2	40.9306	20.4653	19.28*
Error	3	3.1841	1.0614	

*Significant at the 5% level.

The sums of squares in Table 12.17 are computed as follows:

$$\text{SSTot} = (1/6) \sum_i D_i^2 - G^2/6p(p - 1)$$

$$= (1/6)(364.73) - (600.25)/(36) = 44.1147$$

$$\text{SST} = (1/12p) \sum_k Q_k^2 = (1/36)(1473.50) = 40.9306$$

To compute the treatment means $\bar{\bar{y}}$ is obtained as

$$\bar{\bar{y}} = \sum_j P_j/3p(p - 1) = 543.7/18 = 30.20$$

Then

$$\bar{y}_a = \bar{\bar{y}} + Q_a/4p = 30.20 + (21.5/12) = 31.99$$

$$\bar{y}_b = \bar{\bar{y}} + Q_b/4p = 30.20 + (9.0/12) = 30.95$$

$$\bar{y}_c = \bar{\bar{y}} + Q_c/4p = 30.20 + (-30.5/12) = 27.66$$

The associated variances are:
for means

$$V(\bar{y}_k) = 3s^2/4p$$

$$= (3)(1.0614)/(4)(3) = .2654$$

TABLE 12.18 Mean Time (min) Required to Process a
Test Document

Model	a	b	c	Standard error
Time	31.99	30.95	27.66	.52

for differences between means

$$V(\bar{y}_k - \bar{y}_{k'}) = 3s^2/2p$$

$$= (3)(1.0614)/(2)(3) = .5307$$

Report of statistical analysis. An experiment was conducted, using an incomplete block switch-back design, to evaluate three models of word processing machines. The mean times required to process a standard test document are given in Table 12.18.

Processing time was about 3 min faster on model c than on the other machines. There was little to choose between model a and model b.

12.6 LATIN SQUARE CHANGE-OVER, OR ROUND ROBIN, DESIGN

In a Latin square change-over (round robin) design the trial is conducted with the experimental units conforming to the columns of a Latin square, or squares, and the rows of the squares corresponding to periods. Treatments are assigned in such a way that each treatment is tested once on each unit and once in each period within a square. Because each treatment is tested in sequence on every unit the Latin square change-over design has been called a "round robin" design by animal scientists.

Unlike the preceding designs, the estimation of carry-over, or residual, effects is fairly efficient with certain Latin square configurations. These designs include squares or sets of squares which are balanced for residual effects. By this we mean that each treatment is followed by every other treatment an equal number of times. This is done by using either an orthogonal set of squares, or else a single balanced square if the number of treatments is even, or a balanced pair if the number of treatments is odd. The orthogonal set, where available, and a selection of balanced squares for trials with from three to eight treatments are included in the accompanying catalog of designs (Section 12.9).

On considering the design patterns it will be seen that carry-over effects (estimated in the period immediately following the period in which the treatment is administered) form a balanced incomplete block design with respect to units on the one hand and with respect to direct effects on the other. Hence, carry-over effects are partially confounded with both units and direct effects. Direct effects, however, are confounded only with carry-over effects. For this reason, and because of reduced replication, residual effects are estimated with lower precision than are direct effects.

In some cases neither the direct effects nor the carry-over effects are of primary interest. Rather, the "permanent," or equilibrium, effects may be wanted. These are estimated, in the present analysis, as the sum of the direct and carry-over effects. Because of the confounding pattern the direct and carry-over effects are highly positively correlated. As a result, estimates of the permanent effect have a high variance. If pernanent effects are of interest better estimates are obtained from the extra-period Latin square designs to be presented in Section 12.7.

It is theoretically possible to study a large number of treatments with relatively few experimental units by using the Latin square change-over designs. If residual effects are negligible the minimum number of periods and units is equal to the number of treatments. This is also true if carry-over effects are to be estimated and the number of treatments is even. If the number of treatments is odd, however, and carry-over effects are to be estimated, the required number of units is two times the number of treatments. From a practical standpoint, however, periods should be long enough to allow the treatments time to take effect. This places a limit on the number of periods and, hence, on the number of treatments which can be studied. In practice, the most useful size of design will be within the range $3 \leq p \leq 8$. Note also that with small squares ($p = 3, 4$) multiple squares must be used to provide adequate degrees of freedom for error.

If the design being used consists of a single square the experimental units are assigned at random to columns (sequences) in the square. If more than one, say q, squares are being used the qp units should first be grouped into q groups of p units each. The grouping should be done just as is done for a randomized block design. The aim here is to maximize the variation among squares and minimize the variation among units within squares. Units are then assigned at random to columns within squares.

For maximum efficiency the periods should all be of the same length. If this is the case then the analysis may be done on a yield-per-period basis. Periods of different length may be used so long as the variation is not too extreme, say not more than 1 week in five. With periods of different length, however, results must be calculated to a common time basis before analysis.

12.6.1 Analysis

Assume that the design is for p treatments over p periods in q Latin squares with p experimental units per square. Let

$y_{ijk(l)}$ = yield of the jth unit in the ith square subjected to the lth treatment during the kth period

The data may be tabulated as shown in Table 12.19. The sums in Table 12.19 are computed as follows:

$$Q_{ik} = \sum_j y_{ijk(l)} = \text{sum for the } k\text{th period in the } i\text{th square}$$

$$U_{ij} = \sum_k y_{ijk(l)} = \text{sum for the } j\text{th unit in the } i\text{th square}$$

TABLE 12.19 Data Table for a Latin Square Change-Over Design

Square	Period	Unit 11	12	$\cdots$	1p	Sum
1	1	(1) y_{111}	(2) y_{121}	$\cdots$	(p) y_{1p1}	Q_{11}
	2	(2) y_{112}	(3) y_{122}	$\cdots$	(1) y_{1p2}	Q_{12}
	$\vdots$	$\vdots$	$\vdots$		$\vdots$	$\vdots$
	p	(p) y_{11p}	(1) y_{12p}	$\cdots$	(p − 1)y_{1pp}	Q_{1p}
Unit sum		U_{11}	U_{12}	$\cdots$	U_{1p}	S_1
Treatment sum		(1) T_{11}	(2) T_{12}	$\cdots$	(p) T_{1p}	S_1
Residual sum		(2) F_{11}	(2) F_{12}	$\cdots$	(p) F_{1p}	$S_1 - Q_{11}$
		21	22	$\cdots$	2p	
2	1	(2) y_{211}	(3) y_{221}	$\cdots$	(1) y_{2p1}	Q_{21}
	2	(3) y_{212}	(4) y_{222}	$\cdots$	(2) y_{2p2}	Q_{22}
	$\vdots$	$\vdots$	$\vdots$		$\vdots$	$\vdots$
	p	(1) y_{21p}	(2) y_{22p}	$\cdots$	(p) y_{2pp}	Q_{2p}
Unit sum		U_{21}	U_{22}	$\cdots$	U_{2p}	S_2
Treatment sum		(1) T_{21}	(2) T_{22}	$\cdots$	(p) T_{2p}	S_2
Residual sum		(1) F_{21}	(2) F_{22}	$\cdots$	(p) F_{2p}	$S_2 - Q_{21}$
$\vdots$	$\vdots$	$\vdots$	$\vdots$	$\vdots$	$\vdots$	$\vdots$
		q1	q2	$\cdots$	qp	
q	1	(p) y_{q11}	(1) y_{q21}	$\cdots$	(p − 1)y_{qp1}	Q_{q1}
	2	(1) y_{q12}	(2) y_{q22}	$\cdots$	(p) y_{qp2}	Q_{q2}
	$\vdots$	$\vdots$	$\vdots$		$\vdots$	$\vdots$
	p	(p − 1)y_{q1p}	(p) y_{q2p}	$\cdots$	(p − 2)y_{qpp}	Q_{qp}
Unit sum		U_{q1}	U_{q2}	$\cdots$	U_{qp}	S_q
Treatment sum		(1) T_{q1}	(2) T_{q2}	$\cdots$	(p) T_{qp}	S_q
Residual sum		(1) F_{q1}	(2) F_{q2}	$\cdots$	(p) F_{qp}	$S_q - Q_{q1}$

Note: Number in parentheses is the treatment number.

T_{il} = sum for the lth treatment in the ith square

F_{il} = sum of the observations following the lth treatment in the ith square

$S_i = \sum_k Q_{ik} = \sum_j U_{ij}^2 = \sum_l T_{il}$ = sum for the ith square

To begin the analysis two tables of totals, a table of period totals, Table 11.20, and a table of treatment totals, Table 12.21, are computed. The entries in Table 12.21 are computed in the following way:

$D_l = \sum_i T_{il}$ = sum for the lth treatment

$R_l = \sum_i F_{il}$ = sum of the observations following the lth treatment

C_l = sum of the U_{ij} for all units in which treatment l was applied during the last (pth) period

$D_l^* = (p^2 - p - 1)D_l + pR_l + C_l + P_1 - pG$

= adjusted direct effect

$R_l^* = pD_l + p^2R_l + pC_l + pP_1 - (p + 2)G$

= adjusted carry-over effect

$T_l^* = D_l^* + R_l^*$ = adjusted permanent effect

TABLE 12.20   Period Totals

Period	1	2	$\cdots$	p	Sum
Sum	P_1	P_2	$\cdots$	P_p	G

Note: $P_k = \sum_i Q_{ik}$, $G = \sum_k P_k$ = grand total.

TABLE 12.21   Treatment Totals

Treatment	Totals			Adjusted effects		
	Direct[1]	Residual	Last	Direct	Residual	Permanent
1	D_1	R_1	C_1	D_1^*	R_1^*	T_1^*
2	D_2	R_2	C_2	D_2^*	R_2^*	T_2^*
$\vdots$	$\vdots$	$\vdots$	$\vdots$	$\vdots$	$\vdots$	$\vdots$
p	D_p	R_p	C_p	D_p^*	R_p^*	T_p^*
Sum	G	$G - P_1$	G	0	0	0

[1]This is the only entry which is required if carry-over effects are not being estimated.

The analysis of variance depends on whether or not it is necessary to consider carry-over effects. In either case the analyses have a number of common elements. The best procedure is to start as if carry-over effects are negligible, then add the effect of carry-over if it is wanted. The analysis without carry-over is the same as that for a replicated Latin square. The format for the analysis of variance with carry-over is presented in Table 12.22.

The sums of squares in Tables 12.22 are computed as follows: First compute a correction term, C, where

$$C = G^2/qp^2$$

Then

$$SSTot = \sum_i \sum_j \sum_k y_{ijk(l)}^2 - C$$

$$SSS = (1/p^2) \sum_i S_i^2 - C$$

$$SSU = (1/p) \sum_i \sum_j U_{ij}^2 - C - SSS$$

$$SSP = (1/qp) \sum_k P_k^2 - C$$

$$SSPQ = (1/p) \sum_i \sum_k Q_{ik}^2 - C - SSS - SSP$$

$$SSD = (1/qp) \sum_l D_l^2 - C$$

TABLE 12.22 ANOVA for a Latin Square Change-Over Design

Source	d.f.	SS
Total	$qp^2 - 1$	SSTot
Squares	$q - 1$	SSS
Units in squares	$q(p - 1)$	SSU
Periods	$p - 1$	SSP
Period × square	$(q - 1)(p - 1)$	SSPQ
Treatment (unadjusted)	$p - 1$	SSD
Treatment × square	$(q - 1)(p - 1)$	SSDQ
Residual (adjusted)	$p - 1$	SSR*
Error	$(p - 1)(qp - 2p - 1)$	SSE
Treatment (adjusted)	$p - 1$	SSD*
Permanent (adjusted)	$p - 1$	SST*

Note: If the residual effect is not wanted then SSR*, SSD*, and SST* are not computed. In this case SSE has $q(p - 1)(p - 2)$ d.f.

$$SSDQ = (1/p) \sum_i \sum_l T_{i(l)}^2 - C - SSS - SSD$$

$$SSE = SSTot - SSS - SSU - SSP - SSPQ - SSD - SSR* - \\ SSDQ$$

If carry-over effects are not computed the analysis continues as with an ordinary replicated Latin square design. If carry-over effects are to be estimated and tested some additional sums of squares must be computed:

$$SSR* = \left[\frac{1}{qp^3(p + 1)(p - 2)} \right] \sum_l R_l^{*2}, \qquad p - 1 \text{ d.f.}$$

= adjusted carry-over effect sum of squares

$$SSD* = \left[\frac{1}{qp(p + 1)(p - 2)(p^2 - p - 1)} \right] \sum_l D_l^{*2}, \qquad p - 1 \text{ d.f.}$$

= adjusted treatment effect sum of squares

$$SST* = \left[\frac{1}{qp(p + 1)^2(p - 2)(2p - 1)} \right] \sum_l T_l^{*2}, \qquad p - 1 \text{ d.f.}$$

= adjusted permanent effect sum of squares

The means and their sample variances, after adjustment, are obtained as

1. Adjusted treatment mean, $\bar{y}_{dl}$

$$\bar{y}_{dl} = \left[\frac{D_l^*}{qp(p + 1)(p - 2)} \right] + \frac{G}{qp^2}$$

$$V(\bar{y}_{dl}) = \left[\frac{p^2 - p - 1}{qp(p + 1)(p - 2)} \right] MSE*$$

2. Adjusted residual effect, $\bar{r}_l$

$$\bar{r}_l = \left[\frac{R_l^*}{qp(p + 1)(p - 2)} \right]$$

$$V(\bar{r}_l) = \left[\frac{p}{q(p + 1)(p - 2)} \right] MSE*$$

3. Permanent mean, $\bar{y}_{tl}$

$$\bar{y}_{tl} = \left[\frac{T_l^*}{qp(p + 1)(p - 2)} \right] + \frac{G}{qp^2}$$

$$V(\bar{y}_{tl}) = \left[\frac{2p - 1}{qp(p - 2)} \right] MSE*$$

12.6.2 Numerical Example: Latin Square Change-Over Trial

An experimental psychologist wanted to determine the effect of three new drugs, A, B, C, on the time required by laboratory rats to work their way through a maze. She was concerned about differences among individual rats so she decided to use a change-over trial in which each rat receives, in sequence, all three drugs. She was also concerned about estimating and testing carry-over effects if they existed, so she decided to use a Latin square change-over trial. Because the effects she was interested in would be poorly estimated with only one balanced pair of squares she decided to use two, which required the use of 12 rats as her experimental units. To conduct the trial she administered a drug to a rat, placed it at the beginning of the maze, and measured the time required for it to find its way to the end. She then repeated the process, using, in sequence, the other two drugs.

The sequences in which the drugs were administered to the individual rats and the times they required to run the maze after each administration are presented in Table 12.23. The period totals are shown in Table 12.24, and the treatment totals and effects are given in Table 12.25.

The analysis of variance of the run time data was computed under the assumption that carry-over effects were present. The results are shown in Table 12.26. Note: In Table 12.26 the adjusted sums of squares are computed in the following way:

$$SSR^* = \left[\frac{1}{qp^3(p + 1)(p - 2)} \right] \sum_i R_i^{*2} = \left(\frac{1}{432} \right) \sum_i R_i^{*2} = 296.72$$

$$SSD^* = \left[\frac{1}{qp(p + 1)(p - 2)(p^2 - p - 1)} \right] \sum_i D_i^{*2}$$

$$= \left(\frac{1}{240} \right) \sum_i D_i^{*2} = 5524.63$$

$$SST^* = \left[\frac{1}{qp(p + 1)^2(p - 2)(2p - 1)} \right] \sum_i T_i^{*2}$$

$$= \left(\frac{1}{960} \right) \sum_i T_i^{*2} = 2373.23$$

In this set of data the effect of carry-over is not significant. The data can be summarized using either the adjusted or unadjusted treatment means. For illustration, however, the adjusted treatment means, residual effects, permanent means, and their standard errors are given in Table 12.27.

Report of statistical analysis. A Latin square change-over trial was conducted to estimate the effect of three experimental drugs on the time required

TABLE 12.23 Time Required (sec) for Rats to Run a Maze After the Administration of Three Experimental Drugs

Square	Period	Rat 11	12	13	Sum
	1	A 138	B 209	C 224	571
	2	B 125	C 186	A 172	483
	3	C 115	A 139	B 127	381
Unit sum		378	534	523	1435
Treatment sum		A 449	B 461	C 525	1435
Residual sum		A 252	B 301	C 311	864
		21	22	23	
	1	A 186	B 175	C 201	562
	2	C 176	A 135	B 163	474
	3	B 146	C 134	A 101	381
Unit sum		508	444	465	1417
Treatment sum		A 422	B 484	C 511	1417
Residual sum		A 310	B 236	C 309	855
		31	32	33	
	1	A 166	B 194	C 186	546
	2	B 152	C 180	A 130	462
	3	C 137	A 97	B 123	357
Unit sum		455	471	439	1365
Treatment sum		A 393	B 469	C 503	1365
Residual sum		A 275	B 317	C 227	819
		41	42	43	
	1	A 138	B 164	C 168	470
	2	C 154	A 128	B 150	432
	3	B 129	C 137	A 106	372
Unit sum		421	429	424	1274
Treatment sum		A 372	B 443	C 459	1274
Residual sum		A 291	B 234	C 279	804

Note: Letter designates drug.

TABLE 12.24 Period Totals

Period	1	2	3	Sum
Sum	2149	1851	1491	5491

TABLE 12.25 Treatment Totals

	Totals			Adjusted effects		
Treatment	Direct	Residual	Last	Direct	Residual	Permanent
A	1636	1128	1894	−866	−266	−1132
B	1857	1088	1891	116	28	144
C	1998	1126	1706	750	238	988
Sum	5491	3342	5491	0	0	0

TABLE 12.26 ANOVA of Times (sec) Required by Rats to Run a Maze After Administration of Three Drugs

Source	d.f.	SS	MS	F
Total	35	33,624.97		
Squares	3	1,738.30	579.43	6.88*
Rats in squares	8	5,944.67	743.08	8.82**
Periods	2	18,093.56	9,046.78	107.43**
Period × square	6	1,001.11	166.85	1.98
Drug (unadjusted)	2	5,549.06	2,774.53	32.95**
Drug × square	6	496.28	82.71	.98
Carry-over (adjusted)	2	296.72	148.36	1.76
Error	6	505.27	84.21	
Drug (adjusted)	2	5,524.63	2,762.32	32.80**
Permanent (adjusted)	2	2,373.23	1,186.62	14.09**

*Significant at the 5% level.
**Significant at the 1% level.

TABLE 12.27 Adjusted Mean Run Times (sec) and
Their Standard Errors

Drug	Direct	Residual	Permanent
A	134.48	− 5.54	128.94
B	154.95	.58	155.53
C	168.16	4.96	173.12
Standard error	2.96	3.97	5.92

TABLE 12.28 Mean Time (sec) Required by Rats to Find
Their Way Through a Maze After Administration of Three Drugs

Drug	*A*	*B*	*C*	Standard error
Time	134.48	154.95	168.16	2.96

by rats to find their way through a maze. The results are summarized in Table 12.28.

The differences in run times among the three were highly significant. Drug *A* produced the fastest time while drug *C* produced the slowest. The effect of the drug lasted only for the run following its administration. There was apparently no real residual effect in the succeeding period.

There were significant differences in the rats' inherent ability to solve the maze. Further, there appeared to be a learning effect since the mean run time decreased with each succeeding run. The experimental design used in this trial appeared to give satisfactory control of differences among rats and period-to-period differences.

12.7 EXTRA-PERIOD LATIN SQUARE CHANGE-OVER DESIGN

The extra-period Latin square change-over designs can be most easily described as Latin square change-over designs with the last period repeated. Hence, the number of units required is the same as the number of units needed for a Latin square. To study *p* treatments, however, requires *p* + 1 periods. The extra-period designs are especially adapted for the study of carry-over effects and the efficient estimation of permanent effects.

As with the ordinary Latin square change-over designs, the efficient estimation of carry-over effects is possible only with a limited number of design configurations. The pertinent designs are those formed by repeating the final period of the orthogonal sets of squares or the balanced squares in the catalog of Latin square change-over designs. Hence, no additional catalog will be given for the extra-period designs.

The extra-period designs are better adapted for estimating residual effects than are the regular Latin squares. This is because direct effects rather than residual effects are confounded with the experimental units. This, in part, offsets the greater replication of the direct effects. Further, in the extra-period designs the direct effects are orthogonal to the residual effects and, hence, are independent of each other. This tends to minimize the variance of an estimated permanent effect.

As with the Latin square change-over designs, the most useful range for the extra-period designs is $3 \leq p \leq 8$. If it is certain that carry-over effects are not a problem then it is more efficient to use the Latin square designs, incomplete Latin square designs, or switch-back designs.

12.7.1 Analysis

Assume that the design is for p treatments over $p + 1$ periods in q squares with p units per square. Let

$y_{ijk(l)}$ = yield of the jth unit in the ith square subjected to the lth treatment during the kth period

The data are tabulated as in Table 12.29. The sums in Table 12.29 are computed as follows:

$$Q_{ik} = \sum_j y_{ijk(l)} = \text{sum for the } k\text{th period in the } i\text{th square}$$

$$U_{ij} = \sum_k y_{ijk(l)} = \text{sum for the } j\text{th unit in the } i\text{th square}$$

$$T_{il} = \text{sum for the } l\text{th treatment in the } i\text{th square}$$

$$F_{il} = \text{sum of the observations following the } l\text{th treatment in the } i\text{th square}$$

$$S_i = \sum_k Q_{ik} = \sum_j U_{ij} = \sum_l T_{il} = \text{sum for the } i\text{th square}$$

Two tables of totals, a table of period totals as shown in Table 12.30 and a table of treatment totals, Table 12.31, are now computed. The entries in Table 12.31 are computed as follows:

$$D_l = \sum_i T_{il} = \text{sum for the } l\text{th treatment}$$

$$R_l = \sum_i F_{il} = \text{sum of the observations following the } l\text{th treatment}$$

C_l = sum of the U_{ij} for all units in which treatment l was applied during the last two periods

TABLE 12.29 Data Table for an Extra-Period Latin Square Change-Over Design

Square	Period	Unit 11	12	$\cdots$	$1p$	Sum
1	1	(1) y_{111}	(2) y_{121}	$\cdots$	(p) y_{1p1}	Q_{11}
	2	(2) y_{112}	(3) y_{122}	$\cdots$	(1) y_{1p2}	Q_{12}
	$\vdots$	$\vdots$	$\vdots$		$\vdots$	$\vdots$
	$p+1$	(p) y_{11p+1}	(1) y_{12p+1}	$\cdots$	$(p-1)y_{1pp+1}$	Q_{1p+1}
Unit sum		U_{11}	U_{12}	$\cdots$	U_{1p}	S_1
Treatment sum		(1) T_{11}	(2) T_{12}	$\cdots$	(p) T_{1p}	S_1
Residual sum		(1) F_{11}	(2) F_{12}	$\cdots$	(p) F_{1p}	$S_1 - Q_{11}$
		21	22	$\cdots$	$2p$	
2	1	(2) y_{211}	(3) y_{221}	$\cdots$	(1) y_{2p1}	Q_{21}
	2	(3) y_{212}	(4) y_{222}	$\cdots$	(2) y_{2p2}	Q_{22}
	$\vdots$	$\vdots$	$\vdots$		$\vdots$	$\vdots$
	$p+1$	(1) y_{21p+1}	(2) y_{22p+1}	$\cdots$	(p) y_{2pp+1}	Q_{2p+1}
Unit sum		U_{21}	U_{22}	$\cdots$	U_{2p}	S_2
Treatment sum		(1) T_{21}	(2) T_{22}	$\cdots$	(p) T_{2p}	S_2
Residual sum		(1) F_{21}	(2) F_{22}	$\cdots$	(p) F_{2p}	$S_2 - Q_{21}$
$\vdots$	$\vdots$	$\vdots$	$\vdots$		$\vdots$	$\vdots$
		$q1$	$q2$	$\cdots$	qp	
q	1	(p) y_{q11}	(1) y_{q21}	$\cdots$	$(p-1)y_{qp1}$	Q_{q1}
	2	(1) y_{q12}	(2) y_{q22}	$\cdots$	(p) y_{qp2}	Q_{q2}
	$\vdots$	$\vdots$	$\vdots$		$\vdots$	$\vdots$
	$p+1$	$(p-1)y_{q1p+1}$	(p) y_{q2p+1}	$\cdots$	$(p-2)y_{qpp+1}$	Q_{qp+1}
Unit sum		U_{q1}	U_{q2}	$\cdots$	U_{qp}	S_q
Treatment sum		(1) T_{q1}	(2) T_{q2}	$\cdots$	(p) T_{qp}	S_q
Residual sum		(1) F_{q1}	(2) F_{q2}	$\cdots$	(p) F_{qp}	$S_q - Q_{q1}$

Note: Number is parentheses is the treatment number.

$$D_l^* = (p + 1)D_l - C_l - G$$
$$T_l^* = pD_l^* + p(p + 2)R_l - (p + 2)(G - P_l)$$

The analysis proceeds in essentially the same way whether or not the residuals are to be estimated and tested. The format for the analysis of variance of data from an extra-period Latin square change-over design is shown in Table 12.32.

TABLE 12.30 Period Totals

Period	1	2	$\cdots$	p	$p + 1$	Sum
Sum	P_1	P_2	$\cdots$	P_p	P_{p+1}	G

Note: $P_k = \sum_i Q_{ik}$, $G = \sum_k P_k$.

TABLE 12.31 Treatment Totals

	Totals			Adjusted effects	
Treatment	Direct	Residual	Last	Direct	Permanent
1	D_1	R_1	C_1	D_1^*	T_1^*
2	D_2	R_2	C_2	D_2^*	T_2^*
$\vdots$	$\vdots$	$\vdots$	$\vdots$	$\vdots$	$\vdots$
p	D_p	R_p	C_p	D_p^*	T_p^*
Sum	G	$G - P_1$	G	0	0

TABLE 12.32 ANOVA for an Extra-Period Latin Square Change-Over Trial

Source	d.f.	SS
Total	$qp(p + 1) - 1$	SSTot
Squares	$q - 1$	SSS
Units in squares	$q(p - 1)$	SSU
Periods	p	SSP
Periods $\times$ squares	$p(q - 1)$	SSPQ
Treatment (adjusted)	$p - 1$	SSD
Residual (adjusted)	$p - 1$	SSR
Error	$(qp - 2)(p - 1)$	SSE
Permanent (adjusted)	$p - 1$	SST

Note: If the residual effect is not wanted then SSR and SST are not computed. In this case SSE has $(qp - 1)(p - 1)$ d.f.

The sums of squares in Table 12.32 are computed in the following way: First, compute a correction term, C, where

$$C = G^2/qp(p + 1)$$

Then

$$\text{SSTot} = \sum_i \sum_j \sum_k y_{ijk(l)}^2 - C$$

$$\text{SSS} = [1/p(p + 1)] \sum_i S_i^2 - C$$

$$\text{SSU} = [1/(p + 1)] \sum_i \sum_j U_{ij}^2 - C - \text{SSS}$$

$$\text{SSP} = (1/qp) \sum_k P_k^2 - C$$

$$\text{SSQP} = (1/p) \sum_i \sum_k Q_{ik}^2 - C - \text{SSS} - \text{SSP}$$

$$\text{SSD} = (\sum_l D_l^{*2})/qp(p + 1)(p + 2)$$

$$\text{SSR} = (1/qp) \sum_l R_l^2 - (G - P_1)^2/qp^2$$

$$\text{SSE} = \text{SSTot} - \text{SSS} - \text{SSU} - \text{SSP} - \text{SSPQ} - \text{SSD} - \text{SSR}$$

$$\text{SST} = (\sum_l T_l^{*2})/qp^2(p + 1)(2p + 3)$$

The means and their sample variances are obtained as follows:

1. Treatment mean, $\bar{y}_{dl}$

$$\bar{y}_{dl} = [D_l^*/qp(p + 2)] + [G/qp(p + 1)]$$

$$V(\bar{y}_{dl}) = (p + 1)\text{MSE}/qp(p + 2)$$

2. Residual effect, $\bar{r}_l$

$$\bar{r}_l = (pR_l - G + P_1)/qp^2$$

$$V(\bar{r}_l) = \text{MSE}/qp$$

3. Permanent effect mean, $\bar{y}_{tl}$

$$\bar{y}_{tl} = [T_l^*/qp^2(p + 2)] + [G/qp(p + 1)]$$

$$V(\bar{y}_{tl}) = (2p + 3)\text{MSE}/qp(p + 2)$$

12.7.2 Numerical Example: Extra Period Latin Square Change-Over Trial

A dairy nutritionist wanted to measure the effect of three protein supplements on the production of 4% fat-corrected milk (FCM) by dairy cows. He wanted to use a sequential trial so that he could control differences among different

TABLE 12.33 Average Daily FCM (lb) from Cows Fed Three Protein Supplements

Square	Period	Cow 11		Cow 12		Cow 13		Sum
1	1	A	38.7	B	52.8	C	32.6	124.1
	2	B	41.4	C	44.9	A	32.3	118.6
	3	C	32.4	A	42.0	B	32.5	106.9
	4	C	29.3	A	39.6	B	31.1	100.0
Cow sum			141.8		179.3		128.5	449.6
Treatment sum		A	152.6	B	157.8	C	139.2	449.6
Residual sum		A	113.5	B	108.4	C	103.6	325.5
		21		22		23		
2	1	A	35.2	B	36.9	C	28.4	100.5
	2	C	31.5	A	33.1	B	33.5	98.1
	3	B	32.4	C	25.5	A	26.7	84.6
	4	B	29.1	C	23.1	A	23.1	75.3
Cow sum			128.2		118.6		111.7	358.5
Treatment sum		A	118.1	B	131.9	C	108.5	358.5
Residual sum		A	80.1	B	88.9	C	89.0	258.0
		31		32		33		
3	1	A	25.7	B	34.8	C	23.4	83.9
	2	B	30.1	C	27.3	A	26.0	83.4
	3	C	21.4	A	26.4	B	27.9	75.7
	4	C	16.7	A	23.2	B	23.9	63.8
Cow sum			93.9		111.7		101.2	306.8
Treatment sum		A	101.3	B	116.7	C	88.8	306.8
Residual sum		A	81.2	B	72.6	C	69.1	222.9
		41		42		43		
4	1	A	21.8	B	25.8	C	20.8	68.4
	2	C	21.9	A	21.9	B	25.0	68.8
	3	B	25.7	C	17.4	A	18.6	61.7
	4	B	21.6	C	14.6	A	16.1	52.3
Cow sum			91.0		79.7		80.5	251.2
Treatment sum		A	78.4	B	98.1	C	74.7	251.2
Residual sum		A	55.4	B	62.1	C	65.3	182.8

Note: Letter designates supplement.

cows. He was also concerned about the carry-over effect of a supplement into the period following the period in which it was used. He decided to use two pairs of balanced Latin squares, and to repeat the last period, generating an extra-period design.

He obtained 12 lactating Guernsey cows and randomly assigned them to the sequences shown in Table 12.33. Each cow was fed the assigned protein supplement, *A, B,* or *C,* for a 5-week test period. Average daily FCM production by each cow during each test period was determined. The results are given in Table 12.33.

Before computing the entries in an analysis of variance table the nutritionist first computed a table of period totals, Table 12.34, and a table of treatment totals and effects, Table 12.35. Then the analysis of variance, shown in Table 12.36, was completed.

There were no significant carry-over effects in this trial so we could use the unadjusted treatment means to summarize the results. For illustration, however, we will compute the adjusted treatment means, the residual effect means, and the permanent effect means, along with the appropriate standard errors. These are displayed in Table 12.37.

Report of statistical analysis.   A trial was conducted to determine the effect of three protein supplements on the yield of 4% fat-corrected milk by dairy cattle. The results are summarized in Table 12.38.

Clearly, supplement *B* produced the highest yield while supplement *C* produced the lowest. As expected, average production differed from animal to animal, and average production decreased from the beginning of the trial to the end. The protein supplements affected FCM production only during the periods in which they were fed. There was no evidence of a carry-over effect.

12.8   INCOMPLETE LATIN SQUARE CHANGE-OVER DESIGN

The incomplete Latin squares (Youden squares, or Latin rectangles) are the incomplete block counterpart of the Latin squares. These can be constructed in a number of ways. For our purposes, however, we will restrict consideration to those designs in which the number of units in each square is equal to the number of treatments in the experiment. The number of periods is less than the number of treatments. As with the Latin square and extra-period Latin square change-over trials, columns correspond to experimental units and rows correspond to periods in these designs.

From a practical standpoint we will consider only designs which conform to the following restrictions:

1. $p \geq 5$, at least five treatments
2. $3 \leq w \leq 5$, at least three, but not more than five periods
3. $n \leq 60$, not more than 60 experimental units

TABLE 12.34 Period Totals

Period	1	2	3	4	Sum
Sum	376.9	368.9	328.9	291.4	1366.1

TABLE 12.35 Treatment Totals

	Totals			Adjusted effects	
Treatment	Direct	Residual	Last	Direct	Permanent
A	450.4	330.2	483.2	−47.7	−136.1
B	504.5	332.0	448.9	203.0	643.0
C	411.2	327.0	434.0	−155.3	−506.9
Sum	1366.1	989.2	1366.1	0	0

TABLE 12.36 ANOVA of Average Daily FCM Production with Three Protein Supplements

Source	d.f.	SS	MS	F
Total	47	2923.83		
Squares	3	1777.73	592.58	823.03**
Cows in squares	8	441.23	55.15	76.60**
Periods	3	389.39	129.80	180.28**
Periods × squares	9	18.23	2.02	2.80
Protein (adjusted)	2	281.68	140.84	195.61**
Residual (adjusted)	2	1.07	.53	.74
Error	20	14.50	.72	
Permanent (adjusted)	2	531.57	265.79	369.15**

**Significant at the 1% level.

A catalog of incomplete Latin squares is given in Section 12.10.

In the incomplete Latin squares both direct and residual effects are confounded with the experimental units, with residual effects confounded to a greater degree. Further, direct and residual effects are confounded with each other. Because of this confounding their estimates are positively correlated and the precision of the estimated permanent effects is low. The sensitivity of the incomplete designs is less than that of the full Latin square designs. On the other hand,

TABLE 12.37 Adjusted Mean FCM Yields (lb) and
Their Standard Errors

Protein	Direct	Residual	Permanent
A	27.66	.04	27.70
B	31.84	.19	32.03
C	25.87	-.23	25.64
Standard error	.22	.24	.33

TABLE 12.38 Mean Daily FCM Production (lb) by Dairy
Cattle Fed Three Protein Supplements

Supplement	A	B	C	Standard error
FCM	27.66	31.84	25.87	.22

the incomplete squares permit the study of larger numbers of treatments without
the need for an excessive number of periods.

12.8.1 Analysis

Assume that the design is for p treatments over $w < p$ periods in q squares with
p units per squares. Let

$$y_{ijk(l)} = \text{yield of the } j\text{th unit in the } i\text{th square receiving the } l\text{th treatment}$$
during the kth period

The data are tabulated as in Table 12.39, with the sums computed as follows:

$$Q_{ik} = \sum_j y_{ijk(l)} = \text{sum for the } k\text{th period in the } i\text{th square}$$

$$U_{ij} = \sum_k y_{ijk(l)} = \text{sum for the } j\text{th unit in the } i\text{th square}$$

$T_{il} = $ sum for the lth treatment in the ith square

$F_{il} = $ sum of the observations following the lth treatment in the ith square

$V_{il} = $ sum of the U_{ij} for units receiving the lth treatment in any period

$Z_{il} = $ sum of the U_{ij} for units receiving the lth treatment in any but the
last period

$$S_i = \sum_k Q_{ik} = \sum_j U_{ij} = \sum_l T_{il} = \text{sum for the } i\text{th square}$$

TABLE 12.39 Data Table for an Incomplete Latin Square Change-Over Design

Square	Period	Unit 11		Unit 12		$\cdots$	Unit 1p		Sum
1	1	(1)	y_{111}	(2)	y_{121}	$\cdots$	(p)	y_{1p1}	Q_{11}
	2	(2)	y_{112}	(3)	y_{122}	$\cdots$	(1)	y_{1p2}	Q_{12}
	$\vdots$		$\vdots$		$\vdots$			$\vdots$	$\vdots$
	w	(w)	y_{11w}	(1)	y_{12w}	$\cdots$	$(w-1)$	y_{1pw}	Q_{1w}
U sum			U_{11}		U_{12}	$\cdots$		U_{1p}	S_1
T sum		(1)	T_{11}	(2)	T_{12}	$\cdots$	(p)	T_{1p}	S_1
F sum		(1)	F_{11}	(2)	F_{12}	$\cdots$	(p)	F_{1p}	$S_1 - Q_{11}$
V sum		(1)	V_{11}	(2)	V_{12}	$\cdots$	(p)	V_{1p}	wS_1
Z sum		(1)	Z_{11}	(2)	Z_{12}	$\cdots$	(p)	Z_{1p}	$(w-1)S_1$
		21		22		$\cdots$	2p		
2	1	(2)	y_{211}	(3)	y_{221}	$\cdots$	(1)	y_{2p1}	Q_{21}
	2	(3)	y_{212}	(4)	y_{222}	$\cdots$	(2)	y_{2p2}	Q_{22}
	$\vdots$		$\vdots$		$\vdots$			$\vdots$	$\vdots$
	w	(1)	y_{21w}	(2)	y_{22w}	$\cdots$	(w)	y_{2pw}	Q_{2w}
U sum			U_{21}		U_{22}	$\cdots$		U_{2p}	S_2
T sum		(1)	T_{21}	(2)	T_{22}	$\cdots$	(p)	T_{2p}	S_2
F sum		(1)	F_{21}	(2)	F_{22}	$\cdots$	(p)	F_{2p}	$S_2 - Q_{21}$
V sum		(1)	V_{21}	(2)	V_{22}	$\cdots$	(p)	V_{2p}	wS_2
Z sum		(1)	Z_{21}	(2)	Z_{22}	$\cdots$	(p)	Z_{2p}	$(w-1)S_2$
$\vdots$		$\vdots$		$\vdots$			$\vdots$		$\vdots$
		$q1$		$q2$		$\cdots$	qp		
q	1	(p)	y_{q11}	(1)	y_{q21}	$\cdots$	$(p-1)$	y_{qp1}	Q_{q1}
	2	(1)	y_{q12}	(2)	y_{q22}	$\cdots$	(p)	y_{qp2}	Q_{q2}
	$\vdots$		$\vdots$		$\vdots$			$\vdots$	$\vdots$
	w	$(w-1)$	y_{q1w}	(w)	y_{q2w}	$\cdots$	$(w-2)$	y_{qpw}	Q_{qw}
U sum			U_{q1}		U_{q2}	$\cdots$		U_{qp}	S_q
T sum		(1)	T_{q1}	(2)	T_{q2}	$\cdots$	(p)	T_{qp}	S_q
F sum		(1)	F_{q1}	(2)	F_{q2}	$\cdots$	(p)	F_{qp}	$S_q - Q_{q1}$
V sum		(1)	V_{q1}	(2)	V_{q2}	$\cdots$	(p)	V_{qp}	wS_q
Z sum		(1)	Z_{q1}	(2)	Z_{q2}	$\cdots$	(p)	Z_{qp}	$(w-1)S_q$

Note: Number in parentheses designates the treatment.

TABLE 12.40   Period Totals

Period	1	2	$\cdots$	w	Sum
Sum	P_1	P_2	$\cdots$	P_w	G

Note: $P_k = \sum_i Q_{ik}$, $G = \sum_k P_k$.

TABLE 12.41   Treatment Totals

Treatment	Totals				Adjusted effects			
	D	R	V	Z	A^*	D^*	R^*	T^*
1	D_1	R_1	V_1	Z_1	A_1^*	D_1^*	R_1^*	T_1^*
2	D_2	R_2	V_2	Z_2	A_2^*	D_2^*	R_2^*	T_2^*
$\vdots$	$\vdots$	$\vdots$	$\vdots$	$\vdots$	$\vdots$	$\vdots$	$\vdots$	$\vdots$
p	D_p	R_p	V_p	Z_p	A_p^*	D_p^*	R_p^*	T_p^*
Sum	G	$G - P_1$	wG	$(w-1)G$	0	0	0	0

The next step is to compute a table of period totals, Table 12.40, and a table of treatment totals, Table 12.41.

The entries in Table 12.41 are computed as follows:

$$D_l = \sum_i T_{il} = \text{sum for the } l\text{th treatment}$$

$$R_l = \sum_i F_{il} = \text{sum of the observations following the } l\text{th treatment}$$

$$V_l = \sum_i V_{il} = \text{sum of all units receiving the } l\text{th treatment in any period}$$

$$Z_l = \sum_i Z_{il} = \text{sum of all units receiving the } l\text{th treatment in any but the}$$

last period

$$A_l^* = wD_l - V_l$$

$$D_l^* = (pw - p - 1)A_l^* + pwR_l - pZ_l + wP_1 - G$$

$$R_l^* = wD_l^* - (pw^2 - pw - p - w)A_l^*$$

$$T_l^* = D_l^* + R_l^*$$

The format for the analysis of variance of data from an incomplete Latin square change-over design is shown in Table 12.42.

TABLE 12.42 ANOVA for an Incomplete Latin Square Change-Over Design

Source	d.f.	SS
Total	$qpw - 1$	SSTot
Square	$q - 1$	SSS
Unit in square	$q(p - 1)$	SSU
Period	$w - 1$	SSP
Period × square	$(q - 1)(w - 1)$	SSQP
Treatment (unadjusted)	$p - 1$	SSD
Residual (adjusted)	$p - 1$	SSR*
Error	$q(p - 1)(w - 1) - 2(p - 1)$	SSE
Treatment (adjusted)	$p - 1$	SSD*
Permanent (adjusted)	$p - 1$	SST*

The sums of squares in Table 12.42 are computed as follows: First compute a correction term, C, as

$$C = G^2/qpw$$

Then

$$\text{SSTot} = \sum_i \sum_j \sum_k y^2_{ijk(l)} - C$$

$$\text{SSS} = (1/pw) \sum_i S_i^2 - C$$

$$\text{SSU} = (1/w) \sum_i \sum_j U_{ij}^2 - C - \text{SSS}$$

$$\text{SSP} = (1/qp) \sum_k P_k^2 - C$$

$$\text{SSQP} = (1/p) \sum_i \sum_k Q_{ik}^2 - C - \text{SSS} - \text{SSP}$$

$$\text{SSD} = [(p - 1)/(w - 1)qpw^2] \sum_l A_l^{*2}$$

$$\text{SSR*} = \left[\frac{p - 1}{qp^2 w^2(w - 1)(pw^2 - pw - p - w)} \right] \sum_l R_l^{*2}$$

$$\text{SSE} = \text{SSTot} - \text{SSS} - \text{SSU} - \text{SSP} - \text{SSQP} - \text{SSD} - \text{SSR*}$$

$$\text{SSD*} = \left[\frac{p - 1}{qpw(w - 1)(pw - p - 1)(pw^2 - pw - p - w)} \right] \sum_l D_l^{*2}$$

$$\text{SST*} = \left[\frac{p - 1}{qpw(w - 1)(2pw + p - 1)(pw^2 - pw - p - w)} \right] \sum_i T_i^{*2}$$

Note: If carry-over effects are not wanted it is not necessary to compute the F sums, V sums, and Z sums in Table 12.39, nor is it necessary to compute any but the D totals in Table 12.41. In this case SSR*, SSD*, and SST* in the ANOVA of Table 12.42 are not computed. The treatment sum of squares, SSD, is obtained from

$$\text{SSD} = (1/qw) \sum_i D_i^2 - C, \qquad p - 1 \text{ d.f.}$$

$$\text{SSE} = \text{SSTot} - \text{SSS} - \text{SSU} - \text{SSP} - \text{SSQP} - \text{SSD}, \qquad q(p - 1)(w - 2) \text{ d.f.}$$

The means and their sample variances are obtained in the following way:

1. Adjusted treatment mean, $\bar{y}_{dl}$

$$\bar{y}_{dl} = \frac{(p - 1)D_l^*}{qp(w - 1)(pw^2 - pw - p - w)} + \frac{G}{qpw}$$

$$V(\bar{y}_{dl}) = \left[\frac{w(p - 1)(pw - p - 1)}{qp(w - 1)(pw^2 - pw - p - w)} \right] \text{MSE}$$

2. Adjusted residual effect, $\bar{r}_l$

$$\bar{r}_l = \frac{(p - 1)R_l^*}{pq(w - 1)(pw^2 - pw - p - w)}$$

$$V(\bar{r}_l) = \left[\frac{w^2(p - 1)}{q(w - 1)(pw^2 - pw - p - w)} \right] \text{MSE}$$

3. Adjusted permanent effect mean, $\bar{y}_{tl}$

$$\bar{y}_{tl} = \bar{y}_{dl} + \bar{r}_l$$

$$V(\bar{y}_{tl}) = \left[\frac{w(p - 1)(2pw + p - 1)}{qp(w - 1)(pw^2 - pw - p - w)} \right] \text{MSE}$$

12.8.2 Numerical Example: Incomplete Latin Square Change-Over Trial

A client of a national advertising agency had developed five package designs, A, B, C, D, E, for one of its products. The client asked the agency to determine which of the five designs was most effective in promoting sales of the product. The account executive decided to run an in-store test in which the designs would be displayed on successive weeks. Product sales would be used to measure effectiveness of the design. The account executive selected five stores in each of four cities in which to run the test. He wanted to limit the trial to a total of 3 weeks. He was also concerned about differences among stores, and about the lasting appeal of the designs. Accordingly, he decided to use an incomplete Latin square change-over design in which different designs would be displayed in a store on each of three successive weeks.

The display sequence in the several stores and the sales, in dollars to the nearest tenth, are presented in Table 12.43.

To complete the analysis of the resulting data the weekly total sales were computed and are shown in Table 12.44. Additionally, various treatment totals and effects were computed. These are given in Table 12.45.

The analysis of variance of the sales data is summarized in Table 12.46.

The adjusted means and their associated sample variances are given in Table 12.47.

Report of statistical analysis. A study was conducted to evaluate the effect of five different package designs on the sales of a product sold in supermarkets. Four cities were selected for the test, and five markets in each city were chosen to evaluate the product. The test consisted of displaying different designs in each store during three successive test periods of a week each. Total sales for the week were taken as an indicator of design effectiveness.

Mean sales during the week of display, lasting impact of the design, as measured by the residual effect in the week following display, and the long-time effect (permanent effect) of the five designs are presented in Table 12.48.

Consideration of Table 12.48 leads to the conclusion that design *C* was clearly more effective than any of the others. It had the highest mean sales during the week of display, had the highest residual impact in the week following its display, and the highest adjusted permanent effect mean sales. There was little real difference among the other designs as regards direct sales, carry-over effects, or permanent effects.

Mean sales from city to city, averaged over design, were substantially different and there was significant variation in sales among stores within cities. Mean sales decreased from the first week of the test to the last test week. The rate of decrease, however, differed from city to city. The incomplete Latin square design used for this study was effective in controlling variation among cities, stores within cities, and sales period. The precision with which the effects were estimated appeared to be quite high with the design used here.

TABLE 12.43 Treatment Sequences and Sales ($) for an Experiment Designed to Estimate the Effect of Five Package Designs

City	Week	Store 11		Store 12		Store 13		Store 14		Store 15		Sum
1	1	B	38.7	C	50.9	E	34.6	D	35.2	A	37.4	196.8
	2	C	39.4	D	47.9	A	32.3	E	33.5	B	34.3	187.4
	3	D	35.3	E	42.0	B	28.5	A	28.4	C	33.3	167.5
Store Sum			113.4		140.8		95.4		97.1		105.0	551.7
Treatment sum		A	98.1	B	101.5	C	123.6	D	118.4	E	110.1	551.7
F sum		A	62.8	B	72.7	C	83.2	D	75.5	E	60.7	354.9
V sum		A	297.5	B	313.8	C	359.2	D	351.3	E	333.3	1,655.1
Z sum		A	200.4	B	218.4	C	254.2	D	237.9	E	192.5	1,103.4
		21		22		23		24		25		
2	1	C	34.9	D	30.4	E	30.8	A	25.7	B	46.9	168.7
	2	E	34.1	A	29.5	B	29.3	C	28.6	D	42.0	163.5
	3	B	27.5	C	28.7	D	26.4	E	24.4	A	39.6	146.6
Store sum			96.5		88.6		86.5		78.7		128.5	478.8
Treatment sum		A	94.8	B	103.7	C	92.2	D	98.8	E	89.3	478.8
F sum		A	57.3	B	68.4	C	58.5	D	69.1	E	56.8	310.1
V sum		A	295.8	B	311.5	C	263.8	D	303.6	E	261.7	1,436.4
Z sum		A	167.3	B	215.0	C	175.2	D	217.1	E	183.0	957.6
		31		32		33		34		35		
3	1	D	25.4	E	21.8	A	21.4	B	22.8	C	28.0	119.4
	2	B	26.0	C	25.9	D	22.0	E	21.0	A	24.9	119.8
	3	E	23.9	A	22.7	B	19.4	C	20.6	D	19.9	106.5
Store sum			75.3		70.4		62.8		64.4		72.8	345.7
Treatment sum		A	69.0	B	68.2	C	74.5	D	67.3	E	66.7	345.7
F sum		A	41.9	B	44.9	C	47.6	D	45.4	E	46.5	226.3
V Sum		A	206.0	B	202.5	C	207.6	D	210.9	E	210.1	1,037.1
Z sum		A	135.6	B	139.7	C	143.2	D	138.1	E	134.8	691.4
		41		42		43		44		45		
4	1	E	32.3	A	33.5	B	33.1	C	31.3	D	26.1	156.3
	2	D	28.5	E	28.4	A	27.5	B	27.4	C	25.4	137.2
	3	C	30.1	D	25.1	E	25.1	A	23.2	B	19.7	123.2
Store sum			90.9		87.0		85.7		81.9		71.2	416.7
Treatment sum		A	84.2	B	80.2	C	86.8	D	79.7	E	85.8	416.7
F sum		A	53.5	B	50.7	C	47.1	D	55.5	E	53.6	260.4
V sum		A	254.6	B	238.8	C	244.0	D	249.1	E	263.6	1,250.1
Z sum		A	172.7	B	167.6	C	153.1	D	162.1	E	177.9	833.4

TABLE 12.44 Weekly Total Sales

Week	1	2	3	Sum
Sum	641.2	607.9	543.8	1792.9

TABLE 12.45 Treatment Totals

	Totals				Adjusted effects			
Treatment	D	R	V	Z	A*	D*	R*	T*
A	346.1	215.5	1053.9	676.0	−15.6	−157.2	−128.4	−285.6
B	353.6	236.7	1066.6	740.7	− 5.8	− 74.5	− 95.9	−170.4
C	377.1	236.4	1074.6	725.7	56.7	558.5	428.1	986.6
D	364.2	245.5	1114.9	755.2	−22.3	−163.5	.1	−163.4
E	351.9	217.6	1068.7	688.2	−13.0	−163.3	−203.9	−367.2
Sum	1792.9	1151.7	5378.7	3585.8	0	0	0	0

TABLE 12.46 ANOVA of Weekly Sales ($) for Five Package Designs

Source	d.f.	SS	MS	F
Total	59	2982.6898		
City	3	1543.1405	514.3802	532.93**
Store in city	16	1068.6227	66.7889	69.20**
Week	2	245.0743	122.5372	126.96**
Week × city	6	31.1790	5.1965	5.38**
Design (unadjusted)	4	46.2020	11.5505	11.97**
Residual (adjusted)	4	25.3059	6.3265	6.55**
Error	24	23.1654	.9652	
Design (adjusted)	4	66.5965	16.6491	17.25**
Permanent (adjusted)	4	55.5044	13.8761	14.38**

**Significant at the 1% level.

TABLE 12.47 Adjusted Mean Sales ($)

Package design	Adjusted mean		
	Direct	Residual effect	Permanent
A	29.16	− .58	28.58
B	29.54	− .44	29.10
C	32.42	1.94	34.36
D	29.14	.00	29.14
E	29.14	− .93	28.21
Sample variance	.12	.20	.45

TABLE 12.48 Mean Sales ($) During the Week of Display, Mean Residual
Effect ($), and Mean Permanent Effect ($) for Five Package Designs (Adjusted
for Store, City, Week and Display Sequence)

Package design	Adjusted means		
	Week of sale	Carry-over	Permanent
A	$29.16	$ − .58	$28.58
B	29.54	− .44	29.10
C	32.42	1.94	34.36
D	29.14	.00	29.14
E	29.14	− .93	28.21
Standard error	.35	.45	.67

12.9 CATALOG OF BALANCED LATIN SQUARE DESIGNS

12.9.1 Characteristics

1. All designs are balanced for the estimation of first-order residual effects.
2. The orthogonal set of squares of k treatments is balanced for the estimation of all residual effects up to, and including, order $k - 1$.

12.9.2 Designs for Three Treatments

The orthogonal set (also the balanced pair) (use complete set)

```
1 2 3    1 2 3
2 3 1    3 1 2
3 1 2    2 3 1
```

12.9.3 Designs for Four Treatments

The orthogonal set (use complete set)

```
1 2 3 4    1 2 3 4    1 2 3 4
2 1 4 3    3 4 1 2    4 3 2 1
3 4 1 2    4 3 2 1    2 1 4 3
4 3 2 1    2 1 4 3    3 4 1 2
```

The balanced single square

```
1 2 3 4
2 4 1 3
3 1 4 2
4 3 2 1
```

12.9.4 Designs for Five Treatments

The orthogonal set (use complete set)

1 2 3 4 5	1 2 3 4 5	1 2 3 4 5	1 2 3 4 5
2 3 4 5 1	3 4 5 1 2	4 5 1 2 3	5 1 2 3 4
3 4 5 1 2	5 1 2 3 4	2 3 4 5 1	4 5 1 2 3
4 5 1 2 3	2 3 4 5 1	5 1 2 3 4	3 4 5 1 2
5 1 2 3 4	4 5 1 2 3	3 4 5 1 2	2 3 4 5 1

The balanced pairs (select one or more pairs at random)

Pair 1		Pair 2	
1 2 3 4 5	1 2 3 4 5	1 2 3 4 5	1 2 3 4 5
2 5 4 1 3	5 3 1 2 4	2 4 1 5 3	3 1 5 2 4
3 4 2 5 1	2 5 4 1 3	3 1 5 2 4	2 4 1 5 3
4 1 5 3 2	4 1 5 3 2	4 5 2 3 1	5 3 4 1 2
5 3 1 2 4	3 4 2 5 1	5 3 4 1 2	4 5 2 3 1

12.9.5 Designs for Six Treatments

The balanced single squares (select one or more at random)

1 2 3 4 5 6	1 2 3 4 5 6
2 4 6 1 3 5	2 4 1 6 3 5
3 6 2 5 1 4	3 1 5 2 6 4
4 1 5 2 6 3	4 6 2 5 1 3
5 3 1 6 4 2	5 3 6 1 4 2
6 5 4 3 2 1	6 5 4 3 2 1

12.9.6 Designs for Seven Treatments

The orthogonal set (use complete set)

1 2 3 4 5 6 7	1 2 3 4 5 6 7	1 2 3 4 5 6 7
2 3 4 5 6 7 1	3 4 5 6 7 1 2	4 5 6 7 1 2 3
3 4 5 6 7 1 2	5 6 7 1 2 3 4	7 1 2 3 4 5 6
4 5 6 7 1 2 3	7 1 2 3 4 5 6	3 4 5 6 7 1 2
5 6 7 1 2 3 4	2 3 4 5 6 7 1	6 7 1 2 3 4 5
6 7 1 2 3 4 5	4 5 6 7 1 2 3	2 3 4 5 6 7 1
7 1 2 3 4 5 6	6 7 1 2 3 4 5	5 6 7 1 2 3 4
1 2 3 4 5 6 7	1 2 3 4 5 6 7	1 2 3 4 5 6 7
5 6 7 1 2 3 4	6 7 1 2 3 4 5	7 1 2 3 4 5 6
2 3 4 5 6 7 1	4 5 6 7 1 2 3	6 7 1 2 3 4 5
6 7 1 2 3 4 5	2 3 4 5 6 7 1	5 6 7 1 2 3 4
3 4 5 6 7 1 2	7 1 2 3 4 5 6	4 5 6 7 1 2 3
7 1 2 3 4 5 6	5 6 7 1 2 3 4	3 4 5 6 7 1 2
4 5 6 7 1 2 3	3 4 5 6 7 1 2	2 3 4 5 6 7 1

The balanced pairs (select one or more pairs at random)

Pair 1		Pair 2	
1 2 3 4 5 6 7	1 2 3 4 5 6 7	1 2 3 4 5 6 7	1 2 3 4 5 6 7
2 4 6 7 1 5 3	2 4 6 7 1 5 3	2 5 6 1 3 7 4	4 1 5 7 2 3 6
3 6 2 5 7 4 1	7 3 1 6 4 2 5	3 6 4 5 7 1 2	6 7 1 3 4 2 5
4 7 5 3 2 1 6	5 1 7 2 6 3 4	6 7 1 3 4 2 5	2 5 6 1 3 7 4
5 1 7 2 6 3 4	6 5 4 1 3 7 2	4 1 5 7 2 3 6	7 4 2 6 1 5 3
6 5 4 1 3 7 2	4 7 5 3 2 1 6	5 3 7 2 6 4 1	3 6 4 5 7 1 2
7 3 1 6 4 2 5	3 6 2 5 7 4 1	7 4 2 6 1 5 3	5 3 7 2 6 4 1

Pair 3		Pair 4	
1 2 3 4 5 6 7	1 2 3 4 5 6 7	1 2 3 4 5 6 7	1 2 3 4 5 6 7
2 3 6 7 4 5 1	2 3 6 7 4 5 1	4 5 6 3 1 7 2	3 1 7 6 4 2 5
5 4 7 3 2 1 6	7 1 2 5 6 3 4	2 6 1 5 7 4 3	2 6 1 5 7 4 3
7 1 2 5 6 3 4	6 5 4 2 1 7 3	6 4 2 7 3 5 1	5 7 4 1 2 3 6
6 5 4 2 1 7 3	4 7 1 6 3 2 5	3 1 7 6 4 2 5	4 5 6 3 1 7 2
3 6 5 1 7 4 2	5 4 7 3 2 1 6	5 7 4 1 2 3 6	7 3 5 2 6 1 4
4 7 1 6 3 2 5	3 6 5 1 7 4 2	7 3 5 2 6 1 4	6 4 2 7 3 5 1

	Pair 5
1 2 3 4 5 6 7	1 2 3 4 5 6 7
2 4 1 7 6 3 5	3 1 6 2 7 5 4
3 1 6 2 7 5 4	7 5 4 6 1 2 3
4 7 2 5 3 1 6	6 3 5 1 4 7 2
5 6 7 3 2 4 1	4 7 2 5 3 1 6
6 3 5 1 4 7 2	2 4 1 7 6 3 5
7 5 4 6 1 2 3	5 6 7 3 2 4 1

12.9.7 Designs for Eight Treatments

The orthogonal set (use complete set)

1 2 3 4 5 6 7 8	1 2 3 4 5 6 7 8	1 2 3 4 5 6 7 8	1 2 3 4 5 6 7 8
2 1 4 3 6 5 8 7	5 6 7 8 1 2 3 4	7 8 5 6 3 4 1 2	8 7 6 5 4 3 2 1
3 4 1 2 7 8 5 6	2 1 4 3 6 5 8 7	5 6 7 8 1 2 3 4	7 8 5 6 3 4 1 2
4 3 2 1 8 7 6 5	6 5 8 7 2 1 4 3	3 4 1 2 7 8 5 6	2 1 4 3 6 5 8 7
5 6 7 8 1 2 3 4	7 8 5 6 3 4 1 2	8 7 6 5 4 3 2 1	4 3 2 1 8 7 6 5
6 5 8 7 2 1 4 3	3 4 1 2 7 8 5 6	2 1 4 3 6 5 8 7	5 6 7 8 1 2 3 4
7 8 5 6 3 4 1 2	8 7 6 5 4 3 2 1	4 3 2 1 8 7 6 5	6 5 8 7 2 1 4 3
8 7 6 5 4 3 2 1	4 3 2 1 8 7 6 5	6 5 8 7 2 1 4 3	3 4 1 2 7 8 5 6

1 2 3 4 5 6 7 8	1 2 3 4 5 6 7 8	1 2 3 4 5 6 7 8
4 3 2 1 8 7 6 5	6 5 8 7 2 1 4 3	3 4 1 2 7 8 5 6
8 7 6 5 4 3 2 1	4 3 2 1 8 7 6 5	6 5 8 7 2 1 4 3
5 6 7 8 1 2 3 4	7 8 5 6 3 4 1 2	8 7 6 5 4 3 2 1
6 5 8 7 2 1 4 3	3 4 1 2 7 8 5 6	2 1 4 3 6 5 8 7
7 8 5 6 3 4 1 2	8 7 6 5 4 3 2 1	4 3 2 1 8 7 6 5
3 4 1 2 7 8 5 6	2 1 4 3 6 5 8 7	5 6 7 8 1 2 3 4
2 1 4 3 6 5 8 7	5 6 7 8 1 2 3 4	7 8 5 6 3 4 1 2

The balanced single squares (select one or more at random)

```
1 2 3 4 5 6 7 8     1 2 3 4 5 6 7 8     1 2 3 4 5 6 7 8
2 7 1 8 3 5 4 6     2 4 1 6 3 8 5 7     2 4 8 6 3 1 5 7
3 1 5 7 6 8 2 4     3 8 4 2 7 5 1 6     3 1 4 2 7 5 8 6
4 8 7 5 2 1 6 3     4 3 7 8 1 2 6 5     4 3 7 8 1 2 6 5
5 3 6 2 8 4 1 7     5 6 2 1 8 7 3 4     5 6 2 1 8 7 3 4
6 5 8 1 4 7 3 2     6 1 5 7 2 4 8 3     6 8 5 7 2 4 1 3
7 4 2 6 1 3 8 5     7 5 8 3 6 1 4 2     7 5 1 3 6 8 4 2
8 6 4 3 7 2 5 1     8 7 6 5 4 3 2 1     8 7 6 5 4 3 2 1

1 2 3 4 5 6 7 8     1 2 3 4 5 6 7 8     1 2 3 4 5 6 7 8
2 5 1 6 3 8 4 7     2 5 8 6 3 1 4 7     2 4 1 6 3 8 5 7
3 8 5 2 7 4 1 6     3 1 5 2 7 4 8 6     3 1 5 2 7 4 8 6
4 3 7 8 1 2 6 5     4 3 7 8 1 2 6 5     4 6 2 8 1 7 3 5
5 6 2 1 8 7 3 4     5 6 2 1 8 7 3 4     5 3 7 1 8 2 6 4
6 1 4 7 2 5 8 3     6 8 4 7 2 5 1 3     6 8 4 7 2 5 1 3
7 4 8 3 6 1 5 2     7 4 1 3 6 8 5 2     7 5 8 3 6 1 4 2
8 7 6 5 4 3 2 1     8 7 6 5 4 3 2 1     8 7 6 5 4 3 2 1

1 2 3 4 5 6 7 8     1 2 3 4 5 6 7 8     1 2 3 4 5 6 7 8
2 4 8 6 3 1 5 7     2 5 1 6 3 8 4 7     2 5 8 6 3 1 4 7
3 8 5 2 7 4 1 6     3 1 4 2 7 5 8 6     3 8 4 2 7 5 1 6
4 6 2 8 1 7 3 5     4 6 2 8 1 7 3 5     4 6 2 8 1 7 3 5
5 3 7 1 8 2 6 4     5 3 7 1 8 2 6 4     5 3 7 1 8 2 6 4
6 1 4 7 2 5 8 3     6 8 5 7 2 4 1 3     6 1 5 7 2 4 8 3
7 5 1 3 6 8 4 2     7 4 8 3 6 1 5 2     7 4 1 3 6 8 5 2
8 7 6 5 4 3 2 1     8 7 6 5 4 3 2 1     8 7 6 5 4 3 2 1
```

12.10   SELECTED INCOMPLETE LATIN SQUARE DESIGNS

Design restrictions:

1. Balanced for first-order carry-over effects
2. At least five treatments
3. At least 10 degrees of freedom for error

4. At least three, but not more than five, periods
5 No more than 60 experimental units

12.10.1 Designs for Five Treatments

3 periods, 20 units: Select any 3 rows from the full orthogonal set of balanced 5 × 5 Latin squares. Example:

2 3 4 5 1	3 4 5 1 2	4 5 1 2 3	5 1 2 3 4
3 4 5 1 2	5 1 2 3 4	2 3 4 5 1	4 5 1 2 3
4 5 1 2 3	2 3 4 5 1	5 1 2 3 4	3 4 5 1 2

4 periods, 20 units: Select any 4 rows from the full orthogonal set of balanced 5 × 5 Latin squares.

1 2 3 4 5	1 2 3 4 5	1 2 3 4 5	1 2 3 4 5
3 4 5 1 2	5 1 2 3 4	2 3 4 5 1	4 5 1 2 3
4 5 1 2 3	2 3 4 5 1	5 1 2 3 4	3 4 5 1 2
5 1 2 3 4	4 5 1 2 3	3 4 5 1 2	2 3 4 5 1

12.10.2 Designs for Six Treatments

4 periods, 30 units

1 2 3 4 5 6	1 2 3 4 5 6	1 2 3 4 5 6
2 3 4 5 6 1	3 4 5 6 1 2	4 5 6 1 2 3
3 4 5 6 1 2	5 6 1 2 3 4	2 3 4 5 6 1
4 5 6 1 2 3	2 3 4 5 6 1	5 6 1 2 3 4

1 2 3 4 5 6	1 2 3 4 5 6
5 6 1 2 3 4	6 1 2 3 4 5
4 5 6 1 2 3	5 6 1 2 3 4
6 1 2 3 4 5	3 4 5 6 1 2

5 periods, 30 units

1 2 3 4 5 6	1 2 3 4 5 6	1 2 3 4 5 6
2 3 4 5 6 1	3 4 5 6 1 2	4 5 6 1 2 3
3 4 5 6 1 2	4 5 6 1 2 3	5 6 1 2 3 4
6 1 2 3 4 5	6 1 2 3 4 5	3 4 5 6 1 2
4 5 6 1 2 3	5 6 1 2 3 4	2 3 4 5 6 1

```
1 2 3 4 5 6      1 2 3 4 5 6
5 6 1 2 3 4      6 1 2 3 4 5
3 4 5 6 1 2      2 3 4 5 6 1
6 1 2 3 4 5      5 6 1 2 3 4
2 3 4 5 6 1      4 5 6 1 2 3
```

12.10.3 Designs for Seven Treatments

3 periods, 21 units (use one or two sets)

	Set I	
1 2 3 4 5 6 7	1 2 3 4 5 6 7	1 2 3 4 5 6 7
6 7 1 2 3 4 5	4 5 6 7 1 2 3	7 1 2 3 4 5 6
7 1 2 3 4 5 6	6 7 1 2 3 4 5	4 5 6 7 1 2 3

	Set II	
1 2 3 4 5 6 7	1 2 3 4 5 6 7	1 2 3 4 5 6 7
2 3 4 5 6 7 1	3 4 5 6 7 1 2	5 6 7 1 2 3 4
5 6 7 1 2 3 4	2 3 4 5 6 7 1	3 4 5 6 7 1 2

4 periods, 14 units (use one or more pairs)

Pair I		Pair II	
1 2 3 4 5 6 7	1 2 3 4 5 6 7	1 2 3 4 5 6 7	1 2 3 4 5 6 7
7 1 2 3 4 5 6	2 3 4 5 6 7 1	2 3 4 5 6 7 1	7 1 2 3 4 5 6
4 5 6 7 1 2 3	5 6 7 1 2 3 4	4 5 6 7 1 2 3	5 6 7 1 2 3 4
6 7 1 2 3 4 5	3 4 5 6 7 1 2	7 1 2 3 4 5 6	2 3 4 5 6 7 1

Pair III	
1 2 3 4 5 6 7	1 2 3 4 5 6 7
2 3 4 5 6 7 1	3 4 5 6 7 1 2
3 4 5 6 7 1 2	2 3 4 5 6 7 1
4 5 6 7 1 2 3	5 6 7 1 2 3 4

4 periods, 42 units: Select any 4 rows from the full orthogonal set of balanced 7×7 Latin squares.

5 periods, 21 units (use one or both sets)

Set I

1 2 3 4 5 6 7	1 2 3 4 5 6 7	1 2 3 4 5 6 7
2 3 4 5 6 7 1	3 4 5 6 7 1 2	5 6 7 1 2 3 4
5 6 7 1 2 3 4	2 3 4 5 6 7 1	3 4 5 6 7 1 2
7 1 2 3 4 5 6	6 7 1 2 3 4 5	4 5 6 7 1 2 3
6 7 1 2 3 4 5	4 5 6 7 1 2 3	7 1 2 3 4 5 6

Set II

1 2 3 4 5 6 7	1 2 3 4 5 6 7	1 2 3 4 5 6 7
6 7 1 2 3 4 5	4 5 6 7 1 2 3	7 1 2 3 4 5 6
7 1 2 3 4 5 6	6 7 1 2 3 4 5	4 5 6 7 1 2 3
3 4 5 6 7 1 2	5 6 7 1 2 3 4	2 3 4 5 6 7 1
5 6 7 1 2 3 4	2 3 4 5 6 7 1	3 4 5 6 7 1 2

5 periods, 42 units: Select any 5 rows from the full orthogonal set of balanced 7×7 Latin squares.

12.10.4 Designs for Eight Treatments

3 periods, 56 units: Select any 3 rows from the full orthogonal set of balanced 8×8 Latin squares.

4 periods, 56 units: Select any 4 rows from the full orthogonal set of balanced 8×8 Latin squares.

5 periods, 56 units: Select any 5 rows from the full orthogonal set of balanced 8×8 Latin squares.

12.10.5 Designs for Eleven Treatments

3 periods, 55 units (use complete set)

Set I

1 2 3 4 5 6 7 8 9 10 11	1 2 3 4 5 6 7 8 9 10 11
2 3 4 5 6 7 8 9 10 11 1	5 6 7 8 9 10 11 1 2 3 4
4 5 6 7 8 9 10 11 1 2 3	2 3 4 5 6 7 8 9 10 11 1
1 2 3 4 5 6 7 8 9 10 11	1 2 3 4 5 6 7 8 9 10 11
6 7 8 9 10 11 1 2 3 4 5	10 11 1 2 3 4 5 6 7 8 9
5 6 7 8 9 10 11 1 2 3 4	6 7 8 9 10 11 1 2 3 4 5
1 2 3 4 5 6 7 8 9 10 11	
4 5 6 7 8 9 10 11 1 2 3	
10 11 1 2 3 4 5 6 7 8 9	

Set II

1 2 3 4 5 6 7 8 9 10 11	1 2 3 4 5 6 7 8 9 10 11
2 3 4 5 6 7 8 9 10 11 1	4 5 6 7 8 9 10 11 1 2 3
10 11 1 2 3 4 5 6 7 8 9	6 7 8 9 10 11 1 2 3 4 5
1 2 3 4 5 6 7 8 9 10 11	1 2 3 4 5 6 7 8 9 10 11
5 6 7 8 9 10 11 1 2 3 4	6 7 8 9 10 11 1 2 3 4 5
4 5 6 7 8 9 10 11 1 2 3	2 3 4 5 6 7 8 9 10 11 1
1 2 3 4 5 6 7 8 9 10 11	
10 11 1 2 3 4 5 6 7 8 9	
5 6 7 8 9 10 11 1 2 3 4	

5 periods, 55 units

```
1 2 3 4 5 6 7 8 9 10 11      1 2 3 4 5 6 7 8 9 10 11
5 6 7 8 9 10 11 1 2 3 4      8 9 10 11 1 2 3 4 5 6 7
6 7 8 9 10 11 1 2 3 4 5      3 4 5 6 7 8 9 10 11 1 2
7 8 9 10 11 1 2 3 4 5 6      5 6 7 8 9 10 11 1 2 3 4
9 10 11 1 2 3 4 5 6 7 8      2 3 4 5 6 7 8 9 10 11 1

1 2 3 4 5 6 7 8 9 10 11      1 2 3 4 5 6 7 8 9 10 11
4 5 6 7 8 9 10 11 1 2 3      5 6 7 8 9 10 11 1 2 3 4
7 8 9 10 11 1 2 3 4 5 6      3 4 5 6 7 8 9 10 11 1 2
2 3 4 5 6 7 8 9 10 11 1      8 9 10 11 1 2 3 4 5 6 7
11 1 2 3 4 5 6 7 8 9 10      2 3 4 5 6 7 8 9 10 11 1

1 2 3 4 5 6 7 8 9 10 11
11 1 2 3 4 5 6 7 8 9 10
10 11 1 2 3 4 5 6 7 8 9
6 7 8 9 10 11 1 2 3 4 5
3 4 5 6 7 8 9 10 11 1 2
```

12.10.6 Designs for Thirteen Treatments

4 periods, 52 units

```
1 2 3 4 5 6 7 8 9 10 11 12 13      1 2 3 4 5 6 7 8 9 10 11 12 13
2 3 4 5 6 7 8 9 10 11 12 13 1      9 10 11 12 13 1 2 3 4 5 6 7 8
4 5 6 7 8 9 10 11 12 13 1 2 3      13 1 2 3 4 5 6 7 8 9 10 11 12
10 11 12 13 1 2 3 4 5 6 7 8 9      3 4 5 6 7 8 9 10 11 12 13 1 2

1 2 3 4 5 6 7 8 9 10 11 12 13      1 2 3 4 5 6 7 8 9 10 11 12 13
11 12 13 1 2 3 4 5 6 7 8 9 10      8 9 10 11 12 13 1 2 3 4 5 6 7
7 8 9 10 11 12 13 1 2 3 4 5 6      6 7 8 9 10 11 12 13 1 2 3 4 5
12 13 1 2 3 4 5 6 7 8 9 10 11      5 6 7 8 9 10 11 12 13 1 2 3 4
```

5 periods, 39 units

```
1 2 3 4 5 6 7 8 9 10 11 12 13      1 2 3 4 5 6 7 8 9 10 11 12 13
2 3 4 5 6 7 8 9 10 11 12 13 1      4 5 6 7 8 9 10 11 12 13 1 2 3
4 5 6 7 8 9 10 11 12 13 1 2 3      10 11 12 13 1 2 3 4 5 6 7 8 9
8 9 10 11 12 13 1 2 3 4 5 6 7      9 10 11 12 13 1 2 3 4 5 6 7 8
3 4 5 6 7 8 9 10 11 12 13 1 2      7 8 9 10 11 12 13 1 2 3 4 5 6

1 2 3 4 5 6 7 8 9 10 11 12 13
10 11 12 13 1 2 3 4 5 6 7 8 9
2 3 4 5 6 7 8 9 10 11 12 13 1
12 13 1 2 3 4 5 6 7 8 9 10 11
6 7 8 9 10 11 12 13 1 2 3 4 5
```

FURTHER READING*

Cochran, W. G., and G. M. Cox (1957), Ch. 4
Patterson, H. D., and H. L. Lucas (1962).

*See Bibliography for complete citation.

13

Incomplete Block Designs

13.1 INTRODUCTION

In our consideration of the principles of experimental design we have stressed the idea that we should group the experimental units, if a criterion for grouping exists, into blocks of homogeneous units. The aim is to remove the among-block variability from experimental error, to permit the comparison of treatments under as nearly uniform conditions as possible, and to increase the precision of these comparisons. In general it is true that precision increases as block size decreases, leading to the conclusion that small blocks are preferable to large ones. As the number of treatments to be studied in an experiment is increased, however, the size of block required to contain a full replication also increases. Hence with large numbers of treatments the effect of blocking is lost unless block size can be made smaller than the number of treatments in a full replication.

We have seen that when the treatments consist of factorial combinations there are a couple of ways we can reduce block size:

1. Confound one or more factorial contrasts with block contrasts.
2. Use a split-plot design, which, in effect, confounds a factorial main effect.

In both cases we sacrifice all or a part of the information on some contrasts in the hope that reduced block size will increase the precision of the other contrasts.

In some types of experiments, however, we may have a large number of treatments and we may want to make all comparisons among pairs of treatments with equal precision. For example, plant breeders are often interested in making comparisons among a large number of selections in a single trial. For these trials we require a procedure other than confounding to reduce the size of the blocks. This is the feature of a class of designs called *incomplete block designs*. As the name implies, the experimental units in these designs are grouped into blocks which are smaller than a complete replication of the treatments.

13.1.1 Balanced Designs

Incomplete block designs fall into two broad classes: *balanced* or *partially balanced* designs. In the balanced designs each treatment occurs together in the same block with every other treatment an equal number of times. Consider, for example, the following design for four treatments in blocks of size two:

Block	Rep I		Rep II		Rep III	
(1)	A	B	(3) A	C	(5) A	D
(2)	C	D	(4) B	D	(6) B	C

Note that treatment A occurs with treatment B in block (1), with treatment C in block (3), and with treatment D in block (5). Similarly, every other pair of treatments is seen to occur once, and only once, in the same block. This property of balance insures that all pairs of treatments will be compared with nearly the same precision even though the differences among blocks may be large. With this particular design the blocks can be grouped into three replicates each with two blocks. Grouping into replicates is not always possible. However, whether or not this grouping can be made does not affect the uniformity with which pairs can be compared.

13.1.2 Partially Balanced Designs

It is possible to construct a balanced design with any number of treatments and any number of units per block. These two variables fix the minimum number of replications, which, in many cases, is too large to be of practical use. For this reason it is often necessary to use designs which lack the symmetry of complete balance. Such designs are called partially balanced incomplete block designs.

The most useful of these partially balanced designs are those with two associate classes. In these designs, which have been tabulated by Bose et al. (1954), some pairs of treatments occur together in the same block λ_1 times, while other pairs occur together λ_2 times, where λ_1 and λ_2 are whole numbers. Partially balanced designs, although useful from a practical standpoint, are not as desirable as balanced designs. They are more difficult to analyze statistically, and several different standard errors may be possible. Further, when the variation among blocks is large comparisons among some pairs of treatments are made with greater precision than among other pairs.

An example of partially balanced incomplete block design for nine treatments in blocks of size three is

Block	Rep I				Rep II		
(1)	A	B	C	(4)	A	D	G
(2)	D	E	F	(5)	B	F	H
(3)	G	H	I	(6)	C	F	I

Note that A occurs together in the same block once with B, C, D, and G, but not in the same block with E, F, H, and I. In this design $\lambda_1 = 1$ and $\lambda_2 = 0$.

In many of the incomplete block designs, both balanced and partially balanced, the incomplete blocks can be grouped together to form complete replications. Such designs can be thought of as randomized block designs which have restricted randomization within each complete block.

Yates (1939) has shown that these designs can be analyzed as if they were ordinary randomized block designs. Hence, blocking restrictions can be ignored in the analysis without destroying the validity of F and t tests. Further, the sample means unadjusted for block effects are still unbiased estimates of the treatment means. Of course, if the blocking restriction is ignored the analysis will be less accurate in the presence of block differences than will the complete analysis.

13.2 LATTICE DESIGNS

There are a number of kinds of incomplete block designs. Most of these are described and catalogued by Cochran and Cox (1957). Among the most useful of these kinds are the lattice designs. These will be the only kind of incomplete block design considered in detail here. The reader is referred to Cochran and Cox (1957) for other incomplete block designs.

In the lattice designs the number of treatments must be an exact square. The number of units in each block is the square root of the number of treatments. These incomplete blocks are combined into groups which form separate, complete replications. Since lattices are composed of two or more complete replications they can be analyzed as randomized block designs if necessary. Both balanced and partially balanced lattices are available and we will consider both kinds in detail.

13.3 BALANCED LATTICES

In the balanced lattices each treatment occurs together in the same block with every other treatment once. As a consequence, the statistical analysis is relatively simple and each pair of treatments is compared with the same precision. To obtain balance, however, places some restrictions on the number of treatments and the number of blocks in the design. In particular, balanced lattices are not available for 36, 100, or 144 treatments. In addition, if k = block size (k^2 = number of treatments) then $k + 1$ replications are required to attain balance. Plans for balanced lattices are given in Section 13.7 for k = 3, 4, 5, 7, 8, and 9.

13.3.1 Field Arrangement and Randomization

The incomplete blocks should be composed of units which are as homogeneous as possible. Further, blocks in the same replication should be made as nearly alike as possible to maximize the variation among replications. This will result in increased precision if the experiment is analyzed as a randomized block design. Randomization involves three steps:

1. Randomize the order of the blocks within replications, using a separate randomization in each replication.
2. Randomize treatment code numbers separately in each block.
3. Randomize the assignment of treatments to the code numbers. (Note: The third step is probably unnecessary, but it provides added insurance.)

13.3.2 Analysis

Assume that the design is for k^2 treatments replicated $k + 1$ times. There are k blocks of k units each in each replication. Let

$y_{ij(l)}$ = yield of the jth treatment applied to a unit in the lth block of the ith replication

The data may be tabulated as shown in Table 13.1. The sums in Table 13.1 are computed as follows:

B_{il} = sum for k units in the lth block of replication i

$R_i = \sum_l B_{il}$ = sum of the yield in the ith replication

To continue the analysis a table of treatment totals and adjustments is computed. This is given in Table 13.2. The entries in Table 13.2 are obtained as

$T_j = \sum_i y_{ij(l)}$ = sum of the yields on treatment j

TABLE 13.1 Data Table for a Balanced Lattice Design

Replication	Block	Yield				Sum
1	1	$y_{1.1.(1)}$	$y_{1.2.(1)}$	$\cdots$	$y_{1.k.(1)}$	B_{11}
	2	$y_{1.k+1.(2)}$	$y_{1.k+2.(2)}$	$\cdots$	$y_{1.k+k.(2)}$	B_{12}
	$\vdots$	$\vdots$			$\vdots$	$\vdots$
	k	$y_{1.k^2-k.(k)}$	$\cdots$		$y_{1.k^2.(k)}$	B_{1k}
Sum						R_1
2	1	$y_{2.1.(1)}$	$y_{2.2.(1)}$	$\cdots$	$y_{2.k.(1)}$	B_{21}
	2	$y_{2.k+1.(2)}$	$y_{2.k+2.(2)}$	$\cdots$	$y_{2.k+k.(2)}$	B_{22}
	$\vdots$	$\vdots$			$\vdots$	$\vdots$
	k	$y_{2.k^2-k(k)}$	$\cdots$		$y_{2.k^2.(k)}$	B_{2k}
Sum						R_2
$\vdots$			$\vdots$			$\vdots$
$k+1$	1	$y_{k+1.1.(1)}$	$y_{k+1.2.(1)}$	$\cdots$	$y_{k+1.k(1)}$	$B_{k+1.1}$
	2	$y_{k+1.k+1.(2)}$	$y_{k+1.k+2.(2)}$	$\cdots$	$y_{k+1.k+k.(2)}$	$B_{k+1.2}$
	$\vdots$	$\vdots$			$\vdots$	$\vdots$
	k	$y_{k+1.k^2-k.(k)}$	$\cdots$		$y_{k+1.k^2.(k)}$	$B_{k+1.k}$
Sum						R_{k+1}

TABLE 13.2 Treatment Totals and Adjustments for a Balanced Lattice Design

Treatment	Total	Block	Weight
1	T_1	B_1	W_1
2	T_2	B_2	W_2
$\vdots$	$\vdots$	$\vdots$	$\vdots$
k^2	T_{k^2}	B_{k^2}	W_{k^2}
Sum	G	G	0

$$G = \sum_j T_j = \sum_i R_i = \text{grand total of all yields}$$

B_j = sum of the B_{il} for all blocks in which the jth treatment occurs

$W = kT_j - (k + 1)B_j + G$

The format for the analysis of variance is presented in Table 13.3. The sums of squares in Table 13.3 are computed using the following equations:

Compute a correction term, C, as

$$C = G^2/(k^3 + k^2)$$

Then

$$\text{SSTot} = \sum_i \sum_j y_{ij(l)}^2 - C$$

$$\text{SSR} = (1/k^2) \sum_i R_i^2 - C$$

$$\text{SST} = [1/(k + 1)] \sum_j T_j^2 - C$$

$$\text{SSB} = [1/k^3(k + 1)] \sum_j W_j^2$$

$$\text{SSE} = \text{SSTot} - \text{SSR} - \text{SST} - \text{SSB}$$

$$E_b = \text{SSB}/(k^2 - 1)$$

$$E_e = \text{SSE}/(k - 1)(k^2 - 1)$$

The next step is to compare E_b with E_e. If $E_b \leq E_e$ the adjustment for blocks will have no effect. In this case the blocking restrictions are ignored and the data are analyzed as if they had come from a randomized block design with replications as blocks.

If $E_b > E_e$ then blocking is effective. In this case adjusted treatment totals and means are computed. These are obtained by first computing an adjustment factor, A, as

$$A = (E_b - E_e)/k^2 E_b$$

TABLE 13.3 ANOVA of a Balanced Lattice Design

Source	d.f.	SS	MS
Total	$k^3 + k^2 + 1$	SSTot	
Replication	k	SSR	
Treatment (unadjusted)	$k^2 - 1$	SST	
Block (adjusted)	$k^2 - 1$	SSB	E_b
Intrablock error	$(k - 1)(k^2 - 1)$	SSE	E_e

The adjusted treatment totals, $\hat{T}_j$, are then obtained as

$$\hat{T}_j = T_j + AW_j$$

For significance tests and interval estimates the next step is to compute an effective error mean square, E'_e, using the equation

$$E'_e = E_e(1 + kA), \qquad \text{with} \qquad (k - 1)(k^2 - 1) \text{ d.f.}$$

To test the significance of the differences among the adjusted treatment means compute an adjusted treatment sum of squares, SST(adj),

$$SST(adj) = [1/(k + 1)] \sum_j \hat{T}_j^2 - C$$

Then, the adjusted treatment mean square, E_t, is

$$E_t = SST(adj)/(k^2 - 1)$$

and an approximate F for testing significance is

$$F = E_t/E'_e, \qquad \text{with} \qquad k^2 - 1, (k - 1)(k^2 - 1) \text{ d.f.}$$

The adjusted treatment means, $\hat{\bar{y}}_j$, are computed from the adjusted treatment totals.

$$\hat{\bar{y}}_j = \hat{T}_j/(k + 1)$$

The variance of an adjusted mean is

$$V(\hat{\bar{y}}_j) = E'_e/(k + 1)$$

and the variance of the difference between adjusted means is

$$V(\hat{d}) = 2E'_e/(k + 1)$$

To determine the relative precision of the balanced lattice first compute the pooled error, E_{rb}, of a randomized block design as

$$E_{rb} = (SSB + SSE)/k(k^2 - 1)$$

Then the precision of the balanced lattice relative to that of a randomized block design is

$$\% \text{ relative precision} = (E_{rb}/E'_e)100$$

13.3.3 Numerical Example: Balanced Lattice

The proprietor of the Gourmet Ice Cream Shoppe wanted to evaluate 16 new ice cream flavors for possible inclusion in the selection available in her store. She wanted to use an organoleptic test but she knew that a judge could evaluate, at most, five samples at a single session. She also wanted to have equal precision on all comparisons, and to eliminate day-to-day differences if possible. She decided to use a balanced lattice design with judges as blocks and days as replications. Twenty volunteers were selected as judges, and four of these were randomly assigned to each of five test days.

On the test day each judge was presented with 4 samples of ice cream from among the 16 to be tested. Each sample was rated, on a scale of one to ten, on each of five characteristics: flavor, aroma, texture, consistency, and palatability. Overall acceptability was then expressed as the sum of the scores for the five characteristics. The plan for the trial and resulting acceptability scores are shown in Table 13.4. Treatment totals and adjustments are summarized in Table 13.5.

The analysis of variance of the ice cream data is summarized in Table 13.6.

A comparison of E_b with E_e shows that $16.46 > 5.20$ so that blocking on judges has been effective in this trial. The adjustment factor, A, is

$$A = (E_b - E_e)/k^2E_b = (16.46 - 5.20)/(16)(16.46) = .0428$$

The effective error mean square, E_e', is found to be

$$E_e' = E_e(1 + kA) = 5.20[1 + (4)(.0428)] = 6.09$$

To test the significance of the differences among adjusted flavor means the adjusted treatment totals are computed. These are shown, along with the adjusted means, in Table 13.7.

The adjusted treatment sum of squares, SST(adj) is now computed using the entries in Table 13.7.

$$\text{SST(adj)} = (1/5) \sum_j \hat{T}_j^2 - C = 1411.95$$

The adjusted treatment mean square, E_t, is

$$E_t = \text{SST(adj)}/15 = 94.13$$

Then

$$F = E_t/E_e' = 94.13/6.09 = 15.46$$

Since $F_{.01(15,45)} = 2.52 < 15.46$ the differences among the adjusted flavor means are significant at the 1% level.

To estimate the relative precision of the balanced lattice design the pooled randomized block error, E_{rb}, is computed as

$$E_{rb} = (\text{SSB} + \text{SSE})/4(15) = 8.01$$

TABLE 13.4 Trial Plan and Acceptability Scores for a 4 × 4 Balanced Lattice Design to Evaluate 16 Ice Cream Flavors

Day	Judge	Acceptability score				Sum
1	1	(14) 38	(13) 36	(16) 29	(15) 37	140
	2	(1) 26	(2) 23	(4) 27	(3) 30	106
	3	(5) 32	(6) 20	(7) 21	(8) 25	98
	4	(11) 33	(12) 39	(10) 34	(9) 22	128
Sum						472
2	1	(6) 24	(10) 37	(2) 26	(14) 40	127
	2	(12) 29	(4) 33	(16) 26	(8) 24	112
	3	(11) 32	(3) 24	(7) 18	(15) 34	108
	4	(1) 30	(9) 25	(5) 38	(13) 34	127
Sum						474
3	1	(16) 24	(1) 29	(11) 30	(6) 24	107
	2	(7) 20	(13) 30	(4) 23	(10) 28	101
	3	(9) 22	(3) 28	(14) 34	(8) 25	109
	4	(5) 35	(2) 23	(12) 29	(15) 33	120
Sum						437
4	1	(6) 20	(4) 24	(9) 20	(15) 24	88
	2	(10) 28	(16) 24	(3) 25	(5) 31	108
	3	(2) 22	(11) 28	(8) 23	(13) 30	103
	4	(14) 33	(12) 27	(7) 19	(1) 26	105
Sum						404
5	1	(9) 19	(2) 16	(7) 19	(16) 19	73
	2	(5) 21	(11) 25	(14) 19	(4) 18	83
	3	(1) 17	(15) 20	(8) 17	(10) 22	76
	4	(12) 23	(13) 26	(3) 20	(6) 14	83
Sum						315

Note: Numbers in parentheses are treatment numbers.

TABLE 13.5 Treatment Totals and
Adjustments for an Ice Cream Taste Test

Treatment	Total	Block	Weight
1	128	521	9
2	110	529	−103
3	127	514	40
4	125	490	152
5	157	536	50
6	102	503	−5
7	97	485	65
8	114	498	68
9	108	525	−91
10	149	540	−2
11	148	529	49
12	147	548	−50
13	156	554	−44
14	164	564	−62
15	148	532	34
16	122	540	−110
Sum	2102	8408	0

Note: $W_i = 4T_i - 5B_j + 2102$.

TABLE 13.6 ANOVA of Ice Cream Acceptability Scores

Source	d.f.	SS	MS
Total	79	2943.95	
Day	4	1074.32	
Flavor (unadjusted)	15	1388.75	
Judge (adjusted)	15	246.84	$16.46 = E_b$
Intrajudge error	45	234.04	$5.20 = E_c$

The relative precision is

$$\% \text{ relative precision } = (E_{rb}/E_c')100 = (8.01/6.09)100 = 131.6\%$$

Report of statistical analysis. A taste test, using 20 judges testing over a 5-day period in a balanced lattice design, was conducted to measure the acceptability of 16 new flavors of ice cream. The adjusted mean acceptability scores, ranked in order of decreasing acceptability, are presented in Table 13.8.

TABLE 13.7 Adjusted Flavor Totals and
Means

Flavor	Adjusted total $\hat{T}_i$	Adjusted mean $\hat{\bar{y}}_i$
1	128.38	25.68
2	105.59	21.12
3	128.71	25.74
4	131.50	26.30
5	159.14	31.83
6	101.79	20.36
7	99.78	19.96
8	116.91	23.38
9	104.10	20.82
10	148.91	29.78
11	150.10	30.02
12	144.86	28.97
13	154.12	30.82
14	161.35	32.27
15	149.46	29.89
16	117.29	23.46

Note: $\hat{T}_i = T_i + AW_i = T_i + .0428W_i.$

The highest acceptability score was received by flavor 14, although it is not significantly higher than the next five highest flavors. Perhaps the top six should be tested further.

The use of a balanced lattice design for this trial increased the precision 32% as compared to that of a randomized block design. Thus it would have taken about seven replications of a randomized block design to attain the precision of five replications with the balanced lattice.

13.4 PARTIALLY BALANCED LATTICES

The number of replications required for balance becomes very large as the number of treatments increases. For this reason it is not usually practical to use balanced lattices for blocks with more than about seven units per block. In the interest of economy, then, the scientist is forced to accept a partially balanced design with fewer replications than would be required for full balance.

There are a number of types of partially balanced lattices. In all of them the number of treatments must be an exact square, and the number of units per incomplete block is the square root of the number of treatments. The types of designs differ from each other primarily in the number of replicates in the design.

TABLE 13.8 Adjusted Mean
Acceptability Scores for 16 New Ice
Cream Flavors

Flavor	Adjusted mean score
14	32.27 *a*
5	31.83 *a b*
13	30.82 *a b*
11	30.02 *a b*
15	29.89 *a b*
10	29.78 *a b*
12	28.97 *b c*
4	26.30 *d c*
3	25.74 *d*
1	25.68 *d*
16	23.46 *d e*
8	23.38 *d e*
2	21.12 *e f*
9	20.82 *e f*
6	20.36 *e f*
7	19.96 *f*
Standard error	1.10

Note: Means designated by the same letter are
not significantly ($\alpha \leq .05$) different
according to the FPLSD test.

13.4.1 Simple Lattices

These designs have two replications. Plans for the simple lattices may be obtained
by taking the first two replications of the plans for the balanced lattices. Note
that simple lattices can also be constructed for 36 and 100 treatments by taking
the first two replications of the 6 × 6 or 10 × 10 triple lattices.

Although they are available, simple lattices for 9 and 16 treatments are
unlikely to be more precise than randomized block designs. This is primarily
because the blocking restriction reduces the error degrees of freedom below an
acceptable level.

13.4.2 Triple Lattices

Triple lattices contain three replications. They are formed by taking the first
three replications from the plans for the balanced lattices. Note, also, that triple
lattices are possible for all squares from 9 to 169 treatments.

13.4.3 Quadruple Lattices

Quadruple lattices contain four replications. They are constructed by using the first four replications of the balanced lattices. Thus, true quadruple lattices are not available for 36 or 100 treatments. Other designs with four replications may be constructed by repeating the simple lattices. Statistical analysis of such repetitions is somewhat different, so consideration of them will be postponed.

13.4.4 Statistical Analysis: Designs Without Repetitions

The pattern of statistical analysis is the same for simple, triple, and quadruple lattices. We will outline this pattern in general, then go through the numerical details for both a simple and a triple lattice.

Assume that the design is for k^2 treatments replicated r times. There are k blocks of k units each in each replication. Let

$y_{ij(l)}$ = yield of the jth treatment applied to a unit in the lth block of the ith replication

The data may be tabulated as shown in Table 13.9. The sums in Table 13.9 are computed in the following way:

B_{il} = sum for the k units in the lth block of replication i

$R_i = \sum_l B_{il}$ = sum of the yields in the ith replication

C_{il} = sum (over all replicates) of all treatments in the lth block of the ith replicate minus rB_{il}

$C_i = \sum_l C_{il}$

AC_{il} = adjusted sum, where A is the adjustment factor computed after completing the analysis of variance

$G = \sum_i R_i$ = grand total of all yields

To continue the analysis it is also necessary to compute the unadjusted treatment totals. These may be tabulated, along with the adjusted treatment totals and means, as indicated in Table 13.10. The entries in Table 13.10 are computed as:

$T_j = \sum_i y_{ij(l)}$ = sum of yields for the jth treatment

$\hat{T}_j = T_j +$ (sum of the AC_{il} for blocks in which the jth treatment appears)

$\hat{\bar{y}}_j = \hat{T}_j / r$

TABLE 13.9 Data Table for Simple, Triple, or Quadruple Lattice Designs

Replication	Block	Yield				Sum B	Sum C	Adjusted Sum C
1	1	$y_{1,1,(1)}$	$y_{1,2,(1)}$	$\cdots$	$y_{1,k,(1)}$	B_{11}	C_{11}	AC_{11}
	2	$y_{1,k+1,(2)}$	$y_{1,k+2,(2)}$	$\cdots$	$y_{1,k+k,(2)}$	B_{12}	C_{12}	AC_{12}
	$\vdots$	$\vdots$			$\vdots$	$\vdots$	$\vdots$	$\vdots$
	k	$y_{1,k^2,k(k)}$	$\cdots$		$y_{1,k^2,(k)}$	B_{1k}	C_{1k}	AC_{1k}
	Sum					R_1	C_1	
2	1	$y_{2,1,(1)}$	$y_{2,2,(1)}$	$\cdots$	$y_{2,k,(1)}$	B_{21}	C_{21}	AC_{21}
	2	$y_{2,k+1,(2)}$	$y_{2,k+2,(2)}$	$\cdots$	$y_{2,k+k,(2)}$	B_{22}	C_{22}	AC_{21}
	$\vdots$	$\vdots$			$\vdots$	$\vdots$	$\vdots$	$\vdots$
	k	$y_{2,k^2-k(k)}$	$\cdots$		$y_{2,k^2,(k)}$	B_{2k}	C_{2k}	AC_{2k}
	Sum					R_2	C_2	
	$\vdots$		$\vdots$			$\vdots$	$\vdots$	$\vdots$
r	1	$y_{r,1,(1)}$	$y_{r,2,(1)}$	$\cdots$	$y_{r,k,(1)}$	B_{r1}	C_{r1}	AC_{r1}
	2	$y_{r,k+1,(2)}$	$y_{r,k+2,(2)}$	$\cdots$	$y_{r,k+k,(2)}$	B_{r2}	C_{r2}	AC_{r2}
	$\vdots$	$\vdots$			$\vdots$	$\vdots$	$\vdots$	$\vdots$
	k	$y_{r,k^2-k,(k)}$	$\cdots$		$y_{r,k^2,(k)}$	B_{rk}	C_{rk}	AC_{rk}
	Sum					R_r	C_r	
					Sum	G	0	0

TABLE 13.10 Unadjusted Totals, Adjusted Totals, and Adjusted Means for Simple, Triple, and Quadruple Lattices

Treatment	Total (unadjusted)	Total (adjusted)	Mean (adjusted)
1	T_1	$\hat{T}_1$	$\hat{\bar{y}}_1$
2	T_2	$\hat{T}_2$	$\hat{\bar{y}}_2$
$\vdots$	$\vdots$	$\vdots$	$\vdots$
k^2	T_{k^2}	$\hat{T}_{k^2}$	$\hat{\bar{y}}_{k^2}$
Sum	G	G	

TABLE 13.11 ANOVA of Simple, Triple, and Quadruple Lattice Designs

Source	d.f.	SS	MS
Total	$rk^2 - 1$	SSTot	
Replications	$r - 1$	SSR	
Treatments (unadjusted)	$k^2 - 1$	SST	
Blocks (adjusted)	$r(k - 1)$	SSB	E_b
Intrablock error	$(k - 1)(rk - k - 1)$	SSE	E_e

The format for the analysis of variance of these designs is presented in Table 13.11. Except for SSB, the sums of squares in Table 13.11 are computed in the usual way. The adjusted block sum of squares is

$$SSB = [1/kr(r - 1)] \sum_i \sum_l C_{il}^2 - [1/k^2 r(r - 1)] \sum_i C_i^2$$

The mean squares are computed in the usual way:

$$E_b = SSB/r(k - 1)$$

$$E_e = SSE/(k - 1)(rk - k - 1)$$

As with the balanced lattice the next step is to compare E_b with E_e. If $E_b \leq E_e$ blocking is ineffective and the data should be analyzed as if the experiment were conducted as a randomized block design.

If $E_b > E_e$ an adjustment factor, A, is computed. For these designs

$$A = (E_b - E_e)/k(r - 1)E_b$$

This adjustment factor is used to compute the AC_{il} in Table 13.9 and $\hat{T}_j$ in Table 13.10. Note that the AC_{il} values should sum to zero.

The effective error mean square, E_e', is now computed as

$$E_e' = [1 + rkA/(k + 1)]E_e$$

Except for small designs ($k = 3, 4$) E_e' may be used in t tests and interval estimates.

To test the significance of the differences among adjusted treatment means an adjusted treatment mean square is computed. To do this it is first necessary to compute SSBu, the unadjusted sum of squares for blocks within replications. This is found to be

$$SSBu = (1/k) \sum_i \sum_l B_{il}^2 - G^2/rk^2 - SSR$$

Then, the adjusted treatment sum of squares, SST(adj) is computed as

$$\text{SST(adj)} = \text{SST} - Ak(r - 1)\{[r\text{SSBu}/(r - 1)(1 + kA)] - \text{SSB}\}$$

The F ratio for testing the significance of the adjusted treatment differences is

$$F = [\text{SST(adj)}/(k^2 - 1)]/E_e$$

with $k^2 - 1$, $(k - 1)(rk - k - 1)$ d.f.

The variance of an adjusted treatment mean, $V(\hat{\bar{y}}_j)$, is

$$V(\hat{\bar{y}}_j) = E'_e/r$$

The variance of the difference between adjusted means depends on whether or not the treatments occur together in the same block. For treatments in the same block the variance of the difference, $V(\bar{d})$, is

$$V(\bar{d}) = (2E_e/r)[1 + (r - 1)A]$$

while for treatments in different blocks the variance of the difference $V(\bar{d})$, is

$$V(\bar{d}) = (2E_e/r)(1 + rA)$$

For $k > 4$ it is sufficient to use the effective error mean square for the variance of any difference. This is

$$V(\bar{d}) = 2E'_e/r$$

The precision of one of these designs relative to that of a randomized block design is obtained by first computing the error mean square of a randomized block design, E_{rb}. This is

$$E_{rb} = (\text{SSB} + \text{SSE})/(k^2 - 1)(r - 1)$$

Then, the precision of the lattice relative to that of the randomized block is found to be

$$\% \text{ relative precision} = (E_{rb}/E'_e)100$$

13.4.5 Numerical Example: Simple Lattice

A cereal breeder wanted to develop a new, high-yielding variety of barley. As a part of the process he selected 25 promising lines for test in a replicated yield trial. Because of soil variability in the field to be used for the trial he decided to use a simple lattice design, a 5×5 lattice with two replications.

The field arrangement of the experiment and the resulting grain yields (kilograms per plot) are presented in Table 13.12. Also included in Table 13.12 are the C_{il} values and the adjustment factors, AC_{il}, for each block.

The unadjusted treatment totals are summarized in Table 13.13.

The analysis of variance of the yield data is shown in Table 13.14.

TABLE 13.12 Field Arrangement and Yields (kg/plot) of 25 Barley Selections Grown in a Simple Lattice (Selection Numbers in Parentheses)

Replication	Block						Sum B	C	AC
1	1	18.2 (19)	13.0 (16)	9.5 (18)	6.7 (17)	10.1 (20)	57.5	17.6	1.54
	2	13.3 (12)	11.4 (13)	14.2 (15)	11.9 (14)	13.4 (11)	64.2	4.2	.37
	3	15.0 (1)	12.4 (2)	17.3 (3)	20.5 (4)	13.0 (5)	78.2	5.3	.46
	4	7.0 (22)	5.9 (24)	14.1 (21)	19.2 (25)	7.8 (23)	54.0	− .5	− .04
	5	11.9 (9)	15.2 (7)	17.2 (10)	16.3 (8)	16.0 (6)	76.6	3.9	.34
Sum							330.5	30.5	2.67
2	1	7.7 (23)	15.2 (18)	19.1 (3)	15.5 (8)	14.7 (13)	72.2	−9.9	− .86
	2	15.8 (5)	18.0 (20)	18.8 (10)	14.4 (15)	20.0 (25)	87.0	− 13.3	− 1.17
	3	10.2 (22)	11.5 (12)	17.0 (2)	11.0 (17)	15.3 (7)	65.0	− 10.4	− .91
	4	10.9 (14)	4.7 (24)	10.9 (9)	16.6 (4)	9.8 (19)	52.9	15.5	1.35
	5	20.0 (6)	21.1 (16)	16.9 (11)	10.9 (21)	15.0 (1)	83.9	− 12.4	− 1.08
Sum							361.0	− 30.5	− 2.67
						Sum	691.5	0	0

Note: $C =$ sum (over both replications) of all treatments in the block $- 2B_{ik}$.

Example: $C_{11} = (18.2 + 9.8) + (13.0 + 21.1) + (9.5 + 15.2) + (6.7 + 11.0) + (10.1 + 18.0) - (2)(57.5) = 17.6$.

TABLE 13.13 Unadjusted Yield Totals for 25 Barley Selections (Selection Numbers in Parentheses)

(1) 30.0	(2) 29.4	(3) 36.4	(4) 37.1	(5) 28.8
(6) 36.0	(7) 30.5	(8) 31.8	(9) 22.8	(10) 36.0
(11) 30.3	(12) 24.8	(13) 26.1	(14) 22.8	(15) 28.6
(16) 34.1	(17) 17.7	(18) 24.7	(19) 28.0	(20) 28.1
(21) 25.0	(22) 17.2	(23) 15.5	(24) 10.6	(25) 39.2
			Sum	691.5

Since $9.70 > 5.46$ blocking is effective. The adjustment factor, A, is computed using

$$A = (E_b - E_e)/k(r - 1)E_b = (9.70 - 5.46)/(5)(1)(9.70)$$

$$= .0874$$

The block adjustments, AC, are now computed for Table 13.12. These adjustments are used in computing the adjusted selection totals and means given in Table 13.15.

TABLE 13.14 ANOVA of Yields (kg/plot) of 25 Barley
Selections

Source	d.f.	SS	MS
Total	49	805.42	
Replication	1	18.60	
Selection (unadjusted)	24	621.82	
Block in replication (adjusted)	8	77.59	$9.70 = E_b$
Intrablock error	16	87.41	$5.46 = E_e$

$$^1SSB = \frac{\Sigma C_{il}^2}{(5)(2)(1)} - \frac{\Sigma C_i^2}{(25)(2)(1)}$$

$$= 114.80 - 37.21 = 77.59.$$

For this trial the effective error mean square, E_e', is

$$E_e' = [1 + rkA/(k + 1)]E_e = [1 + (2)(5)(.0874)/(6)](5.46)$$

$$= 6.26$$

To test the significance of the differences among the adjusted treatment means it is first necessary to compute the unadjusted block within replication sum of squares, SSBu:

$$SSBu = (1/k) \sum_i \sum_l B_{il}^2 - G^2/rk^2 - SSR$$

$$= 9834.32 - 9563.44 - 18.60 = 252.18$$

Then

$$SST(adj) = SST - Ak(r - 1)\{[rSSBu/(r - 1)(1 + kA)] - SSB\}$$

$$= 621.82 - (.0874)(5)(1)\{[(2)(252.18)/(1)(1 + (5)(.0874))]$$

$$- 77.59\}$$

$$= 502.35$$

and

$$F = [SST(adj)/(k^2 - 1)]/E_e = [(502.35)/(24)]/5.46$$

$$= 3.83$$

Since $F_{.01(24,16)} = 3.18 < 3.83$ the differences among selection means are significant at the 1% level.

The variance of a selection mean, $V(\hat{\bar{y}}_j)$, is

$$V(\hat{\bar{y}}_j), = E_e'/r = 6.26/2 = 3.13$$

TABLE 13.15 Adjusted Selection Totals and
Means (kg/plot)

Selection	Total (adjusted)	Mean (adjusted)
1	29.38[1]	14.69
2	28.95	14.48
3	36.00	18.00
4	38.91	19.46
5	28.09	14.04
6	35.26	17.63
7	29.93	14.96
8	31.28	15.64
9	24.49	12.24
10	35.17	17.58
11	29.59	14.80
12	24.26	12.13
13	25.61	12.80
14	24.52	12.26
15	27.80	13.90
16	34.56	17.28
17	18.33	9.16
18	25.38	12.69
19	30.89	15.44
20	28.47	14.23
21	23.88	11.94
22	16.25	8.12
23	14.60	7.30
24	11.91	5.96
25	37.99	19.00
Sum	691.50	

[1] $29.38 = 30.00 + .46 - 1.08$.

Since $k = 5$ the effective error mean square may be used to compute the variance of a difference between selections. This variance, $V(\bar{d})$, is

$$V(\bar{d}) = 2E'_e/r = (2)(6.26)/2 = 6.26$$

To illustrate the more correct variances it is found that the variance of the difference between means of selections in the same block is

$$V(\bar{d}) = (2E_e/r)[1 + (r - 1)A] = (2)(5.46)[1 + (1)(.0874)]/2$$
$$= 5.94$$

The variance of the difference between means of selections in different blocks is

$$V(\bar{d}) = (2E_e/r)(1 + rA) = (2)(5.46)[1 + (2)(.0874)]/2$$
$$= 6.41$$

The precision of the simple lattice design relative to that of a randomized block design is found by first computing E_{rb}, the error mean square of a randomized block design.

$$E_{rb} = (SSB + SSE)/(k^2 - 1)(r - 1) = (77.59 + 87.41)/(24)(1)$$
$$= 6.88$$

Then

$$\% \text{ relative precision} = (E_{rb}/E_e')100 = (6.88/6.26)100$$
$$= 110.0\%$$

Report of statistical analysis. An experiment was run by-a plant breeder to determine the yield potential of 25 new selections of barley. Because of variation in the experimental area and because of economic considerations the experiment was run using a 5 × 5 simple lattice. The resulting mean yields are summarized in Table 13.16.

Selection 4 was the highest yielding selection, followed closely by selection 25. Selections 17, 22, 23, and 24 are clearly inferior to the others. The statistical analysis of the data indicates that the simple lattice increased the precision by about 10% compared to that of a randomized block design.

13.4.6 Numerical Example: Triple Lattice

A manufacturer of agricultural chemicals developed 16 new chemicals for the purpose of controlling broad-leafed weeds in golf course fairways. He arranged to test the effectiveness of the chemicals at a local country club. He chose three

TABLE 13.16 Adjusted Mean Yields (kg/plot) of 25 Selections of Barley (Selection Numbers in Parentheses)

(1) 14.69	(2) 14.48	(3) 18.00	(4) 19.46	(5) 14.04
(6) 17.63	(7) 14.96	(8) 15.64	(9) 12.24	(10) 17.58
(11) 14.80	(12) 12.13	(13) 12.80	(14) 12.26	(15) 13.90
(16) 17.28	(17) 9.16	(18) 12.69	(19) 15.44	(20) 14.23
(21) 11.94	(22) 8.12	(23) 7.30	(24) 5.96	(25) 19.00

Note: Standard error = 1.77.

fairways on the course for the test. In each fairway he established four blocks each containing four 1-square-meter plots. In each plot he counted the number of broad-leafed weeds. He then treated the plots with the new chemicals, using a 4 × 4 triple lattice experimental design with the three fairways as replications.

Four weeks after applying the chemicals he again counted the broad-leafed weeds in each plot. He then computed the percent of the original weeds killed by the chemicals. The resulting data, along with the C_{il} and the block adjustment factors, AC_{il}, are shown in Table 13.17. The unadjusted treatment totals are summarized in Table 13.18, and Table 13.19 gives the analysis of variance of the data.

Since 59.14 > 20.68 blocking has been effective in this herbicide trial. The adjustment factor is

TABLE 13.17 Field Arrangement and Broad-Leafed Weed Kill (%) by 16 Chemical Herbicides Tested in a 4 × 4 Triple Lattice Design (Chemical Numbers in Parentheses)

Replication	Block		% Kill			B	C	AC
1	1	75 (15)	57 (16)	71 (13)	77 (14)	280	−51	−4.15
	2	78 (12)	66 (11)	68 (10)	45 (9)	257	−55	−4.47
	3	40 (6)	64 (5)	49 (8)	42 (7)	195	23	1.87
	4	59 (3)	53 (1)	46 (2)	54 (4)	212	6	.49
Sum						944	−77	−6.26
2	1	53 (16)	66 (4)	57 (12)	47 (8)	223	−6	− .49
	2	80 (14)	48 (6)	73 (10)	52 (2)	253	−59	−4.80
	3	36 (7)	63 (11)	67 (15)	47 (3)	213	37	3.01
	4	68 (13)	60 (1)	50 (9)	76 (5)	254	−46	−3.74
Sum						943	−74	−6.02
3	1	66 (15)	46 (2)	58 (12)	69 (5)	239	37	3.01
	2	46 (4)	40 (7)	59 (13)	55 (10)	200	78	6.34
	3	43 (9)	55 (3)	50 (8)	68 (14)	216	22	1.79
	4	60 (11)	58 (1)	48 (16)	47 (6)	213	14	1.14
Sum						868	151	12.28
					Sum	2755	0	0

Note: C = sum (over all replications) of all treatments in the block − $3B_{il}$.
Example: C_{11} = (75 + 67 + 66) + (57 + 53 + 48) + (71 + 68 + 59)

$\qquad$ + (77 + 80 + 68) − (3)(280)

$\qquad$ = − 51.

TABLE 13.18   Unadjusted % Kill Totals for 16
Chemical Herbicides (Chemical Numbers in
Parentheses)

(1) 171	(2) 144	(3) 161	(4) 166
(5) 209	(6) 135	(7) 118	(8) 146
(9) 138	(10) 196	(11) 189	(12) 193
(13) 198	(14) 225	(15) 208	(16) 158
		Sum	2755

TABLE 13.19   ANOVA of % Kill by 16 Chemical Herbicides

Source	d.f.	SS	MS
Total	47	6107.48	
Replication	2	237.54	
Herbicide (unadjusted)	15	4903.48	
Block in replication (adjusted)	9	532.27[1]	$59.14 = E_b$
Intrablock error	21	434.19	$20.68 = E_e$

$$^1\text{SSB} = \frac{\Sigma C_{il}^2}{(4)(3)(2)} - \frac{\Sigma C_i^2}{(16)(3)(2)}$$

$$= 888.58 - 356.31 = 532.27.$$

$$A = (E_b - E_e)/k(r - 1)E_b = (59.14 - 20.68)/(4)(2)(59.14)$$

$$= .0813$$

The block adjustments can now be computed and entered in Table 13.17.
These are used in computing the adjusted chemical totals, which are given in
Table 13.20 along with the adjusted means.

For this trial the effective error mean square, E'_e is

$$E'_e = [1 + rkA/(k + 1)]E_e = [1 + (3)(4)(.0813)/(5)](20.68)$$

$$= 24.72$$

The unadjusted block within replication sum of squares, SSBu, is computed
as

$$\text{SSBu} = (1/k) \sum_i \sum_l B_{il}^2 - G^2/rk^2 - \text{SSR}$$

$$= 160,046.75 - 158,125.52 - 237.54 = 1,683.69$$

TABLE 13.20 Adjusted % Kill Totals and
Means for 16 Chemical Herbicides

Chemical	Total (adjusted)	Mean (adjusted)
1	168.89[1]	56.30
2	142.70	47.57
3	166.29	55.43
4	172.34	57.45
5	210.14	70.05
6	133.21	44.40
7	129.22	43.07
8	149.17	49.72
9	131.58	43.86
10	193.07	64.36
11	188.68	62.89
12	191.05	63.68
13	196.45	65.48
14	217.84	72.61
15	209.87	69.96
16	154.50	51.50
Sum	2755.00	

[1]$168.89 = 171 + .49 - 3.74 + 1.14$.

The adjusted treatment sum of squares, SST (adj), is then computed as

$$SST(adj) = SST - Ak(r - 1)\{[rSSBu/(r - 1)(1 + kA)] - SSB\}$$
$$= 4903.48$$
$$- (.0813)(4)(2)\{[(3)(1683.69)/$$
$$(2)(1 + (4)(.0813))] - 532.27\}$$
$$= 4010.15$$

Then

$$F = [SST(adj)/(k^2 - 1)]/E_c = [4010.15/15]/20.68$$
$$= 12.93$$

Since $F_{.01(15,21)} = 3.03 < 12.93$ the differences among chemical herbicide means are significant at the 1% level.

The variance of a herbicide mean, $V(\hat{\bar{y}}_j)$ is

$$V(\hat{\bar{y}}_j) = E'_c/r = 24.72/3 = 8.24$$

The variance of the difference between adjusted means for herbicides in the same block, $V(\bar{d})$, is

$$V(\bar{d}) = (2E_e/r)[1 + (r - 1)A] = (2)(20.68)[1 + (2)(.0813)]/3$$

$$= 16.03$$

The variance of the difference between adjusted means for herbicides in different blocks, $V(\bar{d})$, is

$$V(\bar{d}) = (2E_e/r)(1 + rA) = (2)(20.68)[1 + (3)(.0813)]/3$$

$$= 17.15$$

The precision of the triple lattice design relative to that of a randomized block design is obtained by first computing the error mean square, E_{rb}, of a randomized block design:

$$E_{rb} = (SSB + SSE)/(k^2 - 1)(r - 1) = (532.27 + 434.19)/(15)(2) = 32.22$$

Then

$$\% \text{ relative precision } = (E_{rb}/E'_e)100$$

$$= (32.22/24.72)100 = 130.3\%$$

Report of statistical analysis.   An experiment was conducted to measure the effectiveness of 16 new chemical herbicides in controlling broad-leafed weeds in golf course fairways. Because of the possibility of variability in the incidence of broad-leafed weeds in the experimental area, a local country club, a triple lattice experimental design was used with four blocks of four plots replicated on each of three fairways.

The mean percentage kill, adjusted for block effects, by each of the chemical herbicides is presented in Table 13.21. The five most effective chemicals are 5, 10, 13, 14, and 15, while 6, 7, and 9 are clearly inferior to the others.

The precision of the experiment was increased by about 32% by using the triple lattice in place of a randomized block design. The three replications of the triple lattice had a precision which would have required four replications of a randomized block design.

13.4.7   Statistical Analysis: Repetitions of the Basic Design

With a quadruple lattice we obtain a design which has four replications, each of which has a different within-block association scheme among the treatments. This is the preferred way to obtain four replicates because it comes closer to symmetry than other schemes. Designs with four replicates can also be obtained by repeating the simple lattices. With 36 or 100 treatments this is the only way that such a design can be obtained. When a design is constructed by repeating

TABLE 13.21 Mean % Broad-Leafed Weeds Killed in
a Golf Course Fairway by 16 Chemical Herbicides
(Adjusted for Block Effects)

Chemical	% Kill	Chemical	% Kill
1	56.3 *def*	9	43.9 *h*
2	47.6 *gh*	10	64.4 *abcd*
3	55.4 *efg*	11	62.9 *bcde*
4	57.4 *cdef*	12	63.7 *bcde*
5	70.0 *ab*	13	65.5 *abc*
6	44.4 *h*	14	72.6 *a*
7	43.1 *h*	15	70.0 *ab*
8	49.7 *fgh*	16	51.5 *fgh*

Note: Standard error = 2.87.
Note: Means designated by the same letter are not significantly
(α = .05) different by the FPLSD test.

the basic design the statistical analysis of the resulting data is changed to some
extent.

Suppose that the basic design contains n replications and that it is repeated
p times (p cycles) for a total of $r = np$ replications. Assume that the design is
for k^2 treatments with k blocks of k units in each replication. Let

y_{ijlm} = yield of the jth treatment applied to a unit in the lth block of the ith
replication during the mth repetition (cycle)

The data may be tabulated as shown in Table 13.22. The sums in the data
table are

B_{ilm} = sum for the k units in the lth block of the ith replication of the mth
repetition (cycle)

$R_{im} = \sum_l B_{ilm}$ = sum of the yields in the ith replication of the mth cycle

To continue the analysis it is necessary to compute the unadjusted treatment
totals, Table 13.23. The entries in Table 13.23 are computed as:

$T_j = \sum_i \sum_m y_{ijlm}$ = sum of all yields for the jth treatment

$G = \sum_j T_j$ = grand total of all yields

In addition, a set of n auxiliary tables must be constructed from the block
totals. The entries in these tables are totals for blocks containing the same set

TABLE 13.22 Data Table for p Repetitions of a Basic Simple, Triple, or Quadruple Lattice Design

Cycle	Repetition	Block	Yield				Sum
1	1	1	$y_{1,1,1,1}$	$y_{1,2,1,1}$	$\cdots$	$y_{1,k,1,1}$	B_{111}
		2	$y_{1,k+1,2,1}$	$y_{1,k+2,2,1}$	$\cdots$	$y_{1,2k,2,1}$	B_{121}
		$\vdots$	$\vdots$			$\vdots$	$\vdots$
		k	$y_{1,k^2-k,k,1}$	$\cdots$		$y_{1,k^2,k,1}$	B_{1k1}
Sum						R_{11}	
1	2	1	$y_{2,1,1,1}$	$y_{2,2,1,1}$	$\cdots$	$y_{2,k,1,1}$	B_{211}
		2	$y_{2,k+1,2,1}$	$y_{2,k+2,2,1}$	$\cdots$	$y_{2,2k,2,1}$	B_{221}
		$\vdots$	$\vdots$			$\vdots$	$\vdots$
		k	$y_{2,k^2-k,k,1}$	$\cdots$		$y_{2,k^2,k,1}$	B_{2k1}
	Sum						R_{21}
	$\vdots$		$\vdots$				$\vdots$
1	n	1	$y_{n,1,1,1}$	$y_{n,2,1,1}$	$\cdots$	$y_{n,k,1,1}$	B_{n11}
		2	$y_{n,k+1,2,1}$	$y_{n,k+2,2,1}$	$\cdots$	$y_{n,2k,2,1}$	B_{n21}
		$\vdots$	$\vdots$			$\vdots$	$\vdots$
		k	$y_{n,k^2-k,k,1}$	$\cdots$		$y_{n,k^2,k,1}$	B_{nk1}
	Sum			$\vdots$			$\vdots$
2	1	1	$y_{1,1,1,2}$	$y_{1,2,1,2}$	$\cdots$	$y_{1,k,1,2}$	B_{112}
		2	$y_{1,k+1,2,2}$	$y_{1,k+2,2,2}$	$\cdots$	$y_{1,2k,2,2}$	B_{122}
		$\vdots$	$\vdots$			$\vdots$	$\vdots$
		k	$y_{1,k^2-k,k,2}$	$\cdots$		$y_{1,k^2,k,2}$	B_{1k2}
	Sum						R_{12}
	$\vdots$		$\vdots$				$\vdots$
2	n	1	$y_{n1,1,2}$	$y_{n,2,1,2}$	$\cdots$	$y_{n,k,1,2}$	B_{n12}
		2	$y_{n,k+1,2,2}$	$y_{n,k+2,2,2}$	$\cdots$	$y_{n,2k,2,2}$	B_{n22}
		$\vdots$	$\vdots$				B_{nk2}
		k	$y_{n,k^2-k,k,2}$	$\cdots$		$y_{n,k^2,k,2}$	
	Sum						R_{n2}
	$\vdots$		$\vdots$				$\vdots$
p	1	1	$y_{1,1,1,p}$	$y_{n,2,1,p}$	$\cdots$	$y_{1,k,1,p}$	B_{11p}
		2	$y_{1,k+1,2,p}$	$y_{1,k+2,2,p}$	$\cdots$	$y_{1,2k,2,p}$	B_{12p}
		$\vdots$	$\vdots$			$\vdots$	$\vdots$
		k	$y_{1,k^2-k,k,p}$	$\cdots$		$y_{1,k^2,k,p}$	B_{1kp}

TABLE 13.22 (*Continued*)

Cycle	Repetition	Block	Yield				Sum
	Sum						R_{1p}
	$\vdots$		$\vdots$				$\vdots$
p	n	1	$y_{n,1,1,p}$	$y_{n,2,1,p}$	$\cdots$	$y_{n,k,1,p}$	B_{n1p}
		2	$y_{n,k+1,2,p}$	$y_{n,k+2,2,p}$	$\cdots$	$y_{n,2k,2,p}$	B_{n2p}
		$\vdots$	$\vdots$			$\vdots$	$\vdots$
		k	$y_{n,k^2-k,k,p}$		$\cdots$	$y_{n,k^2,k,p}$	B_{nkp}
	Sum						R_{np}

TABLE 13.23 Unadjusted Treatment Totals for p Repetitions of a Basic Lattice Design

Treatment	Total (unadjusted)
1	T_1
2	T_2
$\vdots$	$\vdots$
k^2	T_{k^2}
Sum	G

of k treatments. These are illustrated in Table 13.24. The sums in Table 13.24 are obtained as follows:

$$S_{il} = \sum_m B_{ilm} = \text{sum for the blocks containing the same set of } k \text{ treatments}$$

$$S_i = \sum_l S_{il} = \text{sum for all blocks in replication } i$$

$$R_{im} = \sum_l B_{ilm} = \text{sum of yields in the } i\text{th replication of the } m\text{th cycle}$$

$C_{il} = $ sum (over all replications) of treatment totals for all treatments appearing in the lth block of the ith replication $- nS_{il}$

$$C_i = \sum_l C_i = \text{sum over blocks of the } C \text{ values in the } i\text{th replication.}$$

These should sum to zero.

$AC_{il} = $ block adjustment factors computed, if necessary, after completing the analysis of variance

TABLE 13.24 Auxiliary Tables for p Repetitions of a Basic Lattice Design

Treatment group		Cycle				Sum		
Replication	Block	1	2	$\cdots$	p	C	AC	
1	1	B_{111}	B_{112}	$\cdots$	B_{11p}	S_{11}	C_{11}	AC_{11}
	2	B_{121}	B_{122}	$\cdots$	B_{12p}	S_{12}	C_{12}	AC_{12}
	$\vdots$	$\vdots$	$\vdots$		$\vdots$	$\vdots$	$\vdots$	$\vdots$
	k	B_{1k1}	B_{1k2}	$\cdots$	B_{1kp}	S_{1k}	C_{1k}	AC_{1k}
Sum		R_{11}	R_{12}	$\cdots$	R_{1p}	S_1	C_1	AC_1
2	1	B_{211}	B_{212}	$\cdots$	B_{21p}	S_{21}	C_{21}	AC_{21}
	2	B_{221}	B_{222}	$\cdots$	B_{22p}	S_{22}	C_{22}	AC_{22}
	$\vdots$	$\vdots$	$\vdots$		$\vdots$	$\vdots$	$\vdots$	$\vdots$
	k	B_{2k1}	B_{2k2}	$\cdots$	B_{2kp}	S_{2k}	C_{2k}	AC_{2k}
Sum		R_{21}	R_{22}	$\cdots$	R_{2p}	S_2	C_2	AC_2
$\vdots$		$\vdots$			$\vdots$	$\vdots$	$\vdots$	
n	1	B_{n11}	B_{n12}	$\cdots$	B_{n1p}	S_{n1}	C_{n1}	AC_{n1}
	2	B_{n21}	B_{n22}	$\cdots$	B_{n2p}	S_{n2}	C_{n2}	AC_{n2}
	$\vdots$	$\vdots$	$\vdots$		$\vdots$	$\vdots$	$\vdots$	$\vdots$
	k	B_{nk1}	B_{nk2}	$\cdots$	B_{nkp}	S_{nk}	C_{nk}	AC_{nk}
Sum		R_{n1}	R_{n2}	$\cdots$	R_{np}	S_n	C_n	AC_n
						G	0	0

TABLE 13.25 ANOVA for p Repetitions of a Basic Lattice Design

Source	d.f.	SS	MS
Total	$rk^2 - 1$	SSTot	
Replication	$r - 1$	SSR	
Treatment (unadjusted)	$k^2 - 1$	SST	
Block in replication (adjusted)	$r(k - 1)$	SSBA	E_b
Component (a)	$n(p - 1)(k - 1)$	SSA	
Component (b)	$n(k - 1)$	SSB	
Intrablock error	$(k - 1)(rk - k - 1)$	SSE	E_e

The format for the analysis of variance is presented in Table 13.25. The adjusted block sum of squares, SSBA, is composed of two components, both of which may be obtained from the auxiliary tables, Table 13.24:

Component (a): Using the entries in Table 13.24, compute the following:

$$\text{SSTo} = (1/k) \sum_i \sum_l \sum_m B_{ilm}^2 - (1/pk^2) \sum_i S_i^2$$

$$\text{SSRow} = (1/pk) \sum_i \sum_l S_{il}^2 - (1/pk^2) \sum_i S_i^2$$

$$\text{SSCol} = (1/k^2) \sum_i \sum_m R_{im}^2 - (1/pk^2) \sum_i S_i^2$$

$$\text{SSA} = \text{SSTo} - \text{SSRow} - \text{SSCol}$$

Component (b):

$$\text{SSB} = [1/kr(n-1)] \sum_i \sum_l C_{il}^2 - [1/k^2 r(n-1)] \sum_i C_i^2$$

The other sums of squares, SSTot, SSR, and SST, are obtained in the usual way. Finally

$$\text{SSE} = \text{SSTot} - \text{SSR} - \text{SST} - \text{SSBA}$$

As with the other lattice analyses the next step is to compare E_b with E_e. If $E_b \le E_e$ blocking has no effect, and the design is treated as an ordinary randomized block design with replications as blocks.

If $E_b > E_e$ an adjustment factor, A, is computed as

$$A = \frac{p(E_b - E_e)}{k[(r-p)E_b + (p-1)E_e]}$$

The adjustment factor is used to compute the block adjustments, AC_{il}, in Table 13.24. These, in turn, are used to compute the adjusted treatment totals, $\hat{T}_j$, and adjusted treatment means, $\hat{\bar{y}}_j$. These are computed as

$$\hat{T}_j = T_j + (\text{sum of the } AC_{il} \text{ for all blocks in which the } j\text{th treatment occurs})$$

$$\hat{\bar{y}}_j = \hat{T}_j/r$$

To test the significance of differences among adjusted means an adjusted treatment sum of squares, SST(adj), is computed as

$$\text{SST(adj)} = \text{SST} - k(n-1)A\left[\frac{n\text{SSRow}}{(n-1)(1+kA)} - \text{SSBA}\right]$$

Then, the approximate F for the test is

$$\cdot \quad F = [\text{SST(adj)}/(k^2-1)]/E_e, \quad \text{with}$$

$$k^2 - 1, (k-1)(rk - k - 1) \text{ d.f.}$$

The effective error variance, E'_e, is computed as

$$E'_e = E_e[1 + nkA/(k + 1)]$$

The variance of an adjusted treatment mean, $V(\hat{\bar{y}}_j)$, is

$$V(\hat{\bar{y}}_j) = E'_e/r$$

and the average variance of the difference between two adjusted means $V(\bar{d})$ is

$$V(\bar{d}) = 2E'_e/r$$

The variance of the difference between means of two treatments in the same block is

$$V(\bar{d}) = (2E_e/r)[1 + (n - 1)A]$$

The variance of the difference between means of two treatments in different blocks is

$$V(\bar{d}) = (2E_e/r)[1 + nA]$$

To estimate the precision of the repeated lattice relative to that of a randomized block design the error variance, E_{rb}, of a randomized block design is computed as

$$E_{rb} = [SSBA + SSE]/(r - 1)(k^2 - 1)$$

Then, the precision of the repeated lattice relative to that of a randomized block is

$$\% \text{ relative precision} = (E_{rb}/E'_e)100$$

13.4.8   Numerical Example: Repeated 5 × 5 Simple Lattice

The data in the numerical example used to illustrate the simple lattice (Section 13.4.5) came from a yield trial which was actually run with four replications obtained by using two repetitions of the simple lattice. For illustrating the simple lattice we used the data from the first two replications. We will now use all four replications to illustrate the computations involved with repetitions of the basic design.

The field layout and barley yields are summarized in Table 13.26. The design consists of $p = 2$ repetitions of a 5 × 5 simple lattice each with $n = 2$ replications for a total of $r = np = 4$ replications in the trial. There are $k^2 = 5^2 = 25$ selections of barley grown in blocks of five plots, with five blocks in each replication.

The unadjusted selection totals are given in Table 13.27. The necessary auxiliary tables are shown in Table 13.28.

TABLE 13.26 Field Layout and Yield (kg/plot) of 25 Barley Selections Grown in a Repeated 5 × 5 Simple Lattice (Selection Numbers in Parentheses)

Cycle	Repetition	Block	Yield					Sum
1	1	1	18.2 (19)	13.0 (16)	9.5 (18)	6.7 (17)	10.1 (20)	57.5
		2	13.3 (12)	11.4 (13)	14.2 (15)	11.9 (14)	13.4 (11)	64.2
		3	15.0 (1)	12.4 (2)	17.3 (3)	20.5 (4)	13.0 (5)	78.2
		4	7.0 (22)	5.9 (24)	14.1 (21)	19.2 (25)	7.8 (23)	54.0
		5	11.9 (9)	15.2 (7)	17.2 (10)	16.3 (8)	16.0 (6)	76.6
	Sum							330.5
1	2	1	7.7 (23)	15.2 (18)	19.1 (3)	15.5 (8)	14.7 (13)	72.2
		2	15.8 (5)	18.0 (20)	18.8 (10)	14.4 (15)	20.0 (25)	87.0
		3	10.2 (22)	11.5 (12)	17.0 (2)	11.0 (17)	15.3 (7)	65.0
		4	10.9 (14)	4.7 (24)	10.9 (9)	16.6 (4)	9.8 (19)	52.9
		5	20.0 (6)	21.1 (16)	16.9 (11)	10.9 (21)	15.0 (1)	83.9
	Sum							361.0
2	1	1	17.0 (2)	18.5 (1)	16.3 (3)	13.9 (5)	15.8 (4)	81.5
		2	7.2 (17)	16.5 (16)	15.6 (20)	12.8 (19)	15.6 (18)	67.7
		3	15.1 (14)	14.5 (15)	15.3 (11)	11.9 (13)	15.1 (12)	71.9
		4	6.6 (23)	14.8 (21)	9.4 (22)	20.6 (25)	8.2 (24)	59.6
		5	13.2 (8)	13.8 (6)	13.0 (7)	15.6 (10)	13.2 (9)	68.8
	Sum							349.5
2	2	1	20.3 (25)	18.3 (10)	15.7 (20)	16.3 (5)	13.8 (15)	84.4
		2	12.8 (18)	14.2 (8)	12.6 (13)	6.2 (23)	17.7 (3)	63.5
		3	16.1 (2)	6.6 (17)	11.8 (12)	10.9 (22)	13.3 (7)	58.7
		4	17.5 (1)	18.5 (16)	16.9 (6)	15.2 (21)	14.0 (11)	82.1
		5	18.5 (4)	15.3 (9)	15.5 (14)	8.1 (24)	12.3 (19)	69.7
	Sum							358.4

The analysis of variance of the barley yield data is presented in Table 13.29.

The block in replication component sums of squares in Table 13.29 are computed as follows:

Component (a):

$$SSTo = (1/5) \sum_i \sum_l \sum_m B_{ilm}^2 - [1/(2)(25)] \sum_k S_i^2$$

$$= 20,010.34 - 19,598.73 = 411.61$$

TABLE 13.27 Unadjusted Selection Totals (Selection Numbers in Parentheses)

(1) 66.0	(2) 62.5	(3) 70.4	(4) 71.4	(5) 59.0
(6) 66.7	(7) 56.8	(8) 59.2	(9) 51.3	(10) 69.9
(11) 59.6	(12) 51.7	(13) 50.6	(14) 53.4	(15) 56.9
(16) 69.1	(17) 31.5	(18) 53.1	(19) 53.1	(20) 59.4
(21) 55.0	(22) 37.5	(23) 28.3	(24) 26.9	(25) 80.1
			Sum	1399.4

TABLE 13.28 Auxiliary Tables for Two Repetitions of a 5 × 5 Simple Lattice

Treatment group	Repetition	Block	Cycle 1	Cycle 2	Sum	C	AC
1,2,3,4,5	1	1	78.2	81.5	159.7	9.9	.67
6,7,8,9,10		2	76.6	68.8	145.4	13.1	.89
11,12,13,14,15		3	64.2	71.9	136.1	0.0	.00
16,17,18,19,20		4	57.5	67.7	125.2	15.8	1.08
21,22,23,24,25		5	54.0	59.6	113.6	0.6	.04
		Sum	330.5	349.5	680.0	39.4	2.68
1,6,11,16,21	2	1	83.9	82.1	166.0	− 15.6	− 1.06
2,7,12,17,22		2	65.0	58.7	123.7	− 7.4	− .50
3,8,13,18,23		3	72.2	63.5	135.7	− 9.8	− .67
4,9,14,19,24		4	52.9	69.7	122.6	10.9	.74
5,10,15,20,25		5	87.0	84.4	171.4	− 17.5	− 1.19
		Sum	361.0	358.4	719.4	− 39.4	− 2.68
		Sum			1399.4	0.0	0.00

Note: C = sum of treatment totals for treatment in the group − nS_{il}.
$C_{11} = (66.0 + 62.5 + 70.4 + 71.4 + 59.0) − (2)(159.7) = 9.9.$

$$\text{SSRow} = [1/(2)(5)] \sum_i \sum_l S_{il}^2 - [1/(2)(25)] \sum_i S_i^2$$

$$= 19,942.94 - 19,598.73 = 344.21$$

$$\text{SSCol} = (1/25) \sum_i \sum_m R_{ij}^2 - [1/(2)(25)] \sum_i S_i^2$$

$$= 19,606.08 - 19,598.73 = 7.35$$

TABLE 13.29 ANOVA of Barley Yields from Two
Repetitions of a 5 × 5 Simple Lattice

Source	d.f.	SS	MS
Total	99	1393.84	
Replication	3	22.88	
Selection (unadj)	24	1085.75	
Block in replication (adj)	16	95.94	$6.00 = E_b$
Component (*a*)	8	60.05	
Component (*b*)	8	35.89	
Intrablock error	56	189.27	$3.38 = E_e$

$$SSA = SSTo - SSRow - SSCol$$
$$= 411.61 - 344.21 - 7.35 = 60.05$$

Component (b):

$$SSB = [1/(5)(4)(1)] \sum_i \sum_l C_{il}^2 - [1/(25)(4)(1)] \sum_i C_i^2$$
$$= 66.94 - 31.05 = 35.89$$

Since $6.00 > 3.38$, $E_b > E_e$ and blocking was effective in reducing experimental error in this trial. The adjustment factor, A, is computed as

$$A = \frac{2(E_b - E_e)}{(5)[(2)E_b + (1)E_e]} = \frac{(2)(6.00 - 3.38)}{(5)[(2)(6.00) + 3.38]} = .0681$$

This adjustment factor is used to compute the block adjustments, AC, in Table 13.28, which, in turn, are used to compute the adjusted treatment totals and means given in Table 13.30.

The test of the significance of the differences among the adjusted means is conducted by first computing an adjusted treatment sum of squares, SST(adj),

$$SST(adj) = SST - k(n - 1)A\left[\frac{nSSRow}{(n - 1)(1 + kA)} - SSB\right]$$
$$= 1085.75 - (5)(1)(.0681)\left[\frac{(2)(344.21)}{(1)(1 + (5)(.0681))} - 35.89\right]$$
$$= 923.10$$

Then

$$F = [SST(adj)/(k^2 - 1)]/E_e$$
$$= (923.10/24)/3.38 = 11.38$$

Since $F_{.01(24,56)} = 2.12 < 11.38$ the differences are highly significant.

TABLE 13.30 Adjusted Yield Totals and
Means for 25 Barley Selections

Selection	Adjusted total	Adjusted mean
1	65.61	16.40
2	62.67	15.67
3	70.40	17.60
4	72.81	18.20
5	58.48	14.62
6	66.53	16.63
7	57.19	14.30
8	59.42	14.86
9	52.93	13.23
10	69.60	17.40
11	58.54	14.63
12	51.20	12.80
13	49.93	12.48
14	54.14	13.54
15	55.71	13.93
16	69.12	17.28
17	32.08	8.02
18	53.51	13.38
19	54.92	13.73
20	59.29	14.82
21	53.98	13.50
22	37.04	9.26
23	27.67	6.92
24	27.68	6.92
25	78.95	19.74

The effective error variance, E'_e, for this trial is

$$E'_e = E_e[1 + nkA/(k + 1)] = 3.38[1 + (2)(5)(.0681)/6]$$

$$= 3.76$$

The variance, $V(\hat{y}_j)$, of an adjusted mean is

$$V(\hat{y}) = E'_e/r = 3.76/4 = .94$$

The variance of the difference between means, $V(\overline{d})$, of selections in the same block is

$$V(\overline{d}) = (2E_e/r)[1 + (n - 1)A]$$

$$= [(2)(3.38)/4][1 + (1)(.0681)] = 1.80$$

The variance of the difference between means, $V(\overline{d})$, of selections in different blocks is

$$V(\overline{d}) = (2E_e/r)(1 + nA)$$

$$= [(2)(3.38)/4][1 + (2)(.0681)] = 1.92$$

The average variance, $V(\overline{d})$, of the difference between means is:

$$V(\overline{d}) = 2E_e'/r = (2)(3.76)/4 = 1.88$$

Finally, the efficiency of the repeated simple lattice relative to a randomized block design is computed by first computing the randomized block error mean square, E_{rb}, as

$$E_{rb} = [SSBA + SSE]/(r - 1)(k^2 - 1)$$

$$= [95.94 + 189.27]/(3)(24) = 3.96$$

Then, the relative precision is

$$\% \text{ relative precision} = (E_{rb}/E_e')100$$

$$= (3.96/3.76)100 = 105.4\%$$

Report of statistical analysis. A yield trial was conducted to evaluate 25 new selections of barley. The trial was run using a 5 × 5 simple lattice repeated twice, giving a total of four replications for each selection. Mean yields (kilograms per plot), adjusted for block differences, are presented in Table 13.31.

The highest yielding selections, in descending order, are 25, 4, 3, 10, 16, 6, and 1. The means in this group are not significantly different from each other. Selections 22, 17, 23, and 24 form a group which is distinctly inferior to the rest.

TABLE 13.31 Mean Yields (kg/plot) of 25 New Selections of Barley (Selection Numbers in Parentheses)

abcde	bcdef	abc	ab	bcdef
(1) 16.40	(2) 15.67	(3) 17.60	(4) 18.20	(5) 14.62
abcde	cdef	bcdef	ef	abcd
(6) 16.63	(7) 14.30	(8) 14.86	(9) 13.23	(10) 17.40
bcdef	fg	fg	ef	cdef
(11) 14.63	(12) 12.80	(13) 12.48	(14) 13.54	(15) 13.93
abcd	h	ef	def	bcdef
(16) 17.28	(17) 8.02	(18) 13.38	(19) 13.73	(20) 14.82
ef	gh	h	h	a
(21) 13.50	(22) 9.26	(23) 6.92	(24) 6.92	(25) 19.74

Note: Standard error = .97.
Note: Means included by the same letter are not significantly different at the 1% level by the FPLSD test.

Little was gained by using the repeated lattice design for this trial. Precision was increased by only about 5% relative to that of a randomized block design with the same degree of replication.

13.5 PLANS FOR LATTICE DESIGNS*

In the following plans:

1. t = number of treatments.
2. k = block size.
3. r = number of replications.
4. b = number of blocks.
5. Block number is enclosed in parentheses.
6. Replication number is in Roman numerals.
7. Treatments are indicated by number within the blocks.

13.5.1 3 × 3 Balanced Lattice

$t = 9, k = 3, r = 4, b = 12$

	I		II		III		IV
(1)	1 2 3	(4)	1 4 7	(7)	1 5 9	(10)	1 8 6
(2)	4 5 6	(5)	2 5 8	(8)	7 2 6	(11)	4 2 9
(3)	7 8 9	(6)	3 6 9	(9)	4 8 3	(12)	7 5 3

13.5.2 4 × 4 Balanced Lattice

$t = 16, k = 4, r = 5, b = 20$

	I		II		III
(1)	1 2 3 4	(5)	1 5 9 13	(9)	1 6 11 16
(2)	5 6 7 8	(6)	2 6 10 14	(10)	5 2 15 12
(3)	9 10 11 12	(7)	3 7 11 15	(11)	9 14 3 8
(4)	13 14 15 16	(8)	4 8 12 16	(12)	13 10 7 4

---

*Reproduced from *Experimental Designs*, 2nd ed. (1957), by W. G. Cochran and G. M. Cox, with the permission of the publisher, John Wiley and Sons, Inc., New York.

	IV					V			
(13)	1	14	7	12	(17)	1	10	15	8
(14)	13	2	11	8	(18)	9	2	7	16
(15)	5	10	3	16	(19)	13	6	3	12
(16)	9	6	15	4	(20)	5	14	11	4

13.5.3 5 × 5 Balanced Lattice

$t = 25, k = 5, r = 6, b = 30$

	I						II				
(1)	1	2	3	4	5	(6)	1	6	11	16	21
(2)	6	7	8	9	10	(7)	2	7	12	17	22
(3)	11	12	13	14	15	(8)	3	8	13	18	23
(4)	16	17	18	19	20	(9)	4	9	14	19	24
(5)	21	22	23	24	25	(10)	5	10	15	20	25

	III						IV				
(11)	1	7	13	19	25	(16)	1	12	23	9	20
(12)	21	2	8	14	20	(17)	16	2	13	24	10
(13)	16	22	3	9	15	(18)	6	17	3	14	25
(14)	11	17	23	4	10	(19)	21	7	18	4	15
(15)	6	12	18	24	5	(20)	11	22	8	19	5

	V						VI				
(21)	1	17	8	24	15	(26)	1	22	18	14	10
(22)	11	2	18	9	25	(27)	6	2	23	19	15
(23)	21	12	3	19	10	(28)	11	7	3	24	20
(24)	6	22	13	4	20	(29)	16	12	8	4	25
(25)	16	7	23	14	5	(30)	21	17	13	9	5

13.5.4 6 × 6 Triple Lattice

$t = 36, k = 6, r = 3, b = 18$

	I								II					
(1)	1	2	3	4	5	6	(7)	1	7	13	19	25	31
(2)	7	8	9	10	11	12	(8)	2	8	14	20	26	32
(3)	13	14	15	16	17	18	(9)	3	9	15	21	27	33
(4)	19	20	21	22	23	24	(10)	4	10	16	22	28	34
(5)	25	26	27	28	29	30	(11)	5	11	17	23	29	35
(6)	31	32	33	34	35	36	(12)	6	12	18	24	30	36

	III					
(13)	1	8	15	22	29	36
(14)	31	2	9	16	23	30
(15)	25	32	3	10	17	24
(16)	19	26	33	4	11	18
(17)	13	20	27	34	5	12
(18)	7	14	21	28	35	6

13.5.5 7 × 7 Balanced Lattice

$t = 49, k = 7, r = 8, b = 56$

	I									II						
(1)	1	2	3	4	5	6	7	(8)	1	8	15	22	29	36	43
(2)	8	9	10	11	12	13	14	(9)	2	9	16	23	30	37	44
(3)	15	16	17	18	19	20	21	(10)	3	10	17	24	31	38	45
(4)	22	23	24	25	26	27	28	(11)	4	11	18	25	32	39	46
(5)	29	30	31	32	33	34	35	(12)	5	12	19	26	33	40	47
(6)	36	37	38	39	40	41	42	(13)	6	13	20	27	34	41	48
(7)	43	44	45	46	47	48	49	(14)	7	14	21	28	35	42	49

III

(15)	1	9	17	25	33	41	49
(16)	43	2	10	18	26	34	42
(17)	36	44	3	11	19	27	35
(18)	29	37	45	4	12	20	28
(19)	22	30	38	46	5	13	21
(20)	15	23	31	39	47	6	14
(21)	8	16	24	32	40	48	7

IV

(22)	1	37	24	11	47	34	21
(23)	15	2	38	25	12	48	35
(24)	29	16	3	39	26	13	49
(25)	43	30	17	4	40	27	14
(26)	8	44	31	18	5	41	28
(27)	22	9	45	32	19	6	42
(28)	36	23	10	46	33	20	7

V

(29)	1	30	10	39	19	48	28
(30)	22	2	31	11	40	20	49
(31)	43	23	3	32	12	41	21
(32)	15	44	24	4	33	13	42
(33)	36	16	45	25	5	34	14
(34)	8	37	17	46	26	6	35
(35)	29	9	38	18	47	27	7

VI

(36)	1	23	45	18	40	13	35
(37)	29	2	24	46	19	41	14
(38)	8	30	3	25	47	20	42
(39)	36	9	31	4	26	48	21
(40)	15	37	10	32	5	27	49
(41)	43	16	38	11	33	6	28
(42)	22	44	17	39	12	34	7

VII

(43)	1	16	31	46	12	27	42
(44)	36	2	17	32	47	13	28
(45)	22	37	3	18	33	48	14
(46)	8	23	38	4	19	34	49
(47)	43	9	24	39	5	20	35
(48)	29	44	10	25	40	6	21
(49)	15	30	45	11	26	41	7

VIII

(50)	1	44	38	32	26	20	14
(51)	8	2	45	39	33	27	21
(52)	15	9	3	46	40	34	28
(53)	22	16	10	4	47	41	35
(54)	29	23	17	11	5	48	42
(55)	36	30	24	18	12	6	49
(56)	43	37	31	25	19	13	7

13.5.6 8 × 8 Balanced Lattice

$t = 64, k = 8, r = 9, b = 72$

I

(1)	1	2	3	4	5	6	7	8
(2)	9	10	11	12	13	14	15	16
(3)	17	18	19	20	21	22	23	24
(4)	25	26	27	28	29	30	31	32
(5)	33	34	35	36	37	38	39	40
(6)	41	42	43	44	45	46	47	48
(7)	49	50	51	52	53	54	55	56
(8)	57	58	59	60	61	62	63	64

II

(9)	1	9	17	25	33	41	49	57
(10)	2	10	18	26	34	42	50	58
(11)	3	11	19	27	35	43	51	59
(12)	4	12	20	28	36	44	52	60
(13)	5	13	21	29	37	45	53	61
(14)	6	14	22	30	38	46	54	62
(15)	7	15	23	31	39	47	55	63
(16)	8	16	24	32	40	48	56	64

III

(17)	1	10	19	28	37	46	55	64
(18)	9	2	51	44	61	30	23	40
(19)	17	50	3	36	29	62	15	48
(20)	25	42	35	4	21	14	63	56
(21)	33	58	27	20	5	54	47	16
(22)	41	26	59	12	53	6	39	24
(23)	49	18	11	60	45	38	7	32
(24)	57	34	43	52	13	22	31	8

IV

(25)	1	18	27	44	13	62	39	56
(26)	17	2	35	60	53	46	31	16
(27)	25	34	3	12	45	54	23	64
(28)	41	58	11	4	29	22	55	40
(29)	9	50	43	28	5	38	63	24
(30)	57	42	51	20	37	6	15	32
(31)	33	26	19	52	61	14	7	48
(32)	49	10	59	36	21	30	47	8

V

(33)	1	26	43	60	21	54	15	40
(34)	25	2	11	52	37	62	47	24
(35)	41	10	3	61	20	38	31	56
(36)	57	50	19	4	45	30	39	16
(37)	17	34	59	44	5	14	55	32
(38)	49	58	35	28	13	6	23	48
(39)	9	42	27	36	53	22	7	64
(40)	33	18	51	12	29	46	63	8

VI

(41)	1	34	11	20	53	30	63	48
(42)	33	2	59	28	45	22	15	56
(43)	9	58	3	52	21	46	39	32
(44)	17	26	51	4	13	38	47	64
(45)	49	42	19	12	5	62	31	40
(46)	25	18	43	36	61	6	55	16
(47)	57	10	35	44	29	54	7	24
(48)	41	50	27	60	37	14	23	8

VII

(49)	1	42	59	52	29	38	23	16
(50)	41	2	19	36	13	54	63	32
(51)	57	18	3	28	53	14	47	40
(52)	49	34	27	4	61	46	15	24
(53)	25	10	51	60	5	22	39	48
(54)	33	50	11	44	21	6	31	64
(55)	17	58	43	37	30	12	7	56
(56)	9	26	35	20	45	62	55	8

VIII

(57)	1	50	35	12	61	22	47	32
(58)	49	2	43	20	29	14	39	64
(59)	33	42	3	60	13	30	55	24
(60)	9	18	59	4	37	54	31	48
(61)	57	26	11	36	5	46	23	56
(62)	17	10	27	52	45	6	63	40
(63)	41	34	51	28	21	62	7	16
(64)	25	58	19	44	53	38	15	8

IX

(65)	1	58	51	36	45	14	31	24
(66)	57	2	27	12	21	38	55	48
(67)	49	26	3	44	37	22	63	16
(68)	33	10	43	4	53	62	23	32
(69)	41	18	35	52	5	30	15	64
(70)	9	34	19	60	29	6	47	56
(71)	25	50	59	20	13	46	7	40
(72)	17	42	11	28	61	54	39	8

13.5.7 9 × 9 Balanced Lattice

$t = 81, k = 9, r = 10, b = 90$

I

(1)	1	2	3	4	5	6	7	8	9
(2)	10	11	12	13	14	15	16	17	18
(3)	19	20	21	22	23	24	25	26	27
(4)	28	29	30	31	32	33	34	35	36
(5)	37	38	39	40	41	42	43	44	45
(6)	46	47	48	49	50	51	52	53	54
(7)	55	56	57	58	59	60	61	62	63
(8)	64	65	66	67	68	69	70	71	72
(9)	73	74	75	76	77	78	79	80	81

II

(10)	1	10	19	28	37	46	55	64	73
(11)	2	11	20	29	38	47	56	65	74
(12)	3	12	21	30	39	48	57	66	75
(13)	4	13	22	31	40	49	58	67	76
(14)	5	14	23	32	41	50	59	68	77
(15)	6	15	24	33	42	51	60	69	78
(16)	7	16	25	34	43	52	61	70	79
(17)	8	17	26	35	44	53	62	71	80
(18)	9	18	27	36	45	54	63	72	81

III

(19)	1	20	12	58	77	69	34	53	45
(20)	10	2	21	67	59	78	43	35	54
(21)	19	11	3	76	68	60	52	44	36
(22)	28	47	39	4	23	15	61	80	72
(23)	37	29	48	13	5	24	70	62	81
(24)	46	38	30	22	14	6	79	71	63
(25)	55	74	66	31	50	42	7	26	18
(26)	64	56	75	40	32	51	16	8	27
(27)	73	65	57	49	41	33	25	17	9

IV

(28)	1	11	21	31	41	51	61	71	81
(29)	19	2	12	49	32	42	79	62	72
(30)	10	20	3	40	50	33	70	80	63
(31)	55	65	75	4	14	24	34	44	54
(32)	73	56	66	22	5	15	52	35	45
(33)	64	74	57	13	23	6	43	53	36
(34)	28	38	48	58	68	78	7	17	27
(35)	46	29	39	76	59	69	25	8	18
(36)	37	47	30	67	77	60	16	26	9

V

(37)	1	29	57	22	50	78	16	44	72
(38)	55	2	30	76	23	51	70	17	45
(39)	28	56	3	49	77	24	43	71	18
(40)	10	38	66	4	32	60	25	53	81
(41)	64	11	39	58	5	33	79	26	54
(42)	37	65	12	31	59	6	52	80	27
(43)	19	47	75	13	41	69	7	35	63
(44)	73	20	48	67	14	42	61	8	36
(45)	46	74	21	40	68	15	34	62	9

VI

(46)	1	56	30	13	68	42	25	80	54
(47)	28	2	57	40	14	69	52	26	81
(48)	55	29	3	67	41	15	79	53	27
(49)	19	74	48	4	59	33	16	71	45
(50)	46	20	75	31	5	60	43	17	72
(51)	73	47	21	58	32	6	70	44	18
(52)	10	65	39	22	77	51	7	62	36
(53)	37	11	66	49	23	78	34	8	63
(54)	64	38	12	76	50	24	61	35	9

VII

(55)	1	47	66	76	14	33	43	62	27
(56)	64	2	48	31	77	15	25	44	63
(57)	46	65	3	13	32	78	61	26	45
(58)	37	56	21	4	50	69	79	17	36
(59)	19	38	57	67	5	51	34	80	18
(60)	55	20	39	49	68	6	16	35	81
(61)	73	11	30	40	59	24	7	53	72
(62)	28	74	12	22	41	60	70	8	54
(63)	10	29	75	58	23	42	52	71	9

VIII

(64)	1	74	39	67	32	24	52	17	63
(65)	37	2	75	22	68	33	61	53	18
(66)	73	38	3	31	23	69	16	62	54
(67)	46	11	57	4	77	42	70	35	27
(68)	55	47	12	40	5	78	25	71	36
(69)	10	56	48	76	41	6	34	26	72
(70)	64	29	21	49	14	60	7	80	45
(71)	19	65	30	58	50	15	43	8	81
(72)	28	20	66	13	59	51	79	44	9

IX

(73)	1	65	48	40	23	60	79	35	18
(74)	46	2	66	58	41	24	16	80	36
(75)	64	47	3	22	59	42	34	17	81
(76)	73	29	12	4	68	51	43	26	63
(77)	10	74	30	49	5	69	61	44	27
(78)	28	11	75	67	50	6	25	62	45
(79)	37	20	57	76	32	15	7	71	54
(80)	55	38	21	13	77	33	52	8	72
(81)	19	56	39	31	14	78	70	53	9

					X				
(82)	1	38	75	49	59	15	70	26	36
(83)	73	2	39	13	50	60	34	71	27
(84)	37	74	3	58	14	51	25	35	72
(85)	64	20	30	4	41	78	52	62	18
(86)	28	65	21	76	5	42	16	53	63
(87)	19	29	66	40	77	6	61	17	54
(88)	46	56	12	67	23	33	7	44	81
(89)	10	47	57	31	68	24	79	8	45
(90)	55	11	48	22	32	69	43	80	9

13.5.8   10 × 10 Triple Lattice

$t = 100, k = 10, r = 3, b = 30$

					I					
(1)	1	2	3	4	5	6	7	8	9	10
(2)	11	12	13	14	15	16	17	18	19	20
(3)	21	22	23	24	25	26	27	28	29	30
(4)	31	32	33	34	35	36	37	38	39	40
(5)	41	42	43	44	45	46	47	48	49	50
(6)	51	52	53	54	55	56	57	58	59	60
(7)	61	62	63	64	65	66	67	68	69	70
(8)	71	72	73	74	75	76	77	78	79	80
(9)	81	82	83	84	85	86	87	88	89	90
(10)	91	92	93	94	95	96	97	98	99	100

II

(11)	1	11	21	31	41	51	61	71	81	91
(12)	2	12	22	32	42	52	62	72	82	92
(13)	3	13	23	33	43	53	63	73	83	93
(14)	4	14	24	34	44	54	64	74	84	94
(15)	5	15	25	35	45	55	65	75	85	95
(16)	6	16	26	36	46	56	66	76	86	96
(17)	7	17	27	37	47	57	67	77	87	97
(18)	8	18	28	38	48	58	68	78	88	98
(19)	9	19	29	39	49	59	69	79	89	99
(20)	10	20	30	40	50	60	70	80	90	100

III

(21)	1	12	23	34	45	56	67	78	89	100
(22)	91	2	13	24	35	46	57	68	79	90
(23)	81	92	3	14	25	36	47	58	69	80
(24)	71	82	93	4	15	26	37	48	59	70
(25)	61	72	83	94	5	16	27	38	49	60
(26)	51	62	73	84	95	6	17	28	39	50
(27)	41	52	63	74	85	96	7	18	29	40
(28)	31	42	53	64	75	86	97	8	19	30
(29)	21	32	43	54	65	76	87	98	9	20
(30)	11	22	33	44	55	66	77	88	99	10

FURTHER READING*

Federer, W. T. (1955), Ch. 11–13.
Finney, D. J. (1960), Ch. 6.
Kempthorne, O. (1973), Ch. 23–27.

REFERENCES

Cochran, W. G., and G. M. Cox (1957). *Experimental Designs,* 2nd ed., Wiley, New York, Ch. 9–13.
Bose, R. C., W. H. Clatworthy, and S. S. Shirkhande (1954). Tables of Partially Balanced Designs with Two Associate Classes, *N. C. Agric. Exp. Stn. Tech. Bull. 107.*
Yates, F. (1939). The Recovery of Inter-Block Information in Variety Trials Arranged in Three-Dimensional Lattices, *Ann. Eugen.* 9:136–156.

---

*See Bibliography for complete citation.

Bibliography

Anderson, V. L., and R. A. McLean (1974). *Design of Experiments*, Marcel Dekker, Inc., New York.

Box, G. E. P., and J. S. Hunter (1961). The 2^{k-p} Fractional Factorial Designs, I, *Technometrics* 3:311–351; II, *Technometrics* 3:449–458.

Box, G. E. P., W. G. Hunter, and J. S. Hunter (1978). *Statistics for Experimenters*, Wiley, New York.

Cochran, W. G., and G. M. Cox (1957). *Experimental Designs*, 2nd ed., Wiley, New York.

Connor, W. S., and M. Zelan (1959). Fractional Factorial Experimental Designs for Factors at Three Levels, *Natl. Bur. Stand. Appl. Math. Ser., No. 54*.

Cox, D. R. (1958). *Planning of Experiments*, Wiley, New York.

Cox, D. R., and E. J. Snell (1981). *Applied Statistics*, Chapman & Hall, London.

Daniel, C. (1976). *Applications of Statistics to Industrial Experimentation*, Wiley, New York.

Davies, O. L. (1963). *The Design and Analysis of Industrial Experiments*, 2nd ed., Oliver & Boyd, London.

Federer, W. T. (1955). *Experimental Design*, Macmillan, New York.

Finney, D. J. (1960). *An Introduction to the Theory of Experimental Design*, University of Chicago Press, Chicago.

Kempthorne, O. (1973). *Design and Analysis of Experiments*, reprinted by Robert E. Krieger Publishing Co., Huntington, N.Y.

Mead, R., and D. J. Pike (1975). A Review of Response Surface Methodology from a Biometric Viewpoint, *Biometrics* 31:803–851.

Myers, R. H. (1971). *Response Surface Methodology*, Allyn & Bacon, Boston.

Patterson, H. D., and H. L. Lucas (1962). Change-Over Designs. *N.C. Agric. Exp. Stn. Tech. Bull. 147*.

Snedecor, G. W., and W. G. Cochran (1980). *Statistical Methods*, 7th ed., Iowa State University Press, Ames, Iowa.

Steel, R. G. D., and J. H. Torrie (1980). *Principles and Procedures of Statistics*, 2nd ed., McGraw-Hill, New York.

Yates, F. (1937). The Design and Analysis of Factorial Experiments, *Imp. Bur. Soil Sci. Tech. Comm. 35*.

Appendix

TABLE A1 2500 Randomly Assorted Digits

	00–04	05–09	10–14	15–19	20–24	25–29	30–34	35–39	40–44	45–49
00	70896	44520	64720	49898	48088	76740	47460	83150	78905	59870
01	56809	42909	25853	47624	29486	14196	75841	00494	42390	24847
02	66109	84775	07515	49949	61482	91835	48126	80778	21302	24075
03	18071	36263	14053	52526	44347	04924	68100	57805	19521	15345
04	98732	15120	91754	12657	74675	78500	01247	47919	47635	55514
05	83746	47694	96143	42741	38338	97694	69300	99864	19648	15983
06	27998	42562	63404	10056	81668	48744	08400	83124	19896	18805
07	82685	32323	74625	14510	85927	28017	80588	14756	54937	76369
08	18386	13862	10988	04197	18770	72757	71418	81133	69503	44037
09	21717	13141	68165	58440	19187	08421	23872	03036	22707	34208
10	43674	47103	48614	70823	78252	82406	93424	05236	54588	27757
11	68597	68874	35567	98463	99671	05634	81533	47406	17228	44455
12	91874	70208	06308	40719	02772	69589	79936	07514	44950	35190
13	73854	19470	53014	29375	62256	77488	53949	74388	49607	18916
14	65926	34117	55344	68155	38099	56009	03513	05926	35584	42328
15	93282	62518	17752	53163	63852	44840	02592	88572	03107	90169
16	16215	50809	49326	77232	90155	69955	93892	70445	00906	57002
17	09342	14528	64727	71403	84156	34083	35613	35670	10549	07468
18	38148	79001	03509	79424	39625	73315	18811	86230	99682	82896
19	23689	19997	72382	15247	80205	58090	43804	94548	82693	22799
20	97844	62947	62230	30500	92816	85232	28222	91701	11057	83257
21	07611	71163	82212	20653	21499	51496	40715	78952	33029	64207
22	47744	04603	44522	62783	39347	72310	41460	31052	40814	94297
23	54293	43576	88116	67414	34908	15238	40561	73940	56850	81078
24	67556	93979	73363	00300	11217	74405	18937	79000	68834	48307
25	59391	58030	52098	82718	87024	82848	04190	96754	90464	29065
26	99567	76364	77204	04615	27062	96621	43918	01896	83991	51141
27	10363	97518	51400	25670	98342	61891	27101	37855	06235	33163
28	86859	19558	64432	16706	99612	59798	32803	67708	15297	28612
29	11258	24591	36863	55368	31721	94335	34936	02566	80972	08188
30	31432	96156	89177	75541	81355	24480	77243	76690	42507	84362
31	66890	61505	01240	00660	05873	13568	76082	79172	57913	93448
32	41894	57790	79970	33106	86904	48119	52503	24130	72824	21627
33	11303	87118	81471	52936	08555	28420	49416	44448	04269	27029
34	54374	57325	16947	45356	78371	10563	97191	53798	12693	27928
35	99116	75486	23476	52967	67104	39495	39100	17217	74073	84989
36	15696	10703	65178	90637	63110	17622	53988	71087	84148	11670
37	97720	15369	51269	69620	03388	13699	33423	67453	43269	56720
38	11666	13841	71681	98000	35979	39719	81899	07449	37985	46967
39	71628	73130	78783	73691	41623	09847	61547	18707	85498	69944
40	79788	68243	59732	04257	27084	14743	17520	95401	55811	76099
41	40538	79000	89559	25026	42474	23489	34502	75508	06059	86682
42	64016	73598	18609	73150	62463	33102	45205	87440	96767	67042
43	49767	12691	17903	93871	99721	79109	09425	26904	07419	76013
44	76974	55108	29795	08404	82684	00497	51126	79935	57450	55671
45	57430	82279	10421	00540	43648	75888	66049	21511	47676	33444
46	73528	39559	34434	88596	54086	71693	43132	14414	79949	85193
47	25991	65959	70769	64721	86413	33475	42740	06175	82758	66248
48	78388	16638	09134	59980	63806	48472	39318	35434	24057	74739
49	12477	09965	06657	59439	76330	24596	57994	77515	09577	91871

TABLE A2 Student's t Distribution (Column Headings = Tail Probabilities: A = One-Sided Confidence Interval; B = One-Tailed Significance Test; C = Two-Sided Confidence Interval; D = Two-Tailed Significance Test)

	A .600	.700	.800	.900	.950	.975	.990	.995
	B .400	.300	.200	.100	.050	.025	.010	.005
	C .200	.400	.600	.800	.900	.950	.980	.990
d.f.	D .800	.600	.400	.200	.100	.050	.020	.010
1	.325	.727	1.376	3.078	6.314	12.706	31.821	63.667
2	.289	.617	1.061	1.886	2.920	4.303	6.965	9.925
3	.277	.584	.978	1.638	2.353	3.182	4.541	5.841
4	.271	.569	.941	1.533	2.132	2.776	3.747	4.604
5	.267	.559	.920	1.476	2.015	2.571	3.365	4.032
6	.265	.553	.906	1.440	1.943	2.447	3.143	3.707
7	.263	.549	.896	1.415	1.895	2.365	2.998	3.499
8	.262	.546	.889	1.397	1.860	2.306	2.896	3.355
9	.261	.543	.883	1.383	1.833	2.262	2.821	3.250
10	.260	.542	.879	1.372	1.812	2.228	2.764	3.169
11	.260	.540	.876	1.363	1.796	2.201	2.718	3.106
12	.259	.539	.873	1.356	1.782	2.179	2.681	3.055
13	.259	.538	.870	1.350	1.771	2.160	2.650	3.012
14	.258	.537	.868	1.345	1.761	2.145	2.624	2.977
15	.258	.536	.866	1.341	1.753	2.131	2.602	2.947
16	.258	.535	.865	1.337	1.746	2.120	2.583	2.921
17	.257	.534	.863	1.333	1.740	2.110	2.567	2.898
18	.257	.534	.862	1.330	1.734	2.101	2.552	2.878
19	.257	.533	.861	1.328	1.729	2.093	2.539	2.861
20	.257	.533	.860	1.325	1.725	2.086	2.528	2.845
21	.257	.532	.859	1.323	1.721	2.080	2.518	2.831
22	.256	.532	.858	1.321	1.717	2.074	2.508	2.819
23	.256	.532	.858	1.319	1.714	2.069	2.500	2.807
24	.256	.531	.857	1.318	1.711	2.064	2.492	2.797
25	.256	.531	.856	1.316	1.708	2.060	2.485	2.787
26	.256	.531	.856	1.315	1.706	2.056	2.479	2.779
27	.256	.531	.855	1.314	1.703	2.052	2.473	2.771
28	.256	.530	.855	1.313	1.701	2.048	2.467	2.763
29	.256	.530	.854	1.311	1.699	2.045	2.462	2.756
30	.256	.530	.854	1.310	1.697	2.042	2.457	2.750
40	.255	.529	.851	1.303	1.684	2.021	2.423	2.704
50	.255	.528	.849	1.298	1.676	2.009	2.403	2.678
60	.254	.527	.848	1.296	1.671	2.000	2.390	2.660
80	.254	.527	.846	1.292	1.664	1.990	2.374	2.639
100	.254	.526	.845	1.290	1.660	1.984	2.365	2.626
∞	.253	.524	.842	1.282	1.645	1.960	2.326	2.576

Note: This table is abstracted from Table III of Fisher and Yates: *Statistical Tables for Biological, Agricultural and Medical Research*, published by Longman Group Ltd., London (1974), 6th ed. (previously published by Oliver and Boyd, Ltd., Edinburgh) and by permission of the authors and publishers.

TABLE A3 Percentage Points of the *F* Distribution: 5% Points

Denom. d.f.	1	2	3	4	5	6	7	8	9	10	12	15	20	24	30	40	60	120	∞
1	161.45	199.50	215.71	224.58	230.16	233.99	236.77	238.88	240.54	241.88	243.91	245.95	248.01	249.05	250.09	251.14	252.20	253.25	254.32
2	18.51	19.00	19.16	19.25	19.30	19.33	19.36	19.37	19.38	19.40	19.41	19.43	19.45	19.45	19.46	19.47	19.48	19.49	19.50
3	10.13	9.55	9.28	9.12	9.01	8.94	8.89	8.84	8.81	8.78	8.74	8.70	8.66	8.64	8.62	8.59	8.57	8.55	8.53
4	7.71	6.94	6.59	6.39	6.26	6.16	6.08	6.04	6.00	5.96	5.91	5.86	5.80	5.77	5.74	5.72	5.69	5.66	5.63
5	6.61	5.79	5.41	5.19	5.05	4.95	4.88	4.81	4.77	4.74	4.68	4.62	4.56	4.53	4.50	4.46	4.43	4.40	4.36
6	5.99	5.14	4.76	4.53	4.39	4.28	4.21	4.14	4.10	4.06	4.00	3.94	3.87	3.84	3.81	3.77	3.74	3.70	3.67
7	5.59	4.74	4.35	4.12	3.97	3.87	3.79	3.72	3.68	3.64	3.57	3.51	3.44	3.41	3.38	3.34	3.30	3.27	3.23
8	5.32	4.46	4.07	3.84	3.69	3.58	3.50	3.44	3.39	3.35	3.28	3.22	3.15	3.12	3.08	3.04	3.00	2.97	2.93
9	5.12	4.26	3.86	3.63	3.48	3.37	3.29	3.23	3.18	3.14	3.07	3.01	2.94	2.90	2.86	2.83	2.79	2.75	2.71
10	4.96	4.10	3.71	3.48	3.33	3.22	3.14	3.07	3.02	2.98	2.91	2.85	2.77	2.74	2.70	2.66	2.62	2.58	2.54
11	4.84	3.98	3.59	3.36	3.20	3.09	3.01	2.95	2.90	2.85	2.79	2.72	2.64	2.61	2.57	2.53	2.49	2.45	2.40
12	4.75	3.88	3.49	3.26	3.10	3.00	2.91	2.85	2.80	2.75	2.69	2.62	2.54	2.50	2.47	2.43	2.38	2.34	2.30
13	4.67	3.80	3.41	3.18	3.02	2.92	2.83	2.76	2.71	2.67	2.60	2.53	2.46	2.42	2.38	2.34	2.30	2.25	2.21
14	4.60	3.74	3.34	3.11	2.96	2.85	2.76	2.69	2.64	2.60	2.53	2.46	2.40	2.35	2.31	2.26	2.22	2.18	2.13
15	4.54	3.68	3.29	3.06	2.90	2.79	2.71	2.64	2.59	2.54	2.48	2.40	2.33	2.29	2.25	2.20	2.16	2.11	2.06
16	4.49	3.63	3.24	3.01	2.85	2.74	2.66	2.59	2.54	2.49	2.42	2.35	2.28	2.24	2.19	2.15	2.10	2.06	2.01
17	4.45	3.59	3.20	2.96	2.81	2.70	2.61	2.55	2.49	2.45	2.38	2.31	2.23	2.19	2.15	2.10	2.06	2.01	1.96
18	4.41	3.55	3.16	2.93	2.77	2.66	2.58	2.51	2.46	2.41	2.34	2.27	2.19	2.15	2.11	2.06	2.02	1.97	1.92
19	4.38	3.52	3.13	2.90	2.74	2.63	2.54	2.47	2.42	2.38	2.31	2.23	2.16	2.11	2.07	2.03	1.98	1.93	1.88
20	4.35	3.49	3.10	2.87	2.71	2.60	2.51	2.45	2.39	2.35	2.28	2.20	2.12	2.08	2.04	1.99	1.95	1.90	1.84
21	4.32	3.47	3.07	2.84	2.68	2.57	2.49	2.42	2.37	2.32	2.25	2.18	2.10	2.05	2.01	1.96	1.92	1.87	1.81
22	4.30	3.44	3.05	2.82	2.66	2.55	2.46	2.40	2.34	2.30	2.23	2.15	2.07	2.03	1.98	1.94	1.89	1.84	1.78
23	4.28	3.42	3.03	2.80	2.64	2.53	2.44	2.37	2.32	2.27	2.20	2.13	2.05	2.00	1.96	1.91	1.86	1.81	1.76
24	4.26	3.40	3.01	2.78	2.62	2.51	2.42	2.36	2.30	2.25	2.18	2.11	2.03	1.98	1.94	1.89	1.84	1.79	1.73
25	4.24	3.38	2.99	2.76	2.60	2.49	2.40	2.34	2.28	2.24	2.16	2.09	2.01	1.96	1.92	1.87	1.82	1.77	1.71
26	4.22	3.37	2.98	2.74	2.59	2.47	2.39	2.32	2.26	2.22	2.15	2.07	1.99	1.95	1.90	1.85	1.80	1.75	1.69
27	4.21	3.35	2.96	2.73	2.57	2.46	2.37	2.30	2.25	2.20	2.13	2.06	1.97	1.93	1.88	1.84	1.78	1.73	1.67
28	4.20	3.34	2.95	2.71	2.56	2.45	2.36	2.29	2.24	2.19	2.12	2.04	1.96	1.91	1.87	1.82	1.77	1.71	1.65
29	4.18	3.33	2.93	2.70	2.55	2.43	2.35	2.28	2.22	2.18	2.10	2.03	1.94	1.90	1.85	1.80	1.75	1.70	1.64
30	4.17	3.32	2.92	2.69	2.53	2.42	2.33	2.27	2.21	2.16	2.09	2.01	1.93	1.89	1.84	1.79	1.74	1.68	1.62
40	4.08	3.23	2.84	2.61	2.45	2.34	2.25	2.18	2.12	2.08	2.00	1.92	1.84	1.79	1.74	1.69	1.64	1.58	1.51
60	4.00	3.15	2.76	2.52	2.37	2.25	2.17	2.10	2.04	1.99	1.92	1.84	1.75	1.70	1.65	1.59	1.53	1.47	1.39
120	3.92	3.07	2.68	2.45	2.29	2.18	2.09	2.02	1.96	1.91	1.83	1.75	1.66	1.61	1.55	1.50	1.43	1.35	1.25
∞	3.84	3.00	2.60	2.37	2.21	2.10	2.01	1.94	1.88	1.83	1.75	1.67	1.57	1.52	1.46	1.39	1.32	1.22	1.00

Numerator d.f.

TABLE A3 *(Continued)*: 1% Points

Denom. d.f.	\ Numerator d.f. 1	2	3	4	5	6	7	8	9	10	12	15	20	24	30	40	60	120	∞
1	4052.2	4999.5	5403.3	5624.6	5763.7	5859.0	5928.3	5981.6	6022.5	6055.9	6106.3	6157.3	6208.7	6234.6	6260.7	6286.8	6313.0	6339.4	6366.0
2	98.50	99.00	99.17	99.25	99.30	99.33	99.36	99.37	99.39	99.40	99.42	99.43	99.45	99.46	99.47	99.47	99.48	99.49	99.50
3	34.12	30.82	29.46	28.71	28.24	27.91	27.67	27.49	27.34	27.23	27.05	26.87	26.69	26.60	26.50	26.41	26.32	26.22	26.12
4	21.20	18.00	16.60	15.98	15.52	15.21	14.98	14.80	14.66	14.55	14.37	14.20	14.02	13.93	13.84	13.74	13.65	13.56	13.46
5	16.26	13.27	12.06	11.39	10.97	10.67	10.46	10.29	10.16	10.05	9.89	9.72	9.55	9.47	9.38	9.29	9.20	9.11	9.02
6	13.74	10.92	9.78	9.15	8.74	8.47	8.26	8.10	7.98	7.87	7.72	7.56	7.40	7.31	7.23	7.14	7.06	6.97	6.88
7	12.25	9.55	8.45	7.85	7.46	7.19	6.99	6.84	6.72	6.62	6.47	6.31	6.16	6.07	5.99	5.91	5.82	5.74	5.65
8	11.26	8.65	7.59	7.01	6.63	6.37	6.18	6.03	5.91	5.81	5.67	5.52	5.36	5.28	5.20	5.12	5.03	4.95	4.86
9	10.56	8.02	6.99	6.42	6.06	5.80	5.61	5.47	5.35	5.26	5.11	4.96	4.81	4.73	4.67	4.57	4.48	4.40	4.31
10	10.04	7.56	6.55	5.99	5.64	5.38	5.20	5.06	4.94	4.85	4.70	4.56	4.40	4.32	4.25	4.16	4.08	4.00	3.91
11	9.65	7.20	6.22	5.67	5.32	5.07	4.89	4.74	4.63	4.54	4.40	4.25	4.10	4.02	3.94	3.86	3.78	3.69	3.60
12	9.33	6.93	5.95	5.41	5.06	4.82	4.64	4.50	4.39	4.30	4.16	4.01	3.86	3.78	3.70	3.62	3.54	3.45	3.36
13	9.07	6.70	5.74	5.20	4.86	4.62	4.44	4.30	4.19	4.10	3.96	3.82	3.66	3.59	3.51	3.42	3.34	3.25	3.16
14	8.86	6.51	5.56	5.04	4.70	4.46	4.28	4.14	4.03	3.94	3.80	3.66	3.50	3.43	3.35	3.26	3.18	3.09	3.00
15	8.68	6.36	5.42	4.89	4.56	4.32	4.14	4.00	3.89	3.80	3.67	3.52	3.37	3.29	3.21	3.13	3.05	2.96	2.87
16	8.53	6.23	5.29	4.77	4.44	4.20	4.03	3.89	3.78	3.69	3.55	3.41	3.26	3.18	3.10	3.02	2.93	2.84	2.75
17	8.40	6.11	5.18	4.67	4.34	4.10	3.93	3.79	3.68	3.59	3.46	3.31	3.16	3.08	3.00	2.92	2.83	2.74	2.65
18	8.28	6.01	5.09	4.58	4.25	4.01	3.84	3.71	3.60	3.51	3.37	3.23	3.08	3.00	2.92	2.84	2.75	2.66	2.57
19	8.18	5.93	5.01	4.50	4.17	3.94	3.76	3.63	3.52	3.43	3.30	3.15	3.00	2.92	2.84	2.76	2.67	2.58	2.49
20	8.10	5.85	4.94	4.43	4.10	3.87	3.70	3.56	3.46	3.37	3.23	3.09	2.94	2.86	2.79	2.69	2.61	2.52	2.42
21	8.02	5.78	4.87	4.37	4.04	3.81	3.64	3.51	3.40	3.31	3.17	3.03	2.88	2.80	2.72	2.64	2.55	2.46	2.36
22	7.94	5.72	4.82	4.31	3.99	3.76	3.59	3.45	3.34	3.26	3.12	2.98	2.83	2.75	2.67	2.58	2.50	2.40	2.30
23	7.88	5.66	4.76	4.26	3.94	3.71	3.54	3.41	3.30	3.21	3.07	2.94	2.78	2.70	2.62	2.54	2.45	2.35	2.26
24	7.82	5.61	4.72	4.22	3.90	3.67	3.50	3.36	3.26	3.17	3.03	2.89	2.74	2.66	2.58	2.49	2.40	2.31	2.21
25	7.77	5.57	4.68	4.18	3.86	3.63	3.46	3.32	3.22	3.13	2.99	2.85	2.70	2.62	2.54	2.45	2.36	2.27	2.17
26	7.72	5.53	4.64	4.14	3.82	3.59	3.42	3.29	3.18	3.09	2.96	2.82	2.66	2.58	2.50	2.42	2.33	2.23	2.13
27	7.68	5.49	4.60	4.10	3.78	3.56	3.39	3.26	3.15	3.06	2.92	2.78	2.63	2.55	2.47	2.38	2.29	2.20	2.10
28	7.64	5.45	4.57	4.07	3.75	3.53	3.36	3.23	3.12	3.03	2.90	2.75	2.60	2.52	2.44	2.35	2.26	2.17	2.06
29	7.60	5.42	4.54	4.04	3.72	3.50	3.33	3.20	3.09	3.00	2.87	2.72	2.57	2.49	2.41	2.32	2.23	2.14	2.03
30	7.56	5.39	4.51	4.02	3.70	3.47	3.30	3.17	3.07	2.98	2.84	2.70	2.55	2.47	2.39	2.30	2.21	2.11	2.01
40	7.31	5.18	4.31	3.83	3.51	3.29	3.12	2.99	2.89	2.80	2.66	2.52	2.37	2.29	2.20	2.11	2.02	1.92	1.80
60	7.08	4.98	4.12	3.65	3.34	3.12	2.95	2.82	2.72	2.63	2.50	2.35	2.20	2.11	2.03	1.94	1.84	1.73	1.60
120	6.85	4.79	3.95	3.48	3.17	2.96	2.79	2.66	2.56	2.47	2.34	2.19	2.03	1.95	1.86	1.76	1.66	1.53	1.38
∞	6.63	4.60	3.78	3.32	3.02	2.80	2.64	2.51	2.41	2.32	2.18	2.04	1.88	1.79	1.70	1.59	1.47	1.32	1.00

Note: This table is abstracted from *Tables of Percentage Points of the Inverted Beta (F) Distribution, Biometrika 33* (1943), by M. Merrington and C. M. Thompson, by permission of the authors and editors of *Biometrika*.

TABLE A4 Percentage Points of the Studentized Range (Q): Upper 5% Points of the Studentized Range. Entries are $q_{.05}$ Where $P(q < q_{.05}) = .95$.

ν										n									
	2	3	4	5	6	7	8	9	10	11	12	13	14	15	16	17	18	19	20
1	17.97	26.98	32.82	37.08	40.41	43.12	45.40	47.36	49.07	50.59	51.96	53.20	54.33	55.36	56.32	57.22	58.04	58.83	59.56
2	6.08	8.33	9.80	10.88	11.74	12.44	13.03	13.54	13.99	14.39	14.75	15.08	15.38	15.65	15.91	16.14	16.37	16.57	16.77
3	4.50	5.91	6.82	7.50	8.04	8.48	8.85	9.18	9.46	9.72	9.95	10.15	10.35	10.53	10.69	10.84	10.98	11.11	11.24
4	3.83	5.04	5.78	6.29	6.71	7.05	7.35	7.60	7.83	8.03	8.21	8.37	8.52	8.66	8.79	8.91	9.03	9.13	9.23
5	3.64	4.60	5.22	5.67	6.03	6.33	6.58	6.80	6.99	7.17	7.32	7.47	7.60	7.72	7.83	7.93	8.03	8.12	8.21
6	3.46	4.34	4.90	5.30	5.63	5.90	6.12	6.32	6.49	6.65	6.79	6.92	7.03	7.14	7.24	7.34	7.43	7.51	7.59
7	3.34	4.16	4.68	5.06	5.36	5.61	5.82	6.00	6.16	6.30	6.43	6.55	6.66	6.76	6.85	6.94	7.02	7.10	7.17
8	3.26	4.04	4.53	4.89	5.17	5.40	5.60	5.77	5.92	6.05	6.18	6.29	6.39	6.48	6.57	6.65	6.73	6.80	6.87
9	3.20	3.95	4.41	4.76	5.02	5.24	5.43	5.59	5.74	5.87	5.98	6.09	6.19	6.28	6.36	6.44	6.51	6.58	6.64
10	3.15	3.88	4.33	4.65	4.91	5.12	5.30	5.46	5.60	5.72	5.83	5.93	6.03	6.11	6.19	6.27	6.34	6.40	6.47
11	3.11	3.82	4.26	4.57	4.82	5.03	5.20	5.35	5.49	5.61	5.71	5.81	5.90	5.98	6.06	6.13	6.20	6.27	6.33
12	3.08	3.77	4.20	4.51	4.75	4.95	5.12	5.27	5.39	5.51	5.61	5.71	5.80	5.88	5.95	6.02	6.09	6.15	6.21
13	3.06	3.73	4.15	4.45	4.69	4.88	5.05	5.19	5.32	5.43	5.53	5.63	5.71	5.79	5.86	5.93	5.99	6.05	6.11
14	3.03	3.70	4.11	4.41	4.64	4.83	4.99	5.13	5.25	5.36	5.48	5.63	5.71	5.84	5.79	5.85	5.91	5.97	6.03
15	3.01	3.67	4.08	4.37	4.59	4.78	4.94	5.08	5.20	5.31	5.40	5.49	5.57	5.65	5.72	5.78	5.85	5.90	5.96
16	3.00	3.65	4.05	4.33	4.56	4.74	4.90	5.03	5.15	5.26	5.35	5.44	5.52	5.59	5.66	5.73	5.79	5.84	5.90
17	2.98	3.63	4.02	4.30	4.52	4.70	4.86	4.99	5.11	5.21	5.31	5.39	5.47	5.54	5.61	5.67	5.73	5.79	5.84
18	2.97	3.61	4.00	4.28	4.49	4.67	4.82	4.96	5.07	5.17	5.27	5.35	5.43	5.50	5.57	5.63	5.69	5.74	5.79
19	2.96	3.59	3.98	4.25	4.47	4.65	4.79	4.92	5.04	5.14	5.23	5.31	5.39	5.46	5.53	5.59	5.65	5.70	5.75
20	2.95	3.58	3.96	4.23	4.45	4.62	4.77	4.90	5.01	5.11	5.20	5.28	5.36	5.43	5.49	5.55	5.61	5.66	5.71
24	2.92	3.53	3.90	4.17	4.37	4.54	4.68	4.81	4.92	5.01	5.10	5.18	5.25	5.32	5.38	5.44	5.49	5.55	5.59
30	2.89	3.49	3.85	4.10	4.30	4.46	4.60	4.72	4.82	4.92	5.00	5.08	5.15	5.21	5.27	5.33	5.38	5.43	5.47
40	2.86	3.44	3.79	4.04	4.23	4.39	4.52	4.63	4.73	4.82	4.90	4.98	5.04	5.11	5.16	5.22	5.27	5.31	5.36
60	2.83	3.40	3.74	3.98	4.16	4.31	4.44	4.55	4.65	4.73	4.81	4.88	4.94	5.00	5.06	5.11	5.15	5.20	5.24
120	2.80	3.36	3.68	3.92	4.10	4.24	4.36	4.47	4.58	4.64	4.71	4.78	4.84	4.90	4.95	5.00	5.04	5.09	5.13
∞	2.77	3.31	3.63	3.86	4.03	4.17	4.29	4.39	4.47	4.55	4.62	4.68	4.74	4.80	4.85	4.89	4.93	4.97	5.01

TABLE A4 (Continued): Upper 1% Points of the Studentized Range. Entries are $q_{.01}$ Where $P(q < q_{.01}) = .99$.

ν	n = 2	3	4	5	6	7	8	9	10	11	12	13	14	15	16	17	18	19	20
1	90.03	135.0	164.3	185.6	202.2	215.8	227.2	237.0	245.6	253.2	260.0	266.2	271.8	277.0	281.8	286.3	290.4	294.3	298.0
2	14.04	19.02	22.29	24.72	26.63	28.20	29.53	30.68	31.69	32.59	33.40	34.13	34.81	35.43	36.00	36.53	37.03	37.50	37.95
3	8.26	10.62	12.17	13.33	14.24	15.00	15.64	16.20	16.69	17.13	17.53	17.89	18.22	18.52	18.81	19.07	19.32	19.55	19.77
4	6.51	8.12	9.17	9.96	10.58	11.10	11.55	11.93	12.27	12.57	12.84	13.09	13.32	13.53	13.73	13.91	14.08	14.24	14.40
5	5.70	6.98	7.80	8.42	8.91	9.32	9.67	9.97	10.24	10.48	10.70	10.89	11.08	11.24	11.40	11.55	11.68	11.81	11.93
6	5.24	6.33	7.03	7.56	7.97	8.32	8.61	8.87	9.10	9.30	9.48	9.65	9.81	9.95	10.08	10.21	10.32	10.43	10.54
7	4.95	5.92	6.54	7.01	7.37	7.68	7.94	8.17	8.37	8.55	8.71	8.86	9.00	9.12	9.24	9.35	9.46	9.55	9.65
8	4.75	5.64	6.20	6.62	6.96	7.24	7.47	7.68	7.86	8.03	8.18	8.31	8.44	8.55	8.66	8.76	8.85	8.94	9.03
9	4.60	5.43	5.96	6.35	6.66	6.91	7.13	7.33	7.49	7.65	7.78	7.91	8.03	8.13	8.23	8.33	8.41	8.49	8.57
10	4.48	5.27	5.77	6.14	6.43	6.67	6.87	7.05	7.21	7.36	7.49	7.60	7.71	7.81	7.91	7.99	8.08	8.15	8.23
11	4.39	5.15	5.62	5.97	6.25	6.48	6.67	6.84	6.99	7.13	7.25	7.36	7.46	7.56	7.65	7.73	7.81	7.88	7.95
12	4.32	5.05	5.50	5.84	6.10	6.32	6.51	6.67	6.81	6.94	7.06	7.17	7.26	7.36	7.44	7.52	7.59	7.66	7.73
13	4.26	4.96	5.40	5.73	5.98	6.19	6.37	6.53	6.67	6.79	6.90	7.01	7.10	7.19	7.27	7.35	7.42	7.48	7.55
14	4.21	4.89	5.32	5.63	5.88	6.08	6.26	6.41	6.54	6.66	6.77	6.87	6.96	7.05	7.13	7.20	7.27	7.33	7.39
15	4.17	4.84	5.25	5.56	5.80	5.99	6.16	6.31	6.44	6.55	6.66	6.76	6.84	6.93	7.00	7.07	7.14	7.20	7.26
16	4.13	4.79	5.19	5.49	5.72	5.92	6.08	6.22	6.35	6.46	6.56	6.66	6.74	6.82	6.90	6.97	7.03	7.09	7.15
17	4.10	4.74	5.14	5.43	5.66	5.85	6.01	6.15	6.27	6.38	6.48	6.57	6.66	6.73	6.81	6.87	6.94	7.00	7.05
18	4.07	4.70	5.09	5.38	5.60	5.79	5.94	6.08	6.20	6.31	6.41	6.50	6.58	6.65	6.73	6.79	6.85	6.91	6.97
19	4.05	4.67	5.05	5.33	5.55	5.73	5.89	6.02	6.14	6.25	6.34	6.43	6.51	6.58	6.65	6.72	6.78	6.84	6.89
20	4.02	4.64	5.02	5.29	5.51	5.69	5.84	5.97	6.09	6.19	6.28	6.37	6.45	6.52	6.59	6.65	6.71	6.77	6.82
24	3.96	4.55	4.91	5.17	5.37	5.54	5.69	5.81	5.92	6.02	6.11	6.19	6.26	6.33	6.39	6.45	6.51	6.56	6.61
30	3.89	4.45	4.80	5.05	5.24	5.40	5.54	5.65	5.76	5.85	5.93	6.01	6.08	6.14	6.20	6.26	6.31	6.36	6.41
40	3.82	4.37	4.70	4.93	5.11	5.26	5.39	5.50	5.60	5.69	5.76	5.83	5.90	5.96	6.02	6.07	6.12	6.16	6.21
60	3.76	4.28	4.59	4.82	4.99	5.13	5.25	5.36	5.45	5.53	5.60	5.67	5.73	5.78	5.84	5.89	5.93	5.97	6.01
120	3.70	4.20	4.50	4.71	4.87	5.01	5.12	5.21	5.30	5.37	5.44	5.50	5.56	5.61	5.66	5.71	5.75	5.79	5.83
∞	3.64	4.12	4.40	4.60	4.76	4.88	4.99	5.08	5.16	5.23	5.29	5.35	5.40	5.45	5.49	5.54	5.57	5.61	5.65

Note: Reprinted from *Biometrika Tables for Statisticians*, Vol. 1, 3rd ed. (1966), by permission of the *Biometrika* trustees.

TABLE A5   Minimum-Average-Risk *t* Values

						f						
q	6	8	10	12	14	16	18	20	24	30	40	60	120

(k = 100)

F = 1.2 (a = .913, b = 2.449)

q	6	8	10	12	14	16	18	20	24	30	40	60	120
2-6	*	*	*	*	*	*	*	*	*	*	*	*	*
8	2.91	2.94	2.96	2.97	2.98	2.99	2.99	2.99	3.00	3.00	3.00	3.00	3.00
10	2.93	2.98	3.01	3.04	3.05	3.06	3.07	3.08	3.09	3.10	3.10	3.11	3.12
12	2.95	3.01	3.05	3.08	3.10	3.12	3.13	3.14	3.16	3.17	3.19	3.20	3.21
14	2.96	3.03	3.08	3.12	3.14	3.16	3.18	3.19	3.21	3.23	3.25	3.27	3.29
16	2.97	3.05	3.11	3.15	3.18	3.20	3.22	3.24	3.26	3.28	3.31	3.33	3.36
20	2.99	3.08	3.14	3.19	3.23	3.26	3.28	3.30	3.33	3.37	3.40	3.44	3.47
40	3.02	3.13	3.22	3.29	3.35	3.39	3.43	3.47	3.52	3.58	3.64	3.72	3.79
100	3.04	3.17	3.28	3.36	3.44	3.50	3.55	3.59	3.67	3.76	3.86	3.98	4.11
∞	3.05	3.20	3.32	3.42	3.50	3.58	3.64	3.70	3.80	3.91	4.06	4.24	4.45

F = 1.4 (a = .845, b = 1.871)

q	6	8	10	12	14	16	18	20	24	30	40	60	120
2-4	*	*	*	*	*	*	*	*	*	*	*	*	*
6	2.85	2.84	2.83	2.82	2.82	2.81	2.80	2.80	2.79	2.78	2.77	2.75	2.74
8	2.88	2.89	2.90	2.90	2.90	2.89	2.89	2.89	2.88	2.88	2.87	2.86	2.85
10	2.90	2.93	2.94	2.95	2.95	2.96	2.96	2.96	2.95	2.95	2.95	2.94	2.93
12	2.92	2.95	2.98	2.99	3.00	3.00	3.01	3.01	3.01	3.01	3.01	3.00	2.99
14	2.93	2.97	3.00	3.02	3.03	3.04	3.04	3.05	3.05	3.06	3.06	3.05	3.05
16	2.94	2.99	3.02	3.04	3.06	3.07	3.08	3.08	3.09	3.09	3.10	3.10	3.09
20	2.95	3.01	3.05	3.08	3.10	3.11	3.12	3.13	3.14	3.15	3.16	3.16	3.16
40	2.98	3.06	3.12	3.16	3.19	3.22	3.24	3.25	3.28	3.30	3.31	3.32	3.32
100	2.99	3.09	3.16	3.22	3.26	3.29	3.32	3.34	3.38	3.41	3.43	3.45	3.42
∞	3.01	3.12	3.20	3.26	3.31	3.35	3.39	3.42	3.46	3.50	3.53	3.54	3.46

F = 1.7 (a = .767, b = 1.558)

q	6	8	10	12	14	16	18	20	24	30	40	60	120
2	*	*	*	*	*	*	*	*	*	*	*	*	*
4	*	*	*	*	*	2.61	2.59	2.58	2.56	2.54	2.52	2.50	2.48
6	2.82	2.79	2.76	2.74	2.72	2.71	2.70	2.69	2.67	2.65	2.63	2.61	2.58
8	2.84	2.83	2.81	2.80	2.78	2.77	2.76	2.75	2.74	2.72	2.70	2.68	2.65
10	2.86	2.86	2.85	2.84	2.83	2.82	2.81	2.80	2.79	2.77	2.75	2.73	2.70
12	2.87	2.88	2.88	2.87	2.86	2.85	2.84	2.84	2.82	2.81	2.79	2.76	2.73
14	2.88	2.90	2.90	2.89	2.89	2.88	2.87	2.86	2.85	2.83	2.81	2.79	2.75
16	2.89	2.91	2.91	2.91	2.90	2.90	2.89	2.89	2.87	2.86	2.84	2.81	2.77
20	2.90	2.93	2.93	2.94	2.93	2.93	2.92	2.92	2.91	2.89	2.87	2.84	2.80
40	2.93	2.97	2.99	3.00	3.00	3.00	3.00	2.99	2.98	2.97	2.94	2.89	2.83
100	2.94	2.99	3.02	3.04	3.05	3.05	3.05	3.05	3.04	3.02	2.98	2.92	2.83
∞	2.95	3.01	3.05	3.07	3.08	3.09	3.09	3.08	3.07	3.05	3.01	2.93	2.81

TABLE A5 (*Continued*)

q						f							
	6	8	10	12	14	16	18	20	24	30	40	60	120

F = 2.0 (a = .707, b = 1.414)

q	6	8	10	12	14	16	18	20	24	30	40	60	120
2	*	*	*	*	*	*	*	*	*	*	*	*	*
4	2.74	2.67	2.63	2.59	2.56	2.54	2.52	2.51	2.49	2.46	2.44	2.41	2.39
6	2.79	2.74	2.70	2.67	2.64	2.62	2.60	2.59	2.57	2.54	2.52	2.49	2.46
8	2.81	2.77	2.74	2.71	2.69	2.67	2.65	2.64	2.62	2.59	2.56	2.53	2.49
10	2.83	2.80	2.77	2.74	2.72	2.70	2.69	2.67	2.65	2.62	2.59	2.56	2.52
12	2.84	2.82	2.79	2.77	2.75	2.73	2.71	2.70	2.67	2.64	2.61	2.57	2.53
14	2.85	2.83	2.81	2.79	2.77	2.75	2.73	2.72	2.69	2.66	2.63	2.59	2.54
16	2.85	2.84	2.82	2.80	2.78	2.76	2.74	2.73	2.70	2.67	2.64	2.59	2.54
20	2.86	2.85	2.84	2.82	2.80	2.78	2.77	2.75	2.72	2.69	2.65	2.61	2.55
40	2.88	2.89	2.88	2.86	2.85	2.83	2.81	2.80	2.77	2.73	2.68	2.62	2.55
100	2.89	2.91	2.90	2.89	2.88	2.86	2.84	2.82	2.79	2.75	2.69	2.62	2.53
∞	2.90	2.92	2.92	2.91	2.90	2.88	2.86	2.85	2.81	2.76	2.69	2.61	2.52

F = 2.4 (a = .645, b = 1.309)

q	6	8	10	12	14	16	18	20	24	30	40	60	120
2	*	*	*	*	*	*	*	*	*	*	*	*	2.18
4	2.71	2.63	2.57	2.53	2.49	2.47	2.44	2.43	2.40	2.37	2.34	2.31	2.28
6	2.75	2.68	2.63	2.58	2.55	2.52	2.50	2.48	2.46	2.42	2.39	2.36	2.32
8	2.77	2.71	2.66	2.62	2.59	2.56	2.54	2.52	2.49	2.45	2.42	2.38	2.34
10	2.79	2.73	2.68	2.64	2.61	2.58	2.56	2.54	2.50	2.47	2.43	2.39	2.34
12	2.79	2.74	2.70	2.66	2.62	2.60	2.57	2.55	2.52	2.48	2.44	2.39	2.35
14	2.80	2.75	2.71	2.67	2.64	2.61	2.58	2.56	2.53	2.49	2.44	2.40	2.35
16	2.81	2.76	2.72	2.68	2.65	2.62	2.59	2.57	2.53	2.49	2.45	2.40	2.34
20	2.82	2.77	2.73	2.69	2.66	2.63	2.60	2.58	2.54	2.50	2.45	2.40	2.34
40	2.83	2.80	2.76	2.72	2.69	2.66	2.63	2.60	2.56	2.51	2.46	2.39	2.33
100	2.84	2.81	2.78	2.74	2.71	2.67	2.64	2.62	2.57	2.51	2.45	2.39	2.32
∞	2.85	2.83	2.79	2.76	2.72	2.68	2.65	2.62	2.57	2.51	2.45	2.38	2.31

F = 3.0 (a = .577, b = 1.225)

q	6	8	10	12	14	16	18	20	24	30	40	60	120
2	*	*	2.41	2.36	2.32	2.29	2.27	2.25	2.22	2.20	2.17	2.14	2.11
4	2.68	2.57	2.50	2.45	2.41	2.38	2.35	2.33	2.30	2.27	2.24	2.20	2.17
6	2.71	2.61	2.54	2.49	2.44	2.41	2.39	2.36	2.33	2.29	2.26	2.22	2.18
8	2.72	2.63	2.56	2.51	2.47	2.43	2.40	2.38	2.34	2.31	2.27	2.22	2.18
10	2.74	2.65	2.58	2.52	2.48	2.44	2.41	2.39	2.35	2.31	2.27	2.22	2.18
12	2.74	2.66	2.59	2.53	2.49	2.45	2.42	2.40	2.36	2.31	2.27	2.22	2.18
14	2.75	2.66	2.60	2.54	2.49	2.46	2.43	2.40	2.36	2.32	2.27	2.22	2.17
16	2.75	2.67	2.60	2.55	2.50	2.46	2.43	2.40	2.36	2.32	2.27	2.22	2.17
20	2.76	2.68	2.61	2.55	2.51	2.47	2.43	2.41	2.36	2.32	2.27	2.22	2.17
40	2.77	2.70	2.63	2.57	2.52	2.48	2.44	2.41	2.37	2.32	2.26	2.21	2.16
100	2.78	2.71	2.64	2.58	2.53	2.49	2.45	2.42	2.37	2.31	2.26	2.21	2.16
∞	2.79	2.71	2.65	2.59	2.53	2.49	2.45	2.42	2.37	2.31	2.26	2.20	2.15

TABLE A5 *(Continued)*

q	6	8	10	12	14	16	f 18	20	24	30	40	60	120
F = 4.0 (a = .500, b = 1.155)													
2	2.58	2.44	2.35	2.29	2.25	2.22	2.20	2.18	2.15	2.12	2.09	2.06	2.03
4	2.63	2.50	2.41	2.35	2.30	2.27	2.24	2.22	2.18	2.15	2.12	2.08	2.05
6	2.65	2.52	2.43	2.37	2.32	2.28	2.25	2.23	2.19	2.16	2.12	2.08	2.04
10	2.67	2.55	2.46	2.39	2.34	2.30	2.26	2.24	2.20	2.16	2.12	2.08	2.04
20	2.69	2.57	2.47	2.40	2.35	2.30	2.27	2.24	2.20	2.15	2.11	2.07	2.03
∞	2.71	2.59	2.49	2.42	2.36	2.31	2.27	2.24	2.19	2.15	2.11	2.06	2.02
F = 6.0 (a = .408, b = 1.095)													
2	2.53	2.37	2.27	2.21	2.16	2.13	2.10	2.08	2.05	2.02	1.99	1.96	1.93
4	2.56	2.40	2.30	2.23	2.18	2.14	2.12	2.09	2.06	2.02	1.99	1.96	1.93
6	2.58	2.42	2.31	2.24	2.19	2.15	2.12	2.09	2.06	2.02	1.99	1.95	1.92
10	2.59	2.43	2.32	2.24	2.19	2.15	2.12	2.09	2.06	2.02	1.99	1.95	1.92
20	2.60	2.44	2.32	2.25	2.19	2.15	2.12	2.09	2.05	2.02	1.98	1.95	1.92
∞	2.61	2.44	2.33	2.25	2.19	2.15	2.12	2.09	2.05	2.02	1.98	1.95	1.92
F = 10.0 (a = .316, b = 1.054)													
2	2.48	2.30	2.19	2.12	2.07	2.04	2.01	1.99	1.96	1.93	1.90	1.87	1.85
4	2.49	2.31	2.20	2.13	2.08	2.04	2.01	1.99	1.96	1.93	1.90	1.87	1.84
6	2.50	2.31	2.20	2.13	2.08	2.04	2.01	1.99	1.96	1.93	1.90	1.87	1.84
10-∞	2.51	2.32	2.20	2.13	2.08	2.04	2.01	1.99	1.96	1.93	1.90	1.87	1.84
F = 25.0 (a = .200, b = 1.021)													
2-4	2.40	2.20	2.10	2.03	1.99	1.95	1.93	1.91	1.88	1.86	1.83	1.80	1.78
6-∞	2.41	2.21	2.10	2.03	1.99	1.95	1.93	1.91	1.88	1.86	1.83	1.80	1.78
F = ∞ (a = 0, b = 1)													
2-∞	2.33	2.13	2.03	1.97	1.93	1.90	1.88	1.86	1.84	1.81	1.79	1.76	1.74
							(k = 500)						
F = 1.2 (a = .913, b = 2.449)													
2-16	*	*	*	*	*	*	*	*	*	*	*	*	*
20	4.70	4.82	4.89	*	*	*	*	*	*	*	*	*	*
40	4.75	4.91	5.03	5.12	5.20	5.25	5.30	5.34	5.41	5.48	5.55	5.61	5.67
100	4.79	4.98	5.13	5.25	5.34	5.43	5.50	5.56	5.65	5.76	5.89	6.02	6.13
∞	4.81	5.03	5.20	5.34	5.46	5.56	5.65	5.73	5.86	6.02	6.20	6.41	6.56
F = 1.4 (a = .845, b = 1.871)													
2-14	*	*	*	*	*	*	*	*	*	*	*	*	*
16	4.61	4.66	4.68	4.69	4.69	4.69	4.69	4.68	4.67	4.65	4.62	4.58	4.53
20	4.64	4.70	4.73	4.75	4.76	4.77	4.77	4.76	4.76	4.74	4.72	4.68	4.62
40	4.68	4.78	4.85	4.89	4.92	4.94	4.96	4.96	4.97	4.97	4.95	4.90	4.81
∞	4.74	4.88	4.99	5.06	5.12	5.17	5.20	5.23	5.26	5.28	5.26	5.16	4.82

TABLE A5   (Continued)

q	6	8	10	12	14	16	18	20	24	30	40	60	120

F = 1.7 (a = .767, b = 1.558)

q	6	8	10	12	14	16	18	20	24	30	40	60	120
2-8	*	*	*	*	*	*	*	*	*	*	*	*	*
10	*	*	*	*	*	*	*	*	*	4.08	4.02	3.95	3.87
12	4.50	4.46	4.42	4.38	4.34	4.30	4.27	4.24	4.19	4.14	4.07	3.99	3.90
20	4.55	4.54	4.52	4.49	4.46	4.43	4.40	4.37	4.32	4.26	4.18	4.08	3.95
40	4.59	4.61	4.61	4.60	4.57	4.55	4.52	4.49	4.44	4.36	4.26	4.12	3.93
∞	4.64	4.69	4.71	4.72	4.71	4.69	4.66	4.63	4.57	4.46	4.31	4.07	3.76

F = 2.0 (a = .707, b = 1.414)

q	6	8	10	12	14	16	18	20	24	30	40	60	120
2-6	*	*	*	*	*	*	*	*	*	*	*	*	*
8	*	*	*	*	*	3.98	3.93	3.89	3.83	3.76	3.69	3.60	3.51
10	4.41	4.31	4.22	4.15	4.08	4.03	3.98	3.94	3.88	3.80	3.72	3.63	3.53
20	4.48	4.41	4.34	4.27	4.21	4.16	4.10	4.06	3.98	3.89	3.78	3.65	3.51
40	4.51	4.47	4.41	4.35	4.29	4.23	4.17	4.12	4.03	3.92	3.78	3.62	3.44
∞	4.55	4.53	4.49	4.43	4.37	4.31	4.25	4.19	4.07	3.93	3.75	3.54	3.33

F = 2.4 (a = .645, b = 1.309)

q	6	8	10	12	14	16	18	20	24	30	40	60	120
2-4	*	*	*	*	*	*	*	*	*	*	*	*	*
6	*	*	*	*	3.77	3.71	3.65	3.61	3.54	3.47	3.39	3.30	3.22
8	4.31	4.14	4.01	3.91	3.83	3.76	3.70	3.66	3.58	3.50	3.41	3.32	3.22
10	4.33	4.18	4.05	3.95	3.87	3.79	3.73	3.68	3.60	3.51	3.42	3.31	3.21
20	4.39	4.26	4.14	4.04	3.95	3.87	3.80	3.74	3.64	3.53	3.41	3.28	3.15
∞	4.45	4.35	4.25	4.14	4.03	3.94	3.85	3.78	3.64	3.50	3.34	3.18	3.04

F = 3.0 (a = .577, b = 1.225)

q	6	8	10	12	14	16	18	20	24	30	40	60	120
2	*	*	*	*	*	*	*	*	*	*	*	*	*
4	*	*	*	*	*	3.43	3.38	3.33	3.26	3.19	3.12	3.04	2.97
6	4.19	3.95	3.79	3.66	3.56	3.49	3.43	3.37	3.30	3.21	3.13	3.04	2.95
10	4.24	4.02	3.85	3.72	3.62	3.53	3.46	3.40	3.31	3.21	3.12	3.02	2.92
20	4.28	4.08	3.91	3.77	3.65	3.56	3.48	3.41	3.31	3.20	3.09	2.98	2.87
∞	4.33	4.15	3.97	3.82	3.69	3.57	3.48	3.40	3.28	3.15	3.03	2.92	2.82

F = 4.0 (a = .500, b = 1.155)

q	6	8	10	12	14	16	18	20	24	30	40	60	120
2	*	*	*	*	*	*	*	*	*	*	*	2.81	2.75
4	*	3.74	3.54	3.40	3.30	3.22	3.16	3.11	3.04	2.96	2.89	2.81	2.74
6	4.08	3.78	3.58	3.43	3.32	3.24	3.17	3.12	3.04	2.95	2.87	2.79	2.71
10	4.12	3.83	3.62	3.46	3.34	3.25	3.17	3.11	3.03	2.94	2.85	2.77	2.69
20	4.15	3.86	3.64	3.48	3.35	3.25	3.17	3.10	3.01	2.92	2.83	2.74	2.66
∞	4.19	3.90	3.67	3.49	3.35	3.24	3.15	3.09	2.99	2.89	2.80	2.72	2.65

TABLE A5 *(Continued)*

q	6	8	10	12	14	16	18	20	24	30	40	60	120

F = 6.0 (a = .408, b = 1.095)

q	6	8	10	12	14	16	18	20	24	30	40	60	120
2	*	*	3.28	3.24	3.04	2.97	2.91	2.87	2.81	2.74	2.68	2.62	2.56
4	3.90	3.54	3.32	3.17	3.06	2.98	2.92	2.87	2.80	2.73	2.66	2.60	2.53
6	3.93	3.57	3.33	3.18	3.06	2.98	2.91	2.86	2.79	2.72	2.65	2.58	2.52
10	3.95	3.59	3.34	3.18	3.06	2.97	2. 1	2.85	2.78	2.71	2.64	2.57	2.51
20	3.97	3.60	3.35	3.18	3.06	2.97	2.	2.84	2.77	2.70	2.63	2.56	2.51
∞	3.99	3.62	3.36	3.18	3.05	2.96	2. 9	2.83	2.76	2.69	2.62	2.56	2.50

F = 10.0 (a = .316, b = 1.054)

q	6	8	10	12	14	16	18	20	24	30	40	60	120
2	3.72	3.33	3.10	2.96	2.86	2.79	2.74	2.70	2.64	2.58	2.52	2.47	2.42
4	3.75	3.35	3.11	2.96	2.86	2.79	2.73	2.69	2.63	2.57	2.51	2.46	2.41
10	3.78	3.36	3.11	2.96	2.85	2.78	2.72	2.68	2.62	2.56	2.50	2.45	2.40
20	3.79	3.36	3.11	2.96	2.85	2.78	2.72	2.68	2.62	2.56	2.50	2.45	2.40
∞	3.80	3.37	3.11	2.95	2.85	2.77	2.72	2.67	2.61	2.56	2.50	2.45	2.40

F = 25.0 (a = .200, b = 1.021)

q	6	8	10	12	14	16	18	20	24	30	40	60	120
2	3.55	3.14	2.92	2.79	2.70	2.64	2.59	2.56	2.51	2.46	2.41	2.36	2.32
10	3.57	3.14	2.92	2.79	2.70	2.64	2.59	2.55	2.50	2.45	2.41	2.36	2.32
∞	3.57	3.14	2.92	2.78	2.70	2.63	2.59	2.55	2.50	2.45	2.41	2.36	2.32

F = ∞ (a = 0, b = 1)

q	6	8	10	12	14	16	18	20	24	30	40	60	120
2-∞	3.39	3.00	2.80	2.69	2.61	2.55	2.51	2.48	2.44	2.39	2.35	2.31	2.27

*No differences are significant

$a = 1/\sqrt{F}$ $b = \sqrt{F/(F - 1)}$

Note: Reprinted from A Bayes Rule Rule for the Symmetric Multiple Comparisons Problem, *J. Amer. Stat. Assoc. 67* (1972), by R. A. Waller and D. B. Duncan, by permission of the managing editor of the American Statistical Association.

Author Index

Anderson, R. L., 297, 301
Anderson, V. L., 6, 33, 47, 71, 111,
  145, 202, 229, 250, 401

Bose, R. C., 352, 400
Box, G. E. P., 202, 282, 288, 297,
  301, 401

Carmer, S. G., 84, 85, 111, 159, 165
Chew, V., 85, 111
Clatworthy, W. H., 400
Cochran, W. G., 6, 33, 47, 71, 76,
  111, 145, 189, 202, 219, 222, 224,
  229, 250, 350, 353, 386, 400, 401
Connor, W. S., 189, 202, 222, 229,
  401
Cox, D. R., 6, 165, 401
Cox, G. M., 6, 33, 47, 71, 76, 111,
  145, 189, 202, 219, 222, 224, 229,
  250, 350, 353, 386, 400, 401

Daniel, C., 202, 229, 401
Davies, O. L., 301, 401
Deming, L., 202
Duncan, D. B., 76, 77, 78, 111, 159,
  165

Federer, W. T., 6, 33, 47, 71, 145,
  239, 244, 247, 250, 400, 401
Finney, D. J., 202, 229, 250, 400,
  401
Fisher, R. A., 76, 77, 111, 159
Friedman, M., 297, 301
Gupta, R. P., 247, 250

Harward, M. E., 301
Hunter, J. S., 202, 288, 301, 401
Hunter, W. G., 301, 401

Kempthorne, O., 71, 145, 202, 219,
  222, 229, 250, 400, 401
Kirsch, R. K., 290, 301

Little, T. M., 85, 111
Lucas, H. L., 350, 401

McLean, R. A., 6, 33, 47, 71, 111,
  145, 202, 229, 250, 401
Mead, R., 301, 401
Miller, R., 84, 111
Myers, R. H., 260, 301, 401

Patterson, H. D., 350, 401

Petersen, R. G., 85, 111, 146, 165, 239, 249, 251, 301
Pike, D. J., 301, 401

Satterthwaite, F. E., 140, 145, 237, 238, 241, 251
Savage, L. J., 297, 301
Shirkhande, S. S., 400
Snedecor, G. W., 33, 47, 111, 145, 401
Snell, E. J., 165, 401
Steel, R. G. D., 6, 33, 47, 76, 111, 145, 401
Swanson, M. R., 84, 111, 159, 165

Torrie, J. H., 6, 33, 47, 76, 111, 145, 401
Tukey, J. W., 76, 78, 111

Walker, W. M., 85, 111
Waller, R. A., 76, 77, 78, 111, 159
Wilson, K. B., 282, 297

Yates, F., 145, 169, 192, 202, 353, 400, 401
Young, S., 222, 229

Zelan, M., 202, 401

Subject Index

Accuracy, increasing, 5
Aliases
    in confounded fractional replica-
        tion, 223
    in response surfaces, 268
    3^k factorials, $\frac{1}{3}$ replication, 201
    2^k factorials, $\frac{1}{2}$ replication, 180
Analysis of variance (ANOVA)
    augmented 2^k factorial, 274–275
    balanced lattices, 356
    completely randomized design
        equal replication, 11–12
        subsampling, 25–27
        unequal replication, 17
    cross-over design, 304
    extra-period Latin square change-
        over design, 327
    incomplete block switch-back
        design, 312
    incomplete Latin square change-
        over design, 335
    interblock analysis, 206
    Latin square change-over design,
        319
    Latin square design, 52–53
    partially balanced lattices

[Analysis of variance]
    designs without repetition,
        365
    repetitions of basic designs,
        378
    randomized block design, 38–40
    response surface, 3 × 3 facto-
        rial, 258
    split-plot design, 137–139
    split-split plot design, 237
    strip-plot design, 241
    switch-back design, 308
    three-factor factorial, 125–128
    3^k factorial, confounded, 213–
        214
    two-factor factorial, 118–119
    2^k factorials, 167–168
    2^3 factorial
        completely confounded, 205
        partially confounded, 208–209
Approximate degrees of freedom
    split-plot design, 140
    split-split plot design, 237–238
    strip-plot design, 241
Approximate F test, 40, 44, 55
    balanced lattices, 357

[Approximate *F* test]
  repeated partially balanced lattices, 379
  three-factor factorial, 128
  two-factor factorial, 120
Augmented 2^k factorials, 271–278
  analysis, 273–275
  design characteristics, 272–273
  design construction, 272
  numerical example, 275–278
  pure error, 273
  testing quadratic effects, 273

Balanced Latin squares, catalog, 340–344
Balanced lattices, 354–357
  analysis, 354
  description, 357
  effective error variance, 357
  numerical example, 357–361
  randomization, 354
  relative precision, 357
  significance test, 357
  variance, 357
Basic experimental designs, 68–71
Basic matrix operations, 300–301
Bias, 267–271
  alias matrix, 268
  illustration
    ½ replication of 2^3 factorial, 268–270
    2^2 factorial, 270–271
Block, 6
Blocking, 34–35
  criteria for, 34
  general comments, 35
  reasons for, 4
Blocking in fractional replication, 222–228
  aliases, 223
  confounding contrasts, 223

[Blocking in fractional replication]
  defining contrasts, 222
  generalized interaction, 223
  numerical example, 225–228
BLSD (Waller and Duncan's Bayes LSD), 76, 77, 78–79
  numerical example, 82–84
  risk (*k*) ratio, 78

Carry-over effects, *see* Residual effects
Central composite design, 282–297
  axial points, 283
  center points, 283
  design construction, 282–283
  design matrix, 282–285
  desirable properties, 285
  general model, 283
  numerical example, 290–297
  orthogonal design, 287
    design parameters, 290
    estimators and variances, 292
  rotatable design, 287
  solution, 284
  uniform precision design
    design parameters, 289
    estimators and variances, 291
  variances, 284–285
  $X'X$ matrix, 284–286
Change-over trials, 302–350
  advantages, 302
  characteristics, 302
Coefficient of variation, 13, 40
Compact designs, 247–250
  analysis, 249–250
  characteristics, 247–248
  design construction, 248
  illustration, 248–249
  restrictions, 248
Comparisons, *see* Contrasts
Completely randomized design, 7–33

[Completely randomized design]
  advantages, 7
  design construction, 7
  disadvantages, 7
  equal replication, 11–16
    analysis, 11–12
    numerical example, 14–16
    standard errors, 12–13
  fixed effects model, 21–22
    significance tests, 23
  random effects model, 21–24
    estimation, 23–24
    significance tests, 23
  randomization, 8–9
  subsampling, 24–33
    analysis, 25–27
    estimation, 28–29
    numerical example, 30–33
    sampling plans, 29–30
    significance tests, 27–28
  summary, 68–69
  unequal replication, 17–20
    analysis, 17
    numerical example, 18–20
    standard errors, 18
  uses, 8
Complex designs, Federer's algorithms for, 244–247
Computer analysis
  general comments, 165
  of randomized block design, 163
  of split-plot design, 164
  of two-factor factorial, 163
Confounding, 203–229
  analysis, 205
    interblock estimates, 206
  4^k factorials, 221–222
  illustration in 2^3 factorial, 203–204
  mixed level factorials, 219–220
  principles, 204
  purpose, 203

[Confounding]
  3^k factorials, 213–219
    analysis, 213–214
    further reduction, 213–215
    numerical example, 215–219
  2^k factorials, 203–213
    analysis, 205–206
    further reduction, 207–208
Confounding contrast, 204
  in fractional replication, 233
Contrasts, 85–111
  rules for in 2^k factorials, 168–169
  summary of rules for, 94–96
Contrasts of means, 86–87
  numerical example, 87–89
  orthogonal, 89
Contrasts of totals, 91–93
  rules for, 91–92
Cross-over design, 303–306
  analysis, 303–305
  design construction, 303
  numerical example, 305–306
  uses, 303
  variances, 304–305

Data interpretation examples, 146–162
Defining contrast
  in fractional replication, 222
  3^k factorial, $\frac{1}{3}$ replication, 201
  2^k factorial, $\frac{1}{2}$ replication, 179
Design and analysis example, 96–100
Design construction
  central composite design, 282–283
  compact designs, 248
  completely randomized design, 7
  cross-over design, 303
  extra-period Latin square change-over design, 324

[Design construction]
Graeco-Latin square, 64
incomplete block switch-back
design, 311
incomplete Latin square change-
over design, 330
Latin square, 48
Latin square change-over design,
315–316
½ replication of 2^k factorial,
180–181
⅓ replication of 3^k factorial, 201
partially confounded 2^k factorial,
208
randomized block design, 35
split-plot design, 134–135
split-split plot design, 235
strip-plot design, 239
switch-back design, 306–307
3^k factorial, in blocks of 3^{k-1},
213
2^k factorial, 272
in blocks of 2^{k-1}, 204–205
in blocks of 2^{k-2}, 207
Design parameters for central composite design
orthogonal design, 290
uniform precision design, 289
Double reversal, *see* Switch-back
design

Experiment, 6
Experimental design, 6
purposes, 1
Experimental error, 2, 6
completely randomized design,
subsampling, 27
from pooled high-order interactions, 176–177
Experimental unit, 6
Experimentation
purpose, 1

[Experimentation]
requirements, 4
Extra-period Latin square change-
over design, 324–330
analysis, 325–328
characteristics, 324–325
design construction, 324
numerical example, 328–331
permanent effects, 328
residual effects, 328
restrictions, 325
uses, 324
variances, 328

Factor, 6
Factor level, 6
Factorial experiments, 112–145
advantages, 116
Disadvantages, 116
formal consideration of effects,
189–202
treatment combinations, 114
three factors, 124–134
two factors, 117–124
uses, 117
Federer's algorithms for complex
designs, 244–247
First-order designs, 265–278
augmented 2^k factorials, 271–278
data equation, 265–266
design matrix, 266
linear equation for, 265
orthogonal designs, 266–267
Fisher's protected LSD, *see* FPLSD
Fixed effects model
completely randomized design,
21–22
significance tests, 23
FPLSD (Fisher's protected LSD),
76, 77
numerical example, 79–80

Fractional replication, 179–202
aliases
    ½ replication of 2^k factorial, 180
    ⅓ replication of 3^k factorial, 201
Blocking in, 222–228
Defining contrast
    ½ replication of 2^k factorial, 179
    ⅓ replication of 3^k factorial, 201
Generalized interaction
    in 3^k factorials, 198–199
    in 2^k factorials, 180
sequences of fractions, 187–189
3^k factorials, 201
2^k factorials, 179–189
    analysis, 181–183
    ¼ replication, 186–187

Generalized interaction
in confounded fractional replication, 223
confounded 3^k contrasts, 215
confounded 2^k contrasts, 208
in fractional 3^k factorials, 198–199
in fractional 2^k factorials, 180
Graeco-Latin square design, 64–68
advantages, 64
design construction, 64
disadvantages, 65
numerical example, 65–68
restrictions, 65
uses, 64

Hierarchical classification, 24–33
HSD (Tukey's honestly significant difference), 76, 77, 78

[HSD]
numerical example, 80–81

Incomplete block designs, 351–400
balanced, 352
characteristics, 351–352
partially balanced, 352–353
replicated, analyzed as randomized block designs, 353
Incomplete block switch-back design, 310–315
analysis, 311–312
design construction, 311
numerical example, 313–315
uses, 311
variances, 312
Incomplete Latin square change-over design, 330–340
analysis, 332–336
catalog of designs, 344–350
characteristics, 331
design construction, 330
numerical example, 336–340
permanent effects, 336
residual effects, 336
restrictions, 330
uses, 332
variances, 336
Interaction, 113
first-order (two-factor), 125
high order used to estimate error, 176–177
illustration, 115
second-order (three-factor), 125
Introducing new treatments
alternatives to split-plots, 232–234
split-plots, 233

Latin rectangles, *see* Incomplete Latin square change-over

Latin square change-over design,
        315–324
    analysis, 316–320
    characteristics, 315–316
    design construction, 315–316
    numerical example, 321–324
    permanent effects, 318, 320
    randomization, 316
    residual effects, 318, 320
    restrictions, 316
    uses, 316
    variances, 320
Latin square design, 48–63
    advantages, 49
    analysis, 52–53
    design construction, 48
    disadvantages, 49
    field layout, 50
    missing data, 56
    model, 54
    numerical example, 57–59
        multiple squares, 60–63
    randomization, 51
    relative efficiency, 56–57
    restrictions on use, 59
    significance tests, 54–55
    standard errors, 55
    summary, 70–71
    uses, 49
Lattice designs, 353–400
    catalog of plans, 386–399
    characteristics, 353
Least significant difference, *see*
        LSD
LSD (least significant difference),
        73–75
    for comparisons planned in
        advance, 74
    for gaps between ranked means,
        75
    invalid uses, 74
    valid uses, 74

Main effects, 113
Main plots, *see* Whole plots
Mathematical model
    completely randomized design,
        20–24
        subsampling, 24–25
    Latin square, 54
    randomized block design, 42
    two-factor factorial, 117–118
Means, estimation of
    balanced lattices, 357
    completely randomized design,
        12–13, 18, 28–29
    cross-over design, 304
    extra-period Latin square change-
        over design, 328
    incomplete block switch-back
        design, 312
    incomplete Latin square change-
        over design, 336
    Latin square, 55
    Latin square change-over design,
        320
    partially balanced lattices, 363
    randomized block design, 41
    repeated partially balanced lat-
        tices, 379
    by reverse Yates algorithm, 171
    split-plot design, 139–140
    switch-back design, 307
    three-factor factorial, 129
    two-factor factorial, 120–121
Means, separation of, 72–111
Method of steepest ascent, 298–300
    algorithm, 299–300
Missing data
    Latin square, 56
    randomized block design, 43
Multifactor experiments, 166–202
Multiple comparisons, 75–85
    comments, 84–85
    error rate

[Multiple comparisons]
  comparisonwise, 76
  experimentwise, 77
  numerical example, 79–84

Nested data, *see* Hierarchical
    classification
Numerical example
  augmented 2^k factorial, 275–278
  balanced lattice, 357–361
  blocking in fractional replication,
    225–228
  bush bean spacing, 154–159
  central composite design, 290–
    297
  completely randomized design
    equal replication, 14–16
    subsampling, 30–33
    unequal replication, 18–20
  confounded 3^k factorial, 215–219
  contrasts of means, 87–89
  cross-over design, 305–306
  design and analysis, 96–100
  extra-period Latin square change-
    over design, 328–331
  frozen orange juice, 149–154
  Graeco-Latin square design, 65–
    68
  incomplete block switch-back
    design, 313–315
  incomplete Latin square change-
    over design, 336–340
  Latin square change-over design,
    321–324
  Latin square design, 57–59
  multiple comparisons, 79–84
  multiple Latin squares, 60–63
  ½ replication of 2^k factorial,
    182–185
  partial confounding, 209–213
  partitioning SST, 92–93

[Numerical example]
  randomized block design, 44–47
  regression in the ANOVA, 107–
    110
  repeated 5 × 5 simple lattice,
    380–386
  simple lattice, 366–370
  split-plot design, 141–145
  strip-plot design, 241–244
  switch-back design, 308–310
  3 × 3 factorial, 192–198
  3 × 3 response surface design,
    260–264
  three-factor factorial, 124–134
  triple lattice, 370–375
  two-factor factorial, 122–124
  2^k factorial, 171–175
    single replication, 177–179
  washday products, 147–149
  wheat yield trial, 159–162

Orthogonal contrasts, 89–91
  of means, 89, 94
  polynomial estimation, 106–107
  polynomials, 103–107
  of totals, 91–92, 95

Pairwise comparisons, 73–85
Partial confounding, 208–213
  illustration, 208–209
  numerical example, 209–213
  2^k factorial, 208
Partially balanced lattices, 361–399
  analysis, 363–366
  description, 361
  effective error MS, 365
  relative precision, 366
  significance test, 366
  variances, 366
Partitioning SST, 92–93
  augmented 2^k factorial, 273–275

[Partitioning SST]
  in bush bean spacing trial, 157–
    158
  numerical example, 92–93
  orthogonal polynomials, 105–106
  regression components, 100–111
  response surface analysis, 256–
    257
  3×3 factorial, 195–198
  washday product trial, 148
  Yates algorithm for 2^k factorials,
    169–171
Period effects, 302
Permanent effects
  extra-period Latin square change-
    over design, 328
  incomplete Latin square change-
    over design, 336
  Latin square change-over design,
    318, 320

Quadruple lattices, 363

Random effects model
  completely randomized design,
    21–22
    significance tests, 22
Randomization, 6
  balanced lattices, 354
  completely randomized design,
    8–9
  Latin square change-over design,
    316
  Latin square design, 51
  by lot, 8
  by random number table, 9
  randomized block design, 36–38
  reasons for, 3
  split-plot designs, 134–135
  split-split plot designs, 235

[Randomization]
  strip-plot design, 239
Randomized block design, 34–47
  advantages, 36
  analysis, 38–40
  design construction, 35
  disadvantages, 36
  missing values, 43
  model and assumptions, 42
  numerical example, 44–47
  randomization, 36–38
  relative efficiency, 44
  significance tests, 40
  standard errors, 41
  summary, 69–70
  uses, 36
Regression in the ANOVA, 100–
  110
  estimation, 102–103
  numerical example, 107–110
  partitioning SST, 100–110
  significance tests, 101–102
Relative efficiency
  Latin square design, 56–57
  randomized block design, 44
Repeated partially balanced lattices,
  374–386
  analysis, 374–380
  description, 374–375
  effective error variance, 380
  numerical example, 380–386
  relative precision, 380
  significance tests, 379
  variances, 380
Replication, 6
  purpose, 3
Residual effects, 302–303
  assumptions, 302
  extra-period Latin square change-
    over design, 328
  incomplete Latin square change-
    over design, 336

[Residual effects]
Latin square change-over design, 318, 320
Response surfaces, 252–301
analysis, 255–258
basic assumptions, 253–254
basic ideas, 253
bias, 267–271
examining fitted surface, 258–260
goals, 254
lack of fit, 257
numerical example, 260–264
polynomial approximation, 254–255
3 × 3 factorial illustration, 260–264
Restriction errors, 247
Rotatability, 287–288
Round robin design, *see* Latin square change-over design

Sampling error, 25, 28
Sampling unit, 6
Searching for optima, 297–300
method of steepest ascent, 298–300
strategies, 297
Second order designs, 278–297
central composite design, 282–297
numerical example, 290–297
3 × 3 factorial, 279–282
Second order surfaces, 278–297
model, 278–279
Separation of means, 72–111
ranked in wheat yield trial, 160–161
Sequences of fractional replications, 187–189
Simple effects, 113

Simple lattices, 362
numerical example, 366–370
Simple reversal design, *see* Cross-over design
Single replication
3^k factorial, 200
2^k factorial, 175–179
SNK (Student-Newman-Keuls test), 76, 77, 78
numerical example, 81–82
Split-plot designs, 134–145
advantages, 136
alternatives, 232–234
analysis, 137–139
approximate d.f., 140
design construction, 134–135
disadvantages, 136
interpretation, 140
numerical example, 141–145
randomization, 134–135
significance tests, 139
standard errors, 139–140
subplot errors, 135, 137
uses, 136
variations, 230–251
whole plot errors, 135, 137
Split-plot units, *see* Subplots
Split-split plot designs, 235–238
analysis, 235–237
design construction, 235
randomization, 235
standard errors, 235–238
Standard error
completely randomized design
equal replication, 12–13
unequal replication, 18
Latin square design, 55
randomized block design, 41
split-plot design, 139–140
split-split plot design, 235–238
strip-plot design, 239–241
three-factor factorial, 128–129

[Standard error]
  two-factor factorial, 120–121
Stationary point, 258
Statistical computing packages,
      162–165
Strip-plot designs, 239–244
  analysis, 239–241
  design construction, 239
  numerical example, 241–244
  randomization, 239
  standard errors, 239–241
Student-Newman-Keuls test, *see*
      SNK
Subplots, 135
Switch-back design, 306–310
  analysis, 307–308
  design construction, 306–307
  numerical example, 308–310
  uses, 307
  variances, 307–308

3 × 3 factorial
  equations for factorial effects,
      193–195
  partition of SST, example, 195–
      198
  points in factor space, 192–193
  response surface, 255–258
Three-factor factorial, 124–134
  analysis, 125–128
  interpretation, 129
  numerical example, 129–134
  significance tests, 128
  standard errors, 128–129
3^k factorials, 198–202
  confounding in, 213–219
  design construction, ⅓ replica-
      tion, 201
  equation for effects, 199–200
  fractional replication, 201
  aliases, 201

[3^k factorials]
  defining effects, 201
  restrictions, 201–202
  generalized interaction, 198–199
  single replication, 200
  symbols for effects, 198–199
Treatment, 6
Triple lattices, 362
  numerical example, 370–375
Tukey's honestly significant differ-
      ence, *see* HSD
2 × 2 factorial
  equations for effects, 190–191
  points in factor space, 189–191
Two-factor factorial, 117–124
  analysis, 118–119
  interpretation, 121–122
  model, 117–118
  numerical example, 122–124
  significance tests, 120
  standard errors, 120–121
2^k factorials, 166–192
  analysis, 167–168
  in central composite design,
      289–292
  confounding, 203–213
    blocks of 2^{k-1}, 204–205
    blocks of 2^{k-2}, 207–208
  contrasts, 168–169
  design construction, ½ replica-
      tion, 203–204
  equations for effects, 191–192
  fractional replication, 179–189
  numerical example, 171–175
    ½ replication, 182–185
    ¼ replication, 186–187
    restrictions, 187
  ½ replication in central compos-
      ite design, 289–292
  single replication, 175–179
    numerical example, 177–179
    pooled error, 176–177

[2^k factorials]
standard order, 167
symbols for treatments, 166–167

Uniform precision, 288

Waller-Duncan Bayes LSD, *see* BLSD

Whole plots, 135

Yates algorithm for 2^k factorials, 169–171
fractional replication, 181–183
means computation, 171
Yield, 6
Youden squares, *see* Incomplete Latin square change-over